IX International Symposium on Radiation from Relativistic Electrons in Periodic Structures

(RREPS-2011)

Journal of Physics: Conference Series Volume 357

Egham, United Kingdom
12-16 September 2011

Editors:

Alexander Potylitsyn
Pavel Karataev

ISBN: 978-1-62276-774-8
ISSN: 1742-6588

Printed from e-media with permission by:

Curran Associates, Inc.
57 Morehouse Lane
Red Hook, NY 12571

proceedings.com

Some format issues inherent in the e-media version may also appear in this print version.

Printed by Curran Associates, Inc. (2013)

Institute of Physics
Dirac House, Temple Back
Bristol BS1 6BE UK

Phone: 44 1 17 929 7481
Fax: 44 1 17 920 0979

techtracking@iop.org

Additional copies of this publication are available from:

Curran Associates, Inc.
57 Morehouse Lane
Red Hook, NY 12571 USA
Phone: 845-758-0400
Fax: 845-758-2634
Email: curran@proceedings.com
Web: www.proceedings.com

TABLE OF CONTENTS

GENERAL PROPERTIES OF RADIATION FROM RELATIVISTIC PARTICLES

TRANSITION RADIATION

PARAMETRIC X-RAY RADIATION

DIFFRACTION RADIATION, SMITH–PURCELL EFFECT

COHERENT BREMSSTRAHLUNG AND CHANNELING RADIATION

CRYSTAL ASSISTED PROCESSES

APPLICATIONS OF MONOCHROMATIC X-RAY AND GAMMA BEAMS PRODUCED AT ELECTRON ACCELERATORS

Foreword

This volume contains papers presented at the IX International Symposium on Radiation from Relativistic Electrons in Periodic Structures (RREPS'11) which was held at Royal Holloway, University of London on September 12–16, Egham, United Kingdom. The symposium was organized jointly by Royal Holloway, University of London and Tomsk Polytechnic University, Tomsk, Russia.

RREPS is a biennial series of symposia founded in September 1993 as an initiative of the Nuclear Physics Institute at Tomsk Polytechnic University. The intention was to strengthen the basic and applied research focused on radiation from relativistic electrons in condensed media, particularly from natural and artificial periodic structures, and to review the research activity in this area. Since then, the symposium has developed into a forum attracting young scientists from different areas of research and from many countries. Previous successful symposia were held at Tomsk, Russia (1993, 1995, 1997, 2003), Lake Baikal, Russia (1999), Lake Aiya, Altai, Russia (2001), Czech Technical University in Prague, Czech Republic (2007) and Zvenigorod, Moscow region, Russia (2009). As an outcome of the symposia the conference proceedings have been published in *Nuclear Instruments and Methods in Physics Research*, Section B (Vol. 145 No 1–2, October 1998; Vol. 173 No 1–2, January 2001; Vol. 201 No 1 January 2003; Vol. 227 No 1–2, January 2005; Vol. 266 No 17, September 2008) and *Journal of Physics: Conference Series* (Vol. 236, June 2010).

The purpose of the present RREPS'11 symposium was to review the up-to-date situation in the area of electromagnetic radiation generated by relativistic charged particles in condensed media, and to discuss the research strategy for the near future. Nowadays, electromagnetic radiation studies cover electron energies from a few MeV up to hundreds of GeV in many laboratories throughout the world. The goal is to study the physics of the generation of various kinds of radiation and their interplay or combined effects, and to find successful applications for them. Every kind of radiation reflects specific processes of fundamental atomic physics, classical and quantum electrodynamics with a broad range of applications in accelerator physics, nuclear physics, material science and medicine.

During the symposium the general properties of electromagnetic radiation were discussed. A few reports were devoted to Cherenkov radiation. Such a renewed interest in this problem is related to possible applications in wakefield accelerators and beam diagnostics. Transition radiation appeared as a well-known subject but wide use of it requires a detailed investigation of its characteristics. New prospective schemes for generating intense radiation beams were proposed. During the last few years electromagnetic radiation has been intensively studied as a potential tool for non-invasive charged particle beam diagnostics. In the symposium a few presentations were devoted to both transverse beam size measurements, using optical diffraction radiation and longitudinal beam dynamics monitoring the use of coherent diffraction and synchrotron radiation techniques. The generation of intense THz and soft x-ray beams was a very popular topic. A few presentations were devoted to the development of compact x-ray sources which might be used as an alternative to large central facilities such as third or fourth generation light sources. An application of crystal targets for radiation generation attracted the attention of all RREPS'11 participants. Parametric x-rays may be used for low-emittance beam diagnostics, and channeling radiation and coherent bremsstrahlung are being studied as a possible mechanism for an intense gamma source for positron production.

Traditionally the RREPS symposium includes the following topics:

- General Properties of Radiation from Relativistic Particles; Cherenkov Radiation
- Transition Radiation
- Parametric X-ray Radiation
- Diffraction Radiation and the Smith–Purcell Effect
- Coherent Bremsstrahlung and Channeling Radiation
- Crystal Assisted Processes
- Applications of Monochromatic X-ray and Gamma Beams Produced at Electron Accelerators

We would like to acknowledge the International Program Committee for their suggestions during the preparation of the scientific program. We acknowledge John Adams Institute for Accelerator Science for their financial support of the students, and Royal Holloway, University of London (UK) and Tomsk Polytechnic University (Russia) for their administrative and financial support.

Editors

Prof. Alexander Potylitsyn
Tomsk Polytechnic University, Tomsk, Russia
Dr. Pavel Karataev
Royal Holloway, University of London, Egham, United Kingdom

All papers published in this volume of *Journal of Physics: Conference Series* have been peer reviewed through processes administered by the proceedings Editors. Reviews were conducted by expert referees to the professional and scientific standards expected of a proceedings journal published by IOP Publishing.

Annular Cherenkov High Gradient Wakefield Accelerator: Beam-Breakup Analysis and Energy Transfer Efficiency.

A.M.Altmark[1], A.D.Kanareykin[1, 2]
[1]St.Petersburg Electrotechnical University "LETI", St.Petersburg, Russia
[2]Euclid Techlabs LLC, Gaithersburg MD, USA

E-mail: aaltmark@mail.ru

Abstract. In this paper, we give a complete analytical solution for Cherenkov wakefields generated by an azimuthally asymmetric annular beam propagating in a coaxial two-channel dielectric structure. The transformer ratio of this type of structure is dramatically increased in comparison to a cylindrical wakefield accelerating structure. A particle-Green's function beam dynamics code (BBU-3000) to study beam breakup effects has been upgraded to incorporate annular drive beams and coaxial dielectric wakefield accelerating structures. Beam dynamics simulations of the annular drive beam with asymmetric charge distributions have been carried out to determine the sensitivity of this method to beam imperfections.

1. Introduction

A new application of microwave and THz Cherenkov radiation has been proposed and studied in the last decade to be used for high energy physics colliders and X-ray FELs, the Dielectric Wakefield Accelerator, or DWA [1-3].

In a general sense, a high gradient is desirable for a TeV level linear collider design because it can reduce the total linac length and hence the cost. Recently a high energy linear collider based on a short rf pulse (~22 ns flat top), high gradient (~267 MV/m loaded gradient), high frequency (26 GHz) dielectric two beam accelerator scheme has been proposed. The major parameters of a conceptual 3-TeV linear collider based on a DWA have been developed and are presented in reference [4].

(a) (b)

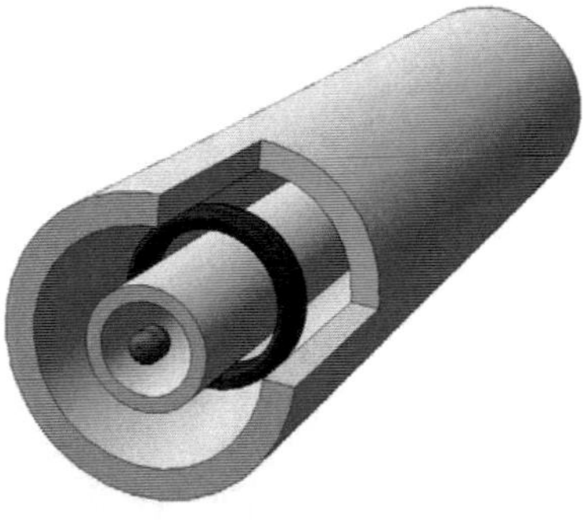

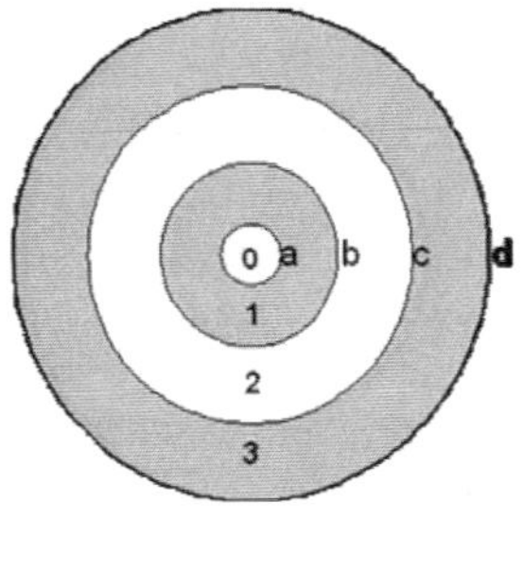

Figure 1. Partially filled dielectric wakefield accelerating structure excited by an annular drive beam. The dielectric tubes have inner radius *a (c)* and outer radius *b (d)*. Regions 0 and 2 are vacuum; layers 1 and 3 are dielectric; and the outermost layer (not shown) is metal.

X-ray free-electron lasers (FELs) are expensive instruments and the accelerator contributes the largest portion of the cost of the entire facility. Using a high-energy gain dielectric wakefield accelerator instead of a conventional accelerator may facilitate reduction of the facility size and significant cost savings. It has been shown that a collinear dielectric wake-field accelerator can accelerate low charge and high peak current electron bunches to a few GeV energy with up to 100 kHz bunch repetition rate [5].

Dielectric loaded accelerator (DLA) structures using various dielectric materials [6] and excited by a high current electron beam or an external high frequency high power RF source have been under extensive study recently [1-3]. The basic wakefield RF structure is very simple - a cylindrical, dielectric loaded waveguide with an axial vacuum channel is inserted into a conductive sleeve. Following at a delay adjusted to catch the accelerating phase of the wakefield is a second electron (witness) beam. The witness beam is accelerated to high energy by the wakefield produced by the drive beam [1]. A series of proof of principle experiments have been successfully performed at Argonne's Advanced Accelerator providing accelerating gradient in the range exceeding 100 MV/m at X-band [1-3,6]. THz wakefields of ~ GV/m magnitude range have been successfully generated by the UCLA-SLAC collaboration as well [7].

Energy transfer efficiency from the drive to witness bunches is a critical issue for wakefield acceleration techniques. The transformer ratio R is defined as the ratio of the maximum energy gain of the witness bunch to the maximum energy loss of the drive bunch. There are two major classes of wakefield accelerator geometries, collinear and two beam. For a collinear wakefield accelerator, R is less than 2 under very general conditions: linear media; a relativistic, longitudinally symmetric drive bunch; and identical paths through the system of both drive and witness beams [8-9]. A number of techniques have been proposed to overcome the transformer ratio limitation. Some of the methods that can be employed to obtain $R>2$ for the dielectric based accelerator include: a triangular longitudinal drive bunch profile [8]; a train of Gaussian drive bunches of progressively increasing charge (ramped bunch train) [10-11]; and use of a proton drive beam so that the particles can change positions within the bunch during deceleration [12].

Another way to achieve high transformer ratio is by designing separate drive and witness beam lines with different shunt impedances. This technique has been proposed in [13] for all-metal structures and initially proposed for dielectric based high gradient accelerator in [14], where a multichannel structure with an annular drive beam propagating through a coaxial outer vacuum channel and a witness beam through a central channel has been considered.

The cited paper [14] is obscure now but it is conceptually important to repeat its basic statements here. The structure geometry consists of a cylindrical cavity filled with a dielectric as presented in figure 1 (a). The structure proposed in [14] anticipates wakefield excitation with an annular drive beam, so the dielectric is separated from the wall by a vacuum region in which this beam is propagated. The metal wall is coated with a dielectric liner, and a small vacuum aperture in the center of the dielectric rod allows for the passage of a low current witness beam there.

In [14] it was correctly noted that because of the dielectric properties the Cherenkov radiation generated by the annular beam passing through the waveguide of this geometry will concentrate near the axis of the device. It was clearly stated that without the central rod a large transformer ratio can only be achieved by proper pulse shaping of the drive beam, and that including the rod results in an enhancement of the transformer ratio due to the geometrical effect.

The analytical solutions for the multizone coaxial geometry was obtained in [14] by solving for the fields in the regions 1–3, figure 1(b), and matching the corresponding boundary conditions. A 10 MeV linearly ramped bunch (100 kA, t = 0.25 ns) has been used in these simulations The maximum accelerated field of 225 MV/m has been simulated providing 50 MeV energy gain of the witness beam in 37 cm with average field of 130 MV/m. It should be noted that the ramped triangular bunch was also considered as a drive beam in reference [14].

Recently the coaxial scheme [14] was considered again for high-gradient wakefield acceleration, and detailed analytical studies for this type of coaxial dielectric-loaded structure have been presented [15–18]. In [15], an analytical solution for wakefields in a coaxial two-channel dielectric structure has been carried out, primarily concentrating on the transformer ratio R optimization in X-band structures. It was found that, by appropriate choice of parameters, R can be much greater than 2, the limiting value for collinear wakefield accelerators. Transformer ratio values as large as $R = 4.5$ were predicted.

In [16], the coaxial scheme initially proposed in [14] has been considered for the THz frequency range and for the SLAC FACET drive bunch parameters: an annular drive beam with the length of 30 μm, 6 nC charge and 5 GeV energy, while the simulated transformer ratio coefficient was found to be close to R~10. The structure parameters correspond to THz frequency range wakefields (~ 0.1 mm witness beam aperture) and can be found in reference [16].

2. Analytic results and comparison with numerical simulations

The coaxial geometry proposed in [14] for transformer ratio enhancement of the dielectric based wakefield structures is considered here for GHz Ka-band [15] and THz acceleration [16–18]. At the same time, dynamics of the beam in structure-based wakefield accelerators leads to beam stability issues not ordinarily found in other machines. In particular, the high current drive beam in an efficient wakefield accelerator loses a large fraction of its energy in the decelerator structure, resulting in physical emittance growth, increased energy spread, and the possibility of head-tail instability for an off axis beam, all of which can lead to severe reduction of beam intensity [20]. Beam breakup (BBU) effects resulting from parasitic wakefields provide a potentially serious limitation to the performance of dielectric structure based wakefield accelerators as well [20-21].

Correspondingly, the transverse stability of the annular driver beam can be a critical issue for a coaxial high transformer ratio DWA [17]. We report on beam breakup study results for the coaxial Ka-band DWA currently planned to be tested at Argonne Wakefield Accelerator (AWA). Recently we have developed a particle-Green's function beam breakup code (BBU-3000) that allows rapid, efficient simulation of beam breakup effects in advanced linear accelerators [21]. The goal of this work is foremost to design mitigation techniques for BBU and to test these concepts as part of an ongoing series of experiments at ANL/AWA, BNL/ATF and SLAC/FACET.

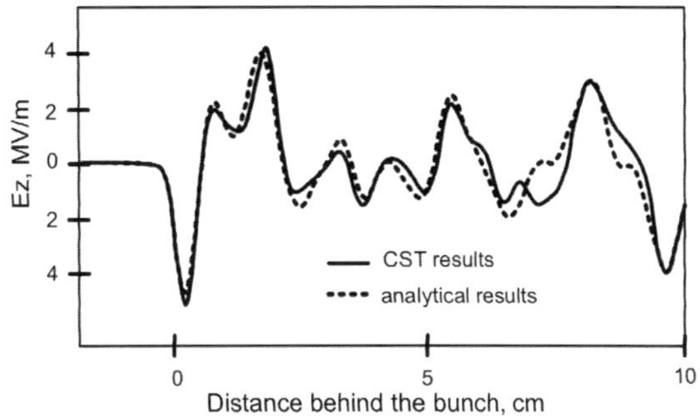

Figure 2. Simulations of wakefields generated by a point-like electron beam (table 2 Beam 1) passing through a coaxial structure (table 1 Structure 1). Solid line is CST simulation, dotted line is the BBU3000 simulation using the analytical Green's functions.

Table 1. Parameters of the coaxial waveguides used in the transverse beam dynamic analysis.

#	a, cm	b, cm	c, cm	d, cm	ε_1	ε_2	f, GHz (TM_{02})
1	0.2	0.60	1.00	1.40	4.76	4.76	13.84
2	0.2	0.32	1.35	1.41	9.80	9.80	28.02

Table 2. Parameters of the annular drive beams used in the transverse beam dynamic analysis

#	beam type	Q, nC	σ_z, cm	σ_r, cm	offset, cm	R, cm	W, MeV
1	transversely point-like beam	10	0.2	0	0.8	-	10
2	symmetrical annular beam	10	0.2	0.01	0.02	0.8	10
3	asymmetrical annular beam	10	0.2	0.01	0	0.8	10
4	transversely point-like beam	50	0.1	0	0.8	-	14
5	symmetrical annular beam	50	0.1	0.05	0.15	0.85	14
6	asymmetrical annular beam	50	0.1	0.05	0.1	0.85	14

The coaxial, two-channel DLA structure of figure 1 has an inner vacuum region "0" (r < a) for the witness beam and an outer vacuum region "2" (b < r < c) for the drive beam. The drive beam used to excite the DLA is azimuthally asymmetric; this is necessary to excite dipole modes which cause beam instabilities. In this paper, we will concentrate on the dipole wakefield excitation due to a non-uniform ring of electrons. For an annular electron beam moving at speed v~c, the corresponding wave equations for the longitudinal field components E_z and H_z can be obtained from

$$\begin{cases} \left(\nabla^2 - \dfrac{1}{c^2}\dfrac{\partial^2}{\partial t^2}\right)E_z = -4\pi e\left(\dfrac{\beta}{c}\dfrac{\partial n}{\partial t} + \dfrac{\partial n}{\partial z}\right), \\ \left(\nabla^2 - \dfrac{1}{c^2}\dfrac{\partial^2}{\partial t^2}\right)H_z = 0, \end{cases} \tag{1}$$

where $n = \dfrac{Q}{e}\delta(z - Vt)\dfrac{\delta(r - r_0)}{r}\delta(\theta - \theta_0)$ is the point-like beam charge density. Consequently one can define transverse components of wakefields by solving system equation (1). Using boundary conditions for the tangential components of electric and magnetic fields at the dielectric surfaces (a,b,c,d) we obtain system equations for unknown coefficients $[X]$:

$$[M][X] = [N].$$

Longitudinal components of the electric field generated by the electron beam passing through vacuum region "2" can be written as

$$Ez = \sum_{\nu=1}^{Nm}\sum_{n=1}^{Nn}\left(A_{\nu,n} I_\nu\left(\chi_2 k_{\nu,n} r\right) + B_{\nu,n} K_\nu\left(\chi_2 k_{\nu,n} r\right)\right)\cos\left(\nu(\theta - \theta_0)\right)\cos\left(k_{\nu,n}\zeta\right) \tag{2}$$

where $A_{\nu,n}$, $B_{\nu,n}$ – coefficients defined by structure geometry, dispersion equation roots, beam charge and transverse position, $k_{\nu,n}$ – dispersion equation roots $\det(M)=0$, ζ – distance behind the bunch. By using Panofsky-Wenzel theorem

$$\frac{1}{r}\frac{\partial}{\partial\theta}E_z=\int F_\theta dz\,,\ \frac{\partial}{\partial r}E_z=\int F_r dz,$$

one can obtain transverse force components F_r and F_θ

$$Fr=\sum_{\nu=1}^{Nm}\sum_{n=1}^{Nn}\chi_2\left(A_{\nu,n}I'_\nu\left(\chi_2 k_{\nu,n}r\right)+B_{\nu,n}K'_\nu\left(\chi_2 k_{\nu,n}r\right)\right)\cos\left(\nu(\theta-\theta_0)\right)\sin\left(k_{\nu,n}\zeta\right) \tag{3}$$

$$F_\theta=\sum_{\nu=1}^{Nm}\sum_{n=1}^{Nn}\frac{\nu}{r}\left(A_{\nu,n}I_\nu\left(\chi_2 k_{\nu,n}r\right)+B_{\nu,n}K_\nu\left(\chi_2 k_{\nu,n}r\right)\right)\sin\left(\nu(\theta-\theta_0)\right)\sin\left(k_{\nu,n}\zeta\right). \tag{4}$$

The wakefield is a superposition of axisymmetric (TM-modes, ν = 0) and non-axisymmetric (HEM-modes, ν = 1, 2, 3 ,...). The magnitude of wake field is determined essentially by the HEM-mode [20], whose contribution increases with the deflection of the beam from the waveguide axis. A comparison of calculations obtained using the expression (1) with the calculations takes into account the CST Particle Studio HEM-mode up to ν = 6 (see Table 1). Comparisons were made for a point-like (in the transverse coordinates) bunch passing the structure along region "2'. figure 2 shows that the results obtained using analytical formulas that are consistent with numerical simulations by CST Particle Studio.

3. Beam dynamics simulations of an annular driving beam in the coaxial DWA structure

BBU-3000 represents an approach to beam dynamics computations which is complementary to the usual electromagnetic PIC (Particle in Cell) approach. Particle pushing is done in the same fashion as a PIC code but the wakefields caused by the charged particles are computed using the known analytic expressions for the Green's functions in a dielectric tube [20]. Details of the scalar code for the Gaussian shaped beam have been published elsewhere [21]. Here we focus on the recent work on the annular electron beam simulation.

It should be noted that for the annular beam the offset value is limited in comparison with ordinary Gaussian beam because of intrinsic large diameter (relatively to a witness beam aperture) of a vacuum channel of the coaxial DWA structure. Consequently, the beam dynamics of the annular drive beam will be dominated by dipole mode fields, or HEM modes with ν=1 with the field structure presented in figure 3. In this paper, the beam breakup of the annular driving bunches has been studied with respect to its both azimuthal and radial asymmetries.

Journal of Physics: Conference Series **357** (2012) 012001 doi:10.1088/1742-6596/357/1/012001

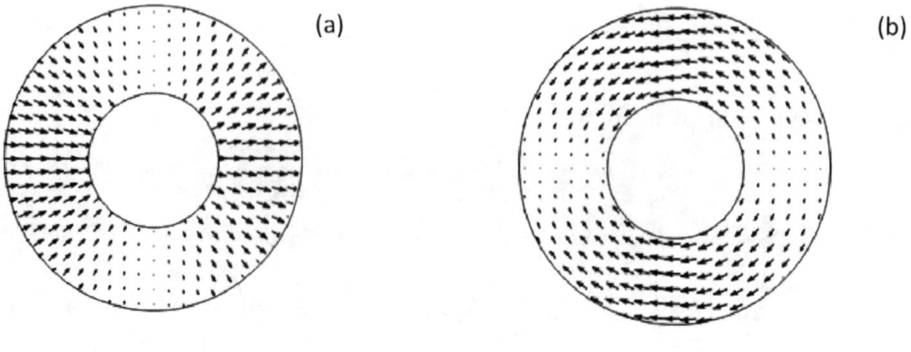

Figure 3. Electric field patterns of the dipole modes generated by an annular beam passing through coaxial dielectric wakefield structure presented in figure 1, region 2.: (a) radial F_r component; (b) azimuthal F_θ component, region 2.

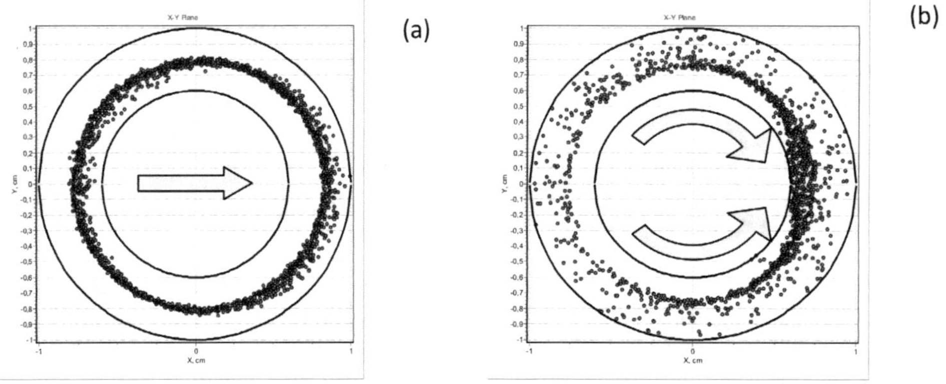

Figure 4. Beam dynamic simulations of an annular beam passing through coaxial dielectric wakefield structure presented in figure 1, region 2: (a) azimuthally symmetric beam with offset (table 1, structure #1; table 2, beam # 2); (b) azimuthally symmetric beam with no offset (table1,structure # 1; table 2, beam # 3).

Figure 4 shows the results of BBU300 simulation for the azimuthally symmetric annular beam with offset and the beam with no offset but with azimuthal charge asymmetry. The beams pass through the coaxial dielectric wakefield structure presented in figure 1. One can see radial deflection for the azimuthally symmetric beam caused by center of mass shift (offset of the beam) while the azimuthal deflection is still compensated by the charge density symmetry, figure 4 (a). At the same time, the azimuthal asymmetry generates F_θ components strong enough to increase dramatically the charge density along the line of structure center – maximum of charge density distortion, figure 4. Figure 5 (a) presents the transverse deflecting field vs. radial position of the beam particle: one can see the flat region where there is no force increase. figure 5 (b) shows the deflecting field of the annular beam vs. its radial offset that shows the same behavior as those of an ordinary Gaussian beam in a cylindrical tube-like DWA structure [20].

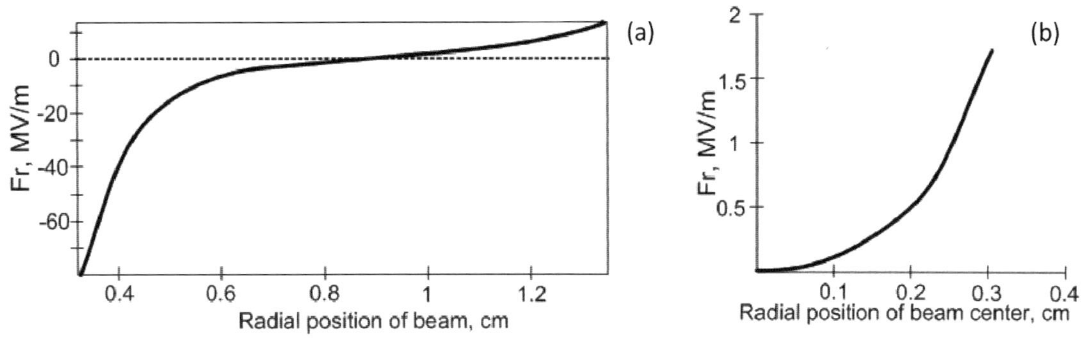

Figure 5. (a) Transverse deflecting field vs. radial position of the beam particle in region 2 for the structure # 1 of table 1, for the beam # 4 of table 2; (b) Maximal deflecting field vs. radial offset for the structure #1 of table 1, beam# 5 of table 2.

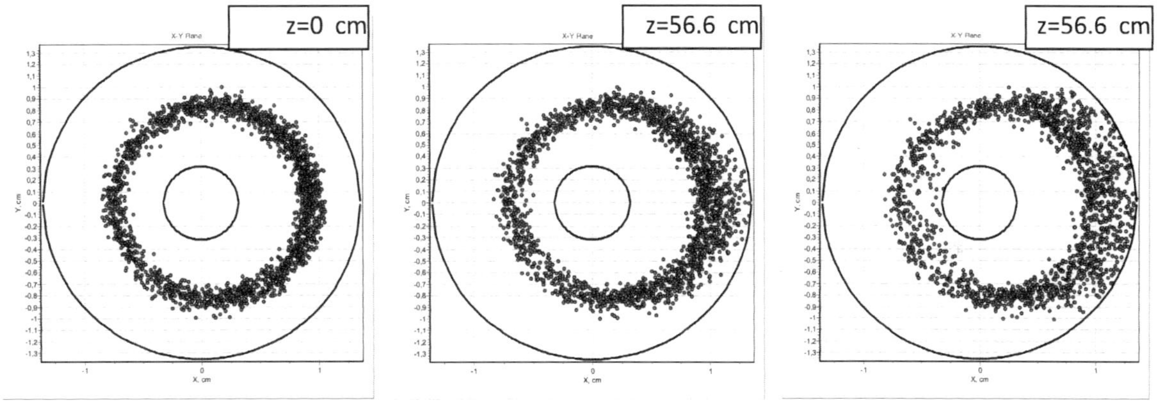

Figure 6. Beam dynamic simulations of an annular beam passing through coaxial structure presented in figure 1, azimuthally asymmetric beam with 0.1 cm offset to the asymmetry direction; structure # 2 of table1, beam # 6 of table 2.

Figure 6 presents simulation results for the annular beam passing through coaxial structure of figure 1. It is azimuthally asymmetric beam with 0.1 cm offset to the asymmetry direction: note strong deflection to both inner and outer dielectric surfaces. The same effect is more clearly seen in figure 7, where an annular beam passing the structure of figure 1 with 0.1 cm offset opposite to the asymmetry direction. It should be noticed azimuthal concentration of a charge density and its strong attraction by the inner dielectric rod. The length of the structure with no focusing is limited to 50 cm for both of the structures, figures 6-7.

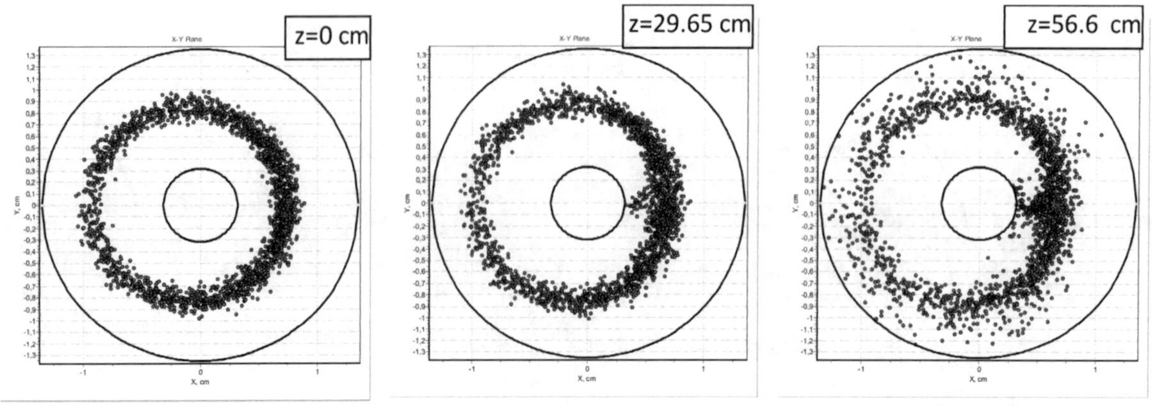

Figure 7. Beam dynamics simulations of an annular beam passing through coaxial dielectric wakefield structure presented in figure 1, azimuthally asymmetric beam with 0.1 cm offset opposite to the asymmetry direction; structure # 2 of table1, beam # 6 of table2.

Acknowledgments. This work is supported by the SBIR Program of the US Department of Energy and RFBR award.

References

[1] Gai W, Schoessow P, Cole B, Konecny R *et al* 1988 *Phys. Rev. Lett.* **61**, 2756
[2] Gai W, 2009 *AIP Conference Proceedings* **1086** 3
[3] Kanareykin A 2010 *Journal of Physics*, Conf. Ser. **236** 012032
[4] Jing C, Kanareykin A, Power J G, Zholents A 2011 *Proc. IPAC'11*, 1485
[5] Jing C, Antipov S, Kanareykin A *et al* 2011 *Proc. PAC'11*, New York, NY, 2279
[6] Kanareykin A 2010 *Proc. AAC'10, AIP Conference Proceedings*, v. 1299, NY, 286
[7] Thompson M C *et al* 2006 *AIP Conf. Proceedings* **877** 903
[8] Bane K L, Chen P, and Wilson P B 1985 *SLAC Report No.* SLAC-PUB-**3662**
[9] Power J G, Gai W, and Schoessow P 1999 *Phys. Rev.* E **60** 6061
[10] Laziev E, Tsakanov V, and Vahanyan S 1988 *Proceedings EPAC'88*. IEEE, 523
[11] Jing C, Kanareykin A, Power J G, Conde M *et al* 2007 *Phys. Rev. Lett.* **98** 144801
[12] Rosenzweig J B, Schoessow P, Cole B, Gai W *et al* 1989 *Phys. Rev.* A **39** 1586
[13] Voss G A and Weiland T 1982 *DESY M-82-10*
[14] Keinigs R, Jones M 1988 *Proc. 7th Int. Conf. High-Power Part. Beams, Beams'88*, Karlsruhe, Germany, **864** (http://library.lanl.gov/cgi-bin/getfile?00261284.pdf)
[15] Liu W and Gai W 2009 *Phys. Rev. ST Accel. Beams* **12** 051301
[16] Sotnikov G V, Marshall T C, and Hirshfield J L 2009 *Phys. Rev. ST-AB* **12** 061302
[17] Sotnikov G V, Marshall T C, Hirshfield J L, and Shchelkunov S V 2010 *Proc. AAC'10, AIP Conference Proceedings*, NY **1299** 342
[18] Sotnikov G V and Marshall T C 2011 *Phys.Rev.ST-Accel.Beams* **14** 031302
[19] Bialowons W, Bieler M, Bremer H D, Decker F J *et al* 1988 *Proc. EPAC'88*, Rome, **55**
[20] Gai W, Kanareykin A D, Kustov A L and Simpson J 1997 *Phys. Rev.* E **55** 3481
[21] Schoessow P, Kanareykin A, Jing C, Altmark A *et al* 2010 *Proc. AAC'10, AIP Conference Proceedings*, N.Y **1299** 262

New peculiarities in angular distribution of Cherenkov radiation from relativistic heavy ions caused by their stopping in radiator: numerical and theoretical research

E I Fiks[1], O V Bogdanov[1, 2] and Yu L Pivovarov[1]
[1] Tomsk Polytechnic University, Tomsk, Russian Federation
[2] Frascati National Laboratory, Frascati, Italy

E-mail: elenafiks@gmail.com

Abstract. A new method and theoretical base for calculation of the Cherenkov Radiation (ChR) angular distribution from relativistic heavy ions (RHI) taking into account stopping in radiator is suggested. Our method is based on the thin radiator approximation and the Bethe-Bloch formula for ionization energy loss (stopping) and provides fast calculations without using any special software. The simple formula for estimation of the ChR angular distribution width in vicinity of the Cherenkov angle taking into account RHI stopping in radiator is obtained. New peculiarities of ChR - dependence on the RHI charge and mass (isotopic effect), emission wavelength (index of refraction) and radiator length - are studied. These new features of ChR from RHI allow for their possible applications to simultaneously reconstruction of both velocity and charge of RHI penetrating through radiator and even to measure the masses of isotopes or radiator stopping power.

1. Introduction

The Cherenkov radiation (ChR) occurs when a charged particle moves in a medium with a velocity greater than the velocity of light in that medium. ChR is emitted into a cone making the Cherenkov angle with particle velocity vector (particle moves through the radiator with a constant velocity v). The detectors based on ChR are widely used to measure the ion velocity. It seems that the theory of ChR is developed in detail and confirmed by experiments. It is true, with respect to ChR from relativistic elementary particles. Nowadays new accelerators of RHI (for example, CERN SPS, GSI Darmstadt, FAIR Darmstadt) allow for the new experiments on ChR. In connection with this, we may ask the question – does the angular distribution of ChR from RHI contain any additional information, for example - on ion charge? To answer this question one needs to reconsider the ChR theory for the case of RHI.

The first theory of ChR was developed by Tamm and Frank (TF) in [1]. TF theory describes the ChR of a charged particle that moves rectilinearly in a medium with a constant velocity. It ignores both bending of particle trajectory due to multiple scattering (MS) and decreasing of its velocity due to ionization energy loss (slowing-down or stopping effect). Dedrick [2] was the first who took into account the influence of MS on the angular distribution of ChR, but he neglected the velocity decreasing due to ionization (and radiation) energy loss. In [3] Kuzmin and Tarasov (KT) considered another limiting case: they calculated the angular distribution of ChR from RHI taking into account their velocity decreasing due to stopping (ionization energy loss) in radiator, but neglecting RHI MS. KT model predicts the complex Fresnel diffraction-like structure of ChR angular distribution, but it is valid only for thin radiators. To avoid this approximation, the authors of [4] and [5] established another approach which uses computer code SRIM [6] to calculate energy loss and velocity depending on RHI penetration depth into a radiator. The obtained velocity is substituted into the formula of classical electrodynamics for the spectral-angular distribution of radiation from a charged particle that moves in a medium along a trajectory r(t):

$$\frac{\mathrm{d}I}{\mathrm{d}\omega\mathrm{d}\Omega}=\frac{z^2e^2\omega^2}{4\pi^2c^3}\sqrt{\varepsilon}\left|\int_{-\infty}^{+\infty}\vec{n}\times[\vec{n}\times\vec{v}]\exp\left\{i\omega\left[t-\sqrt{\varepsilon}(\vec{n}\cdot\vec{r}(t))c^{-1}\right]\right\}]\mathrm{d}t\right|^2. \quad (1)$$

2. Estimation of the ChR angular distribution width

As is well known, the ChR is emitted forward into a cone making an angle with the ion velocity direction (ion moves through the radiator with a constant velocity v). The formula for the Cherenkov angle is: $\cos\theta=c(n\,v)^{-1}$. Here c – is the speed of light in vacuum, n – the radiator refractive index. In the TF theory the width $\Delta\vartheta_{TF}$ of the ChR angular distribution in the vicinity of the Cherenkov angle is determined by the radiation wave length λ and the radiator thickness L:

$$\Delta\vartheta_{TF}\cong\lambda\,L^{-1}.$$

It is easy to obtain the estimation of the ChR angular distribution width in the case of the thin radiator. As in the KT model [3], the inverse ion velocity depends on its penetration depth into a radiator x and stopping (ionization energy loss):

$$\frac{1}{v(x)}\approx\frac{1}{v_0}-\frac{1}{v_0^2}v_0^{'}x,\ v_0^{'}=\left.\frac{\mathrm{d}v(x)}{\mathrm{d}x}\right|_{x=0}<0, \quad (2)$$

$$v_0^{'}=\frac{\mathrm{d}v}{\mathrm{d}E}\frac{\mathrm{d}E}{\mathrm{d}x}=\frac{c^2}{vE}\left(1-\frac{v^2}{c^2}\right)\left.\frac{\mathrm{d}E}{\mathrm{d}x}\right|_{x=0}=-\frac{c^2}{E_0\,\gamma_0^2\,v_0}S(E_0). \quad (3)$$

Here E – is RHI total energy; E_0, γ_0^2, v_0 – are incident energy, relativistic factor and velocity of RHI; $-\frac{\mathrm{d}E}{\mathrm{d}x}=S(E_0)$ – radiator's stopping power (RHI ionization energy loss) which depends on RHI initial velocity. If the thickness of radiator is equal to L ($x=L$) and the approximation (2) is valid, we may easily obtain the RHI velocity v_1 at $x=L$. The ChR angles corresponding to initial v_0 and final v_1 velocities are: $\theta_{0,1}=\arccos\frac{1}{n\beta_{0,1}}$, $\beta_{0,1}=\frac{v_{0,1}}{c}$; since the RHI velocity decreases due to stopping in radiator ($v_1<v_0$), the ChR angle also decreases ($\theta_1<\theta_0$). Now we may estimate the ChR angular distribution width $\Delta\theta_S$ taking into account stopping in a radiator in the following way: $\Delta\theta_S=\theta_0-\theta_1\approx\arccos\frac{1}{n\beta_0}-\arccos\left(\frac{1}{n\beta_0}\left(1-\frac{v_0^{'}L}{v_0}\right)\right)$. If we consider that ChR is located in a narrow angular interval $\theta_1\le\theta\le\theta_0, \theta_0-\theta_1<<1$, $\arccos\left(\frac{1}{n\beta_0}\left(1-\frac{v_0^{'}L}{v_0}\right)\right)$ can be expanded into the Taylor series near $\frac{1}{n\beta_0}$. The expansion reads: $\arccos\left(\frac{1}{n\beta_0}\left(1-\frac{v_0^{'}L}{v_0}\right)\right)\approx\arccos\frac{1}{n\beta_0}+\frac{v_0^{'}L}{v_0}\cot\theta_0$. After some algebra we obtain a very simple formula for $\Delta\theta$:

$$\Delta\theta_S \approx \frac{|v_0'|\,L}{v_0}\cot\theta_0 = \frac{c^2}{E_0\,\gamma_0^2\,v_0} S(E_0)\,L\cot\theta_0, \tag{4a}$$

$$\Delta\theta_S = \frac{c^2}{E_0\,\gamma_0^2\,v_0} S(E_0)\frac{L}{\sqrt{(n(\lambda)\,\beta_0)^2-1}}., \tag{4b}$$

If the thin radiator approximation (2) is not valid, the RHI velocity at $x = L$ can be calculated with the help of the special computer code, e.g. SRIM [6] or ATIMA [7], and the width of ChR angular distribution again can be estimated as $\Delta\theta_S = \theta_0 - \theta_1 \approx \arccos\frac{1}{n\,\beta_0} - \arccos\frac{1}{n\,\beta_1}$.

Expressions (4a) and (4b) demonstrate linear dependence of the ChR angular distribution width $\Delta\theta_S$ on the radiator thickness L, in contrary to the TF theory [1], with angular distribution width $\Delta\vartheta_{TF} \cong \lambda L^{-1}$ being inversely proportional to L. Thus, in a very thin radiator (slowing down is insignificant) the ChR angular distribution width is inversely proportional to radiator thickness L - $\Delta\vartheta_{TF} \cong \lambda L^{-1}$, but if L becomes larger, this dependence transforms to the linear one, i.e $\Delta\theta_S \cong S\,L$. It is clear that if stopping effect is taken into account, the expression for ChR angular distribution radically changes and becomes more complex. Also in (4) we can see an appearance of radiator stopping power $S(E_0)$ (depends on the RHI charge) and wave length dependence, because $n = n(\lambda)$. Moreover, (4) contains an isotopic effect [5]: really, slowing down $S(E_0)$ is proportional to ion charge and is equal for isotopes, but the multiplier $E_0 = \gamma_0 M c^2$ contains the dependence on the isotope mass.

3. Theoretical analysis

According to TF theory [1], the ChR angular distribution in a finite thickness radiator can be written in the following form:

$$\frac{dI}{d\omega d\Omega} = \omega L\left(\frac{z e\sin\theta}{c}\right)^2 f_{TF}(\theta,\omega),\; f_{TF}(\theta,\omega) = \frac{1}{\Delta\theta_{TF}}\left(\frac{\sin x}{x}\right)^2, x = \frac{\pi}{\Delta\theta_{TF}}\left(\cos\theta - \frac{1}{\beta n}\right) \tag{5}$$

Here, $\Delta\theta_{TF} = \lambda (n\,L)^{-1}$ is the ChR angular distribution width in the vicinity of the Cherenkov angle, ze – RHI charge; $\lambda = 2\pi c\omega^{-1}$ – radiation wave length; ω – frequency of the Cherenkov photon. The angular distribution (5) has diffraction-like structure and $\Delta\theta_{TF}$ is proportional to $\lambda \cdot L^{-1}$. In [3] the following expression based on (1) was used to analyze the ChR angular distribution and the ChR angular distribution width:

$$\frac{dW}{d\omega d(\cos\theta)} = \frac{(z e\sin\theta)^2}{2\pi c^2} k\omega \left|\int_0^L e^{i\Phi(x)}dx\right|^2, \tag{6}$$

Here, $\Phi(x) = k\,x\cos\theta - \omega\int_0^x \frac{dx'}{v(x')}$; $k = \frac{\omega\,n}{c}$ is the wave number of the Cherenkov photon. Using (2), the authors of [3] obtained analytic expressions for ChR angular distribution of RHI in a thin radiator taking into account their stopping:

$$\frac{dI}{d\omega d\Omega} = \omega L\left(\frac{z e\sin\theta}{c}\right)^2 f_{KT}(\theta,\omega), \tag{7}$$

Here, the following function appears:

$$f_{KT}(\theta,\omega) = \frac{1}{2\Delta\theta_{KT}\sin\theta_0}\left\{[\mathrm{C}(u_1)-\mathrm{C}(u_0)]^2 + [\mathrm{S}(u_1)-\mathrm{S}(u_0)]^2\right\} \tag{8}$$

In (8) $\mathrm{C}(u) = \sqrt{\frac{2}{\pi}}\int_0^u \cos t^2 dt$ and $\mathrm{S}(u) = \sqrt{\frac{2}{\pi}}\int_0^u \sin t^2 dt$ - are the Fresnel integrals; $u_{0,1} = -\frac{k\cos\theta - \omega/v_{0,1}}{a}$, $a = \sqrt{-\frac{2\omega v_0'}{v_0^2}}$; $\Delta\theta_{KT} = \theta_0 - \theta_1 \approx \frac{|v_0'|L}{v_0}\cot\theta_0$ – ChR angular distribution width, $\theta_{0,1} = \arccos\frac{c}{nv_{0,1}}$; $\frac{1}{v_1} \approx \frac{1}{v_0} - \frac{1}{v_0^2}v_0'L$ – ion velocity on leaving the radiator.

To obtain the expressions for ChR angular distribution we start with the formula of classical electrodynamics (1) and a thin radiator approximation (2) and after some algebra obtained almost the same expressions as in KT model [3]. The difference is in the limits of integration for the Fresnel integrals: we believe that we use the correct expressions for these limits, while the expressions of [3] probably contain misprints. The authors of [3] did not mention, which method was used to calculate the velocity gradient. We did it using the well known Bethe-Bloch formula:

$$-\frac{\mathrm{d}E}{\mathrm{d}x} = \rho 2\pi z^2 N_A r_e^2 m_e c^2 \frac{Z}{A}\frac{1}{\beta^2}\left\{\ln\left(\frac{2m_e c^2\beta^2 T_{\max}}{I^2(1-\beta^2)}\right) - 2\beta^2\right\}. \tag{9}$$

Here, $T_{\max} = \frac{2\beta^2\gamma^2 m_e c^2}{1 + 2\gamma m_e c^2 (Mc^2)^{-1} + (m_e M^{-1})^2}$ is maximal energy that may be transferred to an electron in a radiator, $N_A = 6.022\cdot 10^{23}[\mathrm{mol}^{-1}]$ – Avogadro number; z – ion charge in terms of elementary charge; Z and A – atomic number and atomic weight of the radiator matter; $m_e c^2$ – electron rest energy; $\rho\,[\mathrm{g\,cm^{-3}}]$ – radiator density; $r_e^2 = 2.818\cdot 10^{-13}\,[\mathrm{cm}]$ – classical electron radius; $I = 16\cdot Z^{0.9}\,[\mathrm{eV}]$ – average ionization potential.

Thus, TF theory [1] and KT model [3] differ in functions $f_{TF}(\theta,\omega)$ and $f_{KT}(\theta,\omega)$. The $f_{TF}(\theta,\omega)$ contains dependence on the radiation wave length, refractive index and thickness of the radiator. In addition, in $f_{KT}(\theta,\omega)$ initial ion velocity and radiator stopping power $\mathrm{d}E/\mathrm{d}x$ dependencies appear. The radiator length L dependence is very different: in the first case we see L^{-1} – inverse dependence, and in the second – linear dependence – L.

The schematics of the ChR angular distributions according to TF theory (left) and taking into account stopping in radiator (right) are presented in figure 1.

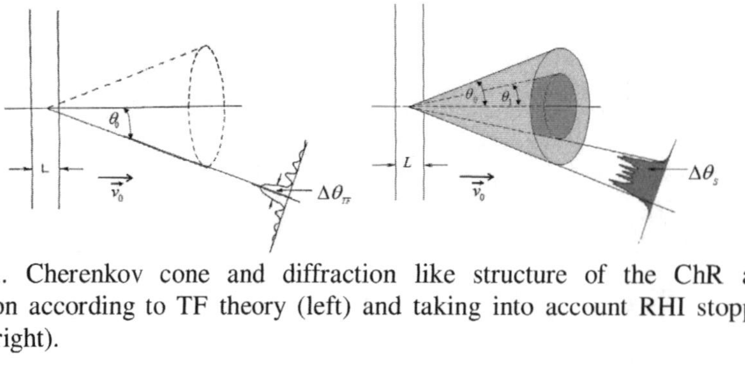

Figure 1. Cherenkov cone and diffraction like structure of the ChR angular distribution according to TF theory (left) and taking into account RHI stopping in radiator (right).

The additional broadening of ChR angular distribution width in the right part accompanied with more complex structure appears (only two Cherenkov cones are presented – for initial (θ_0) and final (θ_1) ChR angles).

The main advantages of the new ChR calculation method based on the Bethe-Bloch formula and a thin radiator approximation are: 1) fast calculation speed; 2) no need to use any special software (like SRIM [6] or ATIMA [7]); 3) no limits in choosing radiators, ions, and energies. The main disadvantage: this method may be used only when the thin radiator approximation (2) is valid.

4. New peculiarities in angular distribution of Cherenkov radiation

Now we will discuss the results of our calculations demonstrating the new features of ChR caused by stopping in radiator. The analytical formula for $\Delta\theta_S$ (4) clearly shows the following dependencies of the ChR angular distribution width $\Delta\theta_S$: 1) L - dependence; 2) λ – dependence; 3) stopping-power $S(E_0)$ dependence; 4) ion charge z – dependence; 5) mass dependence ($E_0 = \gamma_0 M c^2$). To prove it, we applied equation (8) and performed numerical calculations for different targets and ions. Equation (8) connects together ChR angular distribution structure with radiator parameters (length, refractive index, stopping power) and ion beam parameters (charge, mass). Figure 2 shows the thickness dependence (L – dependence) of ChR angular distribution in vicinity of the Cherenkov angle (corresponding zero in the figure 2) for Au ion in LiF radiator (radiator thickness $L = 0.5$ cm; radiation wave length $\lambda = 390$ nm; refractive index $n = 1.4$; initial energy of ion beam 1000 MeV/u). And figure 3 shows z – dependence of ChR angular distribution on RHI charge in LiF radiator (radiator thickness $L = 0.5$ cm; radiation wave length $\lambda = 390$ nm; refractive index $n = 1.4$; initial energy of ion beam 1000 MeV/u). Two methods were used for calculations: the first one – suggested in section 3 – based on the Bethe-Bloch formula and thin radiator approximation (2) (left part of figures 2 and 3), and the second one – based on SRIM [6] (right part of figures 2 and 3).

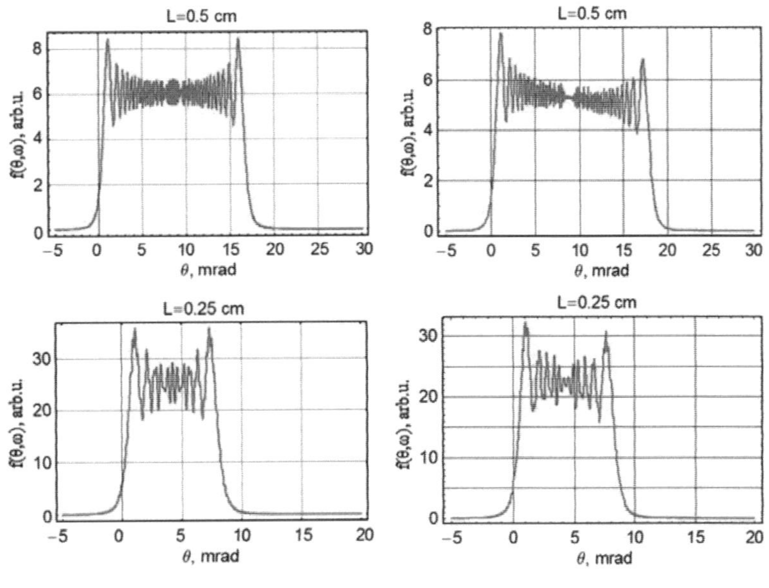

Figure 2. L – dependence of ChR angular distribution.

It is obvious that ChR width is proportional to the radiator thickness and is almost the same in both methods. Also both methods show that the thicker the radiator is the more diffraction maxima appear. Finally, the computer code SRIM allows obtaining more exact values of RHI velocity depending on penetration depth into radiator. The difference in results from two methods is that the graphs obtained using SRIM are not symmetric. Obviously, it is the sequence of the thin radiator approximation used in our suggested method of calculation.

Increasing of isotopic charge z leads to additional broadening of the ChR angular distribution width and also to appearance of diffraction maxima.

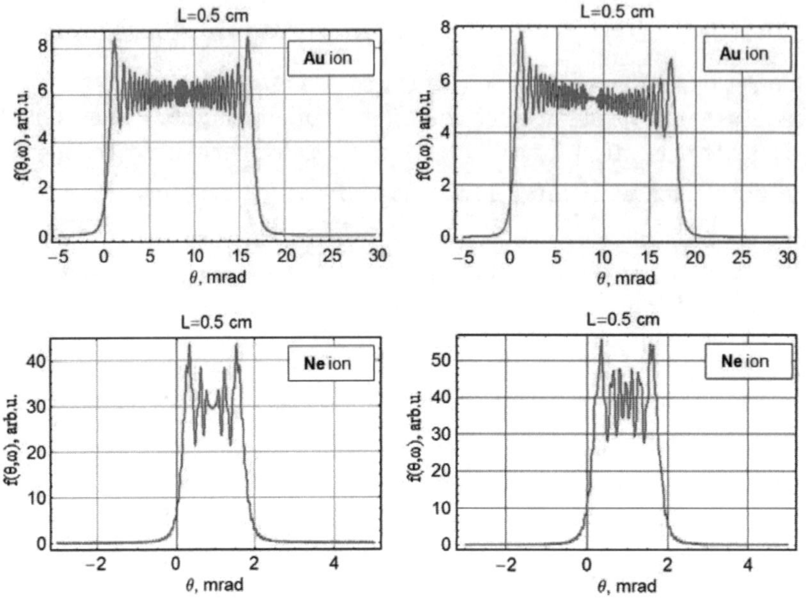

Figure 3. z – dependence of ChR angular distribution.

The newest discovery is that ChR angular distribution width depends on the isotope mass. That's why isotopic [5] effect appears. One may say that 2 isotopes with different masses M_1 and M_2 will lose an equal amount of energy, because the Bethe-Bloch formula for stopping power contains very weak dependence on the mass. But the width and fine structure of the ChR angular distribution in vicinity of the Cherenkov angle are remarkable different for isotopes with different masses, at equal initial relativistic factor (velocity) of isotopes. Indeed, let the isotopes with masses M_1 and M_2 have equal initial relativistic factors $\gamma_{10}=\gamma_{20}=\gamma_0=\dfrac{E_1(0)}{M_1c^2}=\dfrac{E_2(0)}{M_2c^2}$, velocities $v_{10}=v_{20}=v_0$ and initial energies $E_1(0)$ and $E_2(0)$. After a thin layer of radiator Δx both isotopes will lose an equal amount of energy $\Delta E_1=\Delta E_2=-\mathrm{d}E/\mathrm{d}x=S(\gamma_0,v_0)$, according to the Bethe-Bloch formula (9). For isotopes this values are equal at initial layer of radiator Δx. The next layer of radiator they will enter with different relativistic factors, etc.:

$$E_1(\Delta x)=E_1(0)-\Delta E_1=E_1(0)-S(\gamma_0,v_0)\Delta x$$

$$E_2(\Delta x)=E_2(0)-\Delta E_2=E_2(0)-S(\gamma_0,v_0)\Delta x\text{'}$$

$$E_1(0)\neq E_2(0)\Rightarrow E_1(\Delta x)\neq E_2(\Delta x)\Rightarrow \gamma_1(\Delta x)\neq\gamma_2(\Delta x).$$

The change in relativistic factor leads to a change in velocity: $\Delta\beta=\Delta\gamma/\beta\gamma^3$. This, for two isotopes the ratio of changes in velocities is: $\dfrac{\Delta\beta_1}{\Delta\beta_2}=\dfrac{M_2}{M_1}$.

The calculated ChR angular distributions for different isotopes are shown in figure 4:

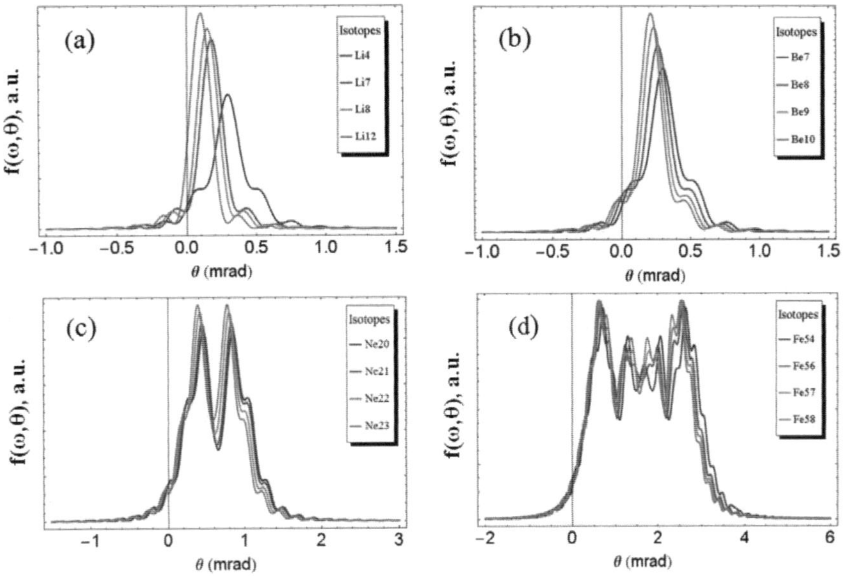

Figure. 4. Angular distribution of ChR from isotopes : a) Li (4, 7, 8, 12), b) Be (7, 8, 9, 10) c) Ne (20, 21, 22, 23), d) Fe (54, 56, 57, 58); radiator – LiF; radiator thickness = 0.25 cm; radiation wave length λ = 390 nm; refractive index n = 1.4; initial energy of isotopes 1000 MeV/u.

Figure 4 clearly shows that the ChR angular distributions for different isotopes of one element are shifted towards each other. This is because the smaller the isotopic mass is the greater changes in velocity occur in radiator; so the final velocity when leaving the radiator for more heavy ions is greater than for smaller ones; and the Cherenkov angle is also greater, since $\theta_c = \arccos\frac{c}{nv}$. That is why the isotopic effect is more noticeable for Li isotopes.

5. Conclusions

In this work a new method for calculation of the ChR angular distribution from (RHI) taking into account stopping in radiator is suggested. Our method is based on the thin radiator approximation [3] and the Bethe-Bloch formula for ionization energy loss. Also the simple formula to estimate the ChR angular distribution width from RHI taking into account energy loss in radiator is obtained. New peculiarities if the ChR angular distribution caused by stopping in radiator were analysed and the results were compared with those calculated using computer code SRIM, which gives more correct values of penetration-dependent velocity. The results of numerical calculations based on equation (8), Bethe-Bloch formula for stopping and thin radiator approximation are very close to those based using SRIM-08.

Now we can answer the question formulated in section 1 of this paper: does the width and structure of the Cherenkov ring contain additional information, e.g. on the ion charge or RHI mass? The answer is – yes, and this statement is proved by numerical calculations.

The experimental studies of predicted effects are possible at existing (GSI) and future (FAIR) accelerators, e.g. using RICH detectors. The predicted new features of ChR allow for: possible applications of Cherenkov detectors (e.g. RICH) to simultaneous reconstruction of the velocity and charge of RHI penetrating through radiator; possible application to measure the masses of isotopes; possible application to measure the stopping power of radiator. For these studies, which should be based on measurements of differential characteristics of ChR, the special optical spectrometer should be designed.

To conclude: only several experiments have been performed until now to observe the ChR from RHI, see [8-10]. An attempt to explain the experimental result was reported in [11], but without detailed analysis. The detailed analysis will be performed by our group and published in separate paper.

Acknowledgements
Yu L Pivovarov thanks Prof. H Geissel, Prof. C Scheidenberger and participants of the NUSTAR Collaboration Seminar at GSI (28 August 2011) for fruitful discussions. The work was partially supported by the Grant of President № MK-2059.2011.2.

References

[1] Tamm I E and Frank I M 1937 *Doklady AN SSSR* **14** 107
[2] Dedrick K G 1952 *Phys. Rev.* **87** 891
[3] Kuzmin E S and Tarasov A V 1993 *JINR Rapid Communications* **4[61]-93** 64
[4] Altapova V R, Bogdanov O V and Pivovarov Yu L 2009 *Nuclear Instruments and Methods in Physics Research.* B **256** 109-113
[5] Bogdanov O V and Pivovarov Yu L 2011 *NuovoCimento.* **V. 034, Issue 04** 1-7
[6] http://www.srim.org
[7] http://www-linux.gsi.de/~weick/atima/
[8] Ruzicka J, Saro S and Zrelov V P, et al. 1996 *Nuclear Instruments and Methods in Physics Research.* A **369** 23-28
[9] Ruzicka J, Saro S, Krupa L, et al. 1999 *Nuclear Instruments and Methods in Physics Research.* A **431** 148-53
[10] Ruzicka J, Hrmo A, Krupa L, et al. 2001 *Vacuum* **63** 591-595
[11] Afanisiev G N and Shilov V M 2000 *J.Phys. D : Appl. Phys* **33** 2931-40

Method of particle energy determination based on measurement of waveguide mode frequencies

A V Tyukhtin
Radiophysics Dep., St Petersburg State University
1 Ul'yanovskaya Str., St Petersburg, 198504, Russia

E-Mail: tyukhtin@bk.ru

Abstract. A new method of determination of charged particle energy is developed. This method consists in use of dependency of waveguide mode frequencies on Lorenz factor of particle. For this method it is principal that the particle bunch generates radiation in waveguide and the mode frequencies depend essentially on the Lorenz factor. Three variants of realization of such technique are considered. The first variant consists in use of a thin dielectric layer in a waveguide. The second variant is based on use of a waveguide containing a system of wires coated with a dielectric material. The third version consists in application of a circular waveguide having a grid wall. For all cases analytical solutions of the problems are obtained (in the case of grid waveguide we use the averaged boundary conditions). Some typical results of computations are given. Advantages and disadvantages of different variants are discussed. It is noted that the grid waveguide can be used as well for generation of microwave radiation.

1. Introduction

Cherenkov radiation is widely used for detection of charged particles [1]. However, as far as we know, there are no Cherenkov detectors using properties of radiation in waveguide structures. Meantime Cherenkov radiation in waveguide is no exotic effect. For example, it is actively used for wakefield acceleration of charged particles [2, 3]. In this technique, a large bunch ("driver") excites Cherenkov radiation (so-called "wakefield") in a waveguide loaded with a dielectric layer, and another (relatively small) bunch is accelerated in this field. Owing to this application, the techniques of Cherenkov radiation generation in waveguides are well-tested in present.

The bunch diagnostics method under consideration consists in use of dependency of the waveguide mode frequencies ω_m on Lorenz factor of the bunch particles $\gamma = \left(1-\beta^2\right)^{-1/2}$ ($\beta = V/c$, V is a bunch velocity, c is a light velocity) [4–11]. Therefore, it is principal that the particle bunch generates Cherenkov radiation in waveguide. This can be achieved with different ways, but it is obvious that the waveguide can not be a simple regular structure with perfectly conductive walls. In the paper [4], the authors have offered to use a cylindrical layer of isotropic nondispersive dielectric in circular waveguide. A disadvantage of such variant is that the dependency of waveguide modes on the Lorenz factor is usually weak, therefore accuracy of Lorenz factor determination will be small especially for the ultra-relativistic case $\gamma >> 1$ (this peculiarity is typical for all known detectors based on Cherenkov phenomenon). However it is possible to overcome partly this imperfection.

We have offered several variants which give essential dependencies $\omega_m(\gamma)$ both for predetermined narrow range of Lorenz factor values and for enough wide range. By our opinion, three versions are perspective for application. One of them is based on use of a thin dielectric layer in a waveguide [9]. Other variant is based on use of a waveguide loading with some metamaterial [5–10], for example, a system of wires coated with a dielectric layer [10]. The third variant is a circular waveguide having a grid wall [11]. Further the peculiarities of these versions are discussed.

2. Thin dielectric layer

In this section we consider the case of a circular waveguide loaded with a simple dielectric layer. This problem is well known in connection with the wakefield acceleration technique [2, 3]. Just this structure has been offered by Poliektov etc. [4] for determination of energy of particles in bunches. Our suggestion consists in using a thin dielectric layer instead a usual relatively thick layer [9].

It is assumed that some dielectric material forms a cylindrical layer in a circular waveguide with a radius a. Vacuum channel has radius b, and thickness of the medium layer is $d = a - b$. The particle bunch possessing a charge q moves along the z-axis with a velocity $\vec{V} = c\beta\vec{e}_z$. The transverse dimension of the bunch is assumed to be negligible, and longitudinal distribution of the charge is determined by the Gaussian function $\exp\left(-\zeta^2/(2\sigma^2)\right)$, where $\zeta = z - Vt$ and σ is much less than the typical wavelength.

It is assumed that the phase velocity of electromagnetic waves in the layer material is less than the charge velocity: $c/\sqrt{\varepsilon} < V$ that is $\varepsilon\beta^2 > 1$, where ε is a permittivity of material (dissipation and dispersion are assumed to be negligible, permeability is $\mu = 1$). Expressions for the components of the wave electromagnetic field (so called "wakefield") are well known [12]. In vacuum channel, they can be written in the following form (cylindrical coordinates ρ, ϕ, ζ are used):

$$E_z = -4q\left(1-\beta^2\right)c^{-2}\beta^{-2}\,\theta(-\zeta)\,\mathrm{Re}\left\{\sum_{m=1}^{\infty}\omega_m \operatorname*{Res}_{\omega_m}\alpha(\omega)\,\mathrm{I}_0\left(k(\omega_m)\rho\right)\exp\left[-\omega_m^2\sigma^2\left(2V^2\right)^{-1} + \mathrm{i}\omega_m\zeta V^{-1}\right]\right\},$$

$$E_\rho = -4qc^{-1}\beta^{-1}\,\theta(-\zeta)\,\mathrm{Im}\left\{\sum_{m=1}^{\infty}k(\omega_m)\operatorname*{Res}_{\omega_m}\alpha(\omega)\,\mathrm{I}_1\left(k(\omega_m)\rho\right)\exp\left[-\omega_m^2\sigma^2\left(2V^2\right)^{-1} + \mathrm{i}\omega_m\zeta V^{-1}\right]\right\},$$

$$H_\phi = \beta E_\rho, \tag{1}$$

where

$$\alpha(\omega) = \frac{s(\omega)\mathrm{K}_1\left(k(\omega)b\right)\psi_1\left(s(\omega)\right) + \varepsilon\, k(\omega)\mathrm{K}_0\left(k(\omega)b\right)\psi_0\left(s(\omega)\right)}{s(\omega)\mathrm{I}_1\left(k(\omega)b\right)\psi_1\left(s(\omega)\right) - \varepsilon\, k(\omega)\mathrm{I}_0\left(k(\omega)b\right)\psi_0\left(s(\omega)\right)}, \tag{2}$$

$$\psi_0(s) = \mathrm{J}_1(sb)\mathrm{N}_0(sa) - \mathrm{J}_0(sa)\mathrm{N}_1(sb), \qquad \psi_1(s) = \mathrm{J}_0(sb)\mathrm{N}_0(sa) - \mathrm{J}_0(sa)\mathrm{N}_0(sb), \tag{3}$$

$$k(\omega) = \omega V^{-1}\sqrt{1-\beta^2}, \qquad s(\omega) = \omega V^{-1}\left(\varepsilon\beta^2 - 1\right), \tag{4}$$

$$\theta(-\zeta) = 1 \text{ for } \zeta < 0, \text{ and } \theta(-\zeta) = 0 \text{ for } \zeta > 0.$$

Here $\operatorname*{Res}_{\omega_m}\alpha(\omega)$ are residues of function $\alpha(\omega)$ in poles $\omega = \omega_m$, $\mathrm{J}_n(\xi)$ and $\mathrm{N}_n(\xi)$ are Bessel and Neumann functions accordingly, $\mathrm{I}_n(\xi)$ and $\mathrm{K}_n(\xi)$ are modified Bessel and Neumann functions correspondingly. The mode frequencies ω_m are determined by the following dispersion equation:

$$s(\omega)\mathrm{I}_1\left(k(\omega)b\right)\psi_1\left(s(\omega)\right) - \varepsilon\, k(\omega)\mathrm{I}_0\left(k(\omega)b\right)\psi_0\left(s(\omega)\right) = 0. \tag{5}$$

Some results of computations of roots of this equation are shown in figures 1, 2. One can see that dependence of $\nu_m = \omega_m/(2\pi)$ on γ becomes stronger with decreasing in the layer thickness d.

For example, one can find from figure 2 that for the range $20 < \gamma < 30$ the relative error of Lorenz factor $\Delta\gamma/\gamma$ has the same order as the relative error of frequency $\Delta\nu_1/\nu_1$ in the case

$d = 2$ mm ($b = 98$ mm). At the same time, these errors differ radically $(\Delta\gamma/\gamma >> \Delta\nu_1/\nu_1)$ for $d > 10$ mm. Thus the thin layer provides essential advantage in comparison of the thick one. Naturally, there is some limitation on value of d because the amplitude of the mode must be enough large for measurement. However computations show that this limitation is not essential for typical bunches with charges > 1 pC.

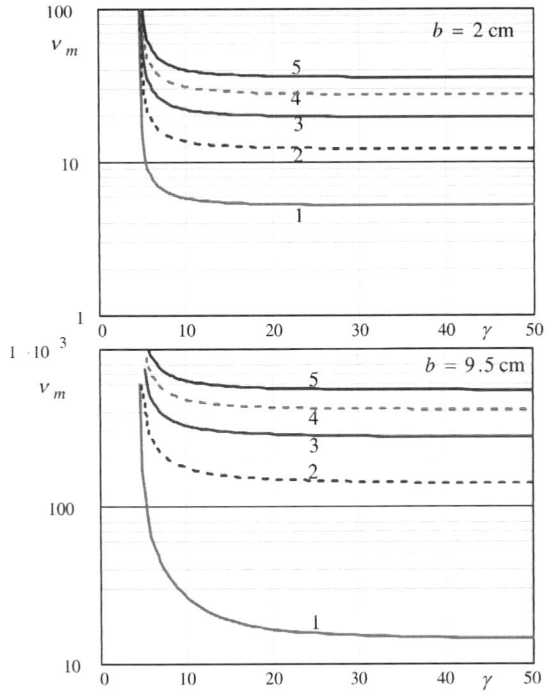

Figure 1. Dependence of frequencies ν_m (GHz) on γ in the case of non-dispersive isotropic dielectric layer for thick layer (above) and thin one (below). Channel radius b and the mode numbers m are indicated in pictures; $a = 10$cm, $\varepsilon = 1.05$.

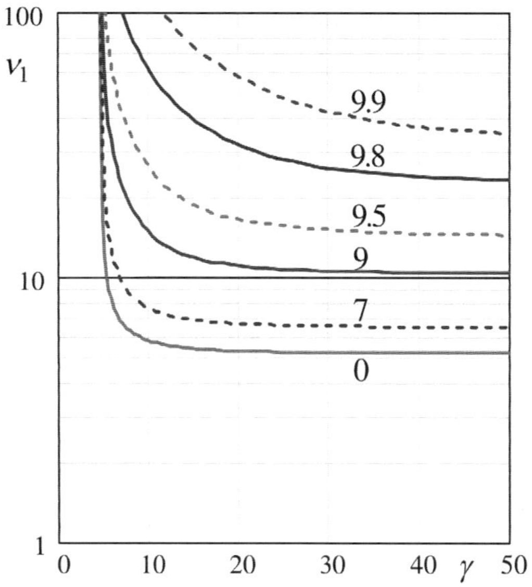

Figure 2. Dependence of the 1st mode frequency ν_1 (GHz) on γ for the channel radius values indicated near the curves (cm); $a = 10$cm, $\varepsilon = 1.05$.

The shortcoming of this technique consists in its limitation on γ. For example, if $b = 98$ mm the range $\gamma \le 10$ corresponds to frequencies $\nu_1 > 50$ GHz (figure 1). The typical bunches with $\sigma \ge 3$mm generate mode with very small amplitude because of the exponential factor in (1). On the other hand for $\gamma > 30$ the precision of γ is essentially lower in comparison with the precision of the frequency measurement. These defects can be partially overcome with the help of different structures.

3. The wire metamaterial

Besides the layer of simple dielectric, different variants of waveguide structures can be used for determination of particle energy. Some of them are connected with use of metamaterials which are artificial periodic structures with small periods. Such structures can be characterized by macroscopic parameters such as permittivity and permeability. One of known metamaterials is a system of parallel wires with coating [13, 14].

Let us consider a periodic system of perfect cylindrical conductors which have a nonconductive cylindrical coating with permittivity ε_1 and permeability μ_1. A radius of conductor is r_0, and a radius of coating is r_{d}. The surrounding medium is characterized by permittivity ε_2 and permeability μ_2. Let the z-axis is parallel to wires. The periods of system along x- and y-axis are equal to d. It is assumed that the following conditions are fulfilled:

$$r_{\mathrm{d}} << d << \Lambda , \tag{6}$$

where $\Lambda \equiv \min(\lambda, \Delta)$, λ is a typical wavelength, Δ is a distance of variation of an "incident" field (i.e. the field in the case of absence of the structure).

For determination of "effective" permittivity of such structures, we have developed a new method [14] which is similar to method of averaged boundary conditions for planar structures [15]. Result of approximate analytical calculation of permittivity tensor can be written in the following form [14]:

$$\hat{\varepsilon} = \begin{pmatrix} \varepsilon_\perp & 0 & 0 \\ 0 & \varepsilon_\perp & 0 \\ 0 & 0 & \varepsilon_\parallel \end{pmatrix}, \qquad \varepsilon_\perp = \varepsilon_{\mathrm{c}}, \qquad \varepsilon_\parallel = \varepsilon_{\mathrm{c}}\left\{1 - \frac{\omega_{\mathrm{p}}^2}{\omega^2 - \chi c^2 k_z^2/(\varepsilon_2\mu_2)}\right\}, \tag{7}$$

where

$$\omega_{\mathrm{p}}^2 = \frac{2\pi c^2}{d^2 \varepsilon_{\mathrm{c}} \mu_2 \left[\ln\left(\frac{d}{r_{\mathrm{d}}}\right) + \frac{\mu_1}{\mu_2}\ln\left(\frac{r_{\mathrm{d}}}{r_0}\right) - 1.0487\right]}, \qquad \chi = 1 + \frac{(\varepsilon_2/\varepsilon_1 - \mu_1/\mu_2)\ln(r_{\mathrm{d}}/r_0)}{\ln(d/r_{\mathrm{d}}) + \mu_1\mu_2^{-1}\ln(r_{\mathrm{d}}/r_0) - 1.0487}, \tag{8}$$

ε_{c} is some constant which is close to ε_2 (it can be estimated more exactly with the Garnett's formula). "Effective" permeability of structure is some constant μ_{c} whish is close to μ_2. One can see that this "medium" has both frequency and spatial dispersion. The role of spatial dispersion is connected with the parameter χ. We can vary this parameter owing to varying magnitudes of $\varepsilon_1/\varepsilon_2$, μ_1/μ_2, and r_{d}/r_0. Increase in any of these parameters results in decreasing the role of spatial dispersion [14].

We consider a circular waveguide containing the structure under consideration (the main axis of structure (z-axis) coincides with the waveguide axis). It is assumed that $\mu_1 = \mu_2 = 1$. As above, a charged particle bunch moves along the z-axis with velocity $\vec{V} = c\beta\vec{e}_z$, the transverse dimension of the bunch is negligible, and longitudinal distribution of the charge is determined by the Gaussian function. Using analysis basing on the mode expansion of the wave field one can obtain the following expression for the mode frequencies:

$$\omega_m \equiv 2\pi\nu_m = \beta\sqrt{\frac{c^2\kappa_m^2}{a^2(\beta^2\varepsilon_{\mathrm{c}} - 1)} + \frac{\varepsilon_{\mathrm{c}}\omega_p^2}{\beta^2\varepsilon_{\mathrm{c}} - \chi}}, \tag{9}$$

where a is a waveguide radius, κ_m are roots of Bessel function ($\mathrm{J}_0(\kappa_m) = 0$).

In the particular case of wires without coating ($\chi = 1$) the modes frequencies are real only for $\beta^2\varepsilon_{\mathrm{c}} > 1$, i.e. Cherenkov radiation is generated only for superlight speed of charge as in an usual medium. In this situation we do not have preferences as compared with the ordinary media. Principally different result takes place in the case of wires with coating. As earlier the ordinary modes are generated for $\beta^2\varepsilon_{\mathrm{c}} > 1$. However there are additional ("anomalous") modes which are generated under condition $\chi < \beta^2\varepsilon_{\mathrm{c}} < 1$. Frequencies of these modes decrease with increase in the mode number and with increase in γ (figures 3, 4). One can see that there are strong dependence $\nu_m(\gamma)$ in some narrow range of γ. In this range the relative error of Lorenz factor $\Delta\gamma/\gamma$ can be much less than the relative error of measurement of frequency $\Delta\nu_m/\nu_m$. Thus, this structure is convenient for high-precision

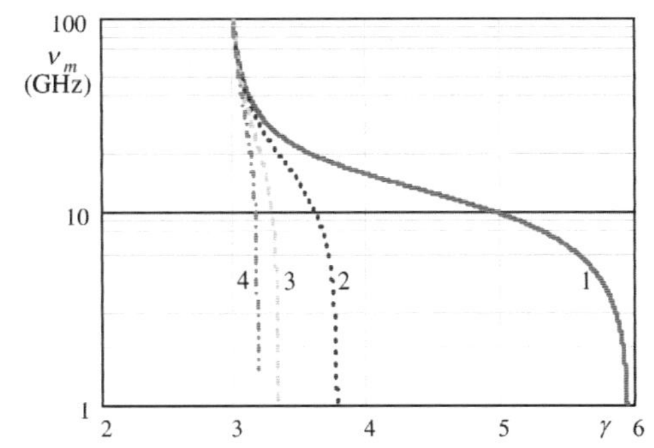

Figure 3. The mode frequencies depending on γ for the medium parameters $\varepsilon_2 = \mu_2 = \mu_1 = 1$, $\varepsilon_c = 1.0027$, $\nu_p = 4.17\,\text{GHz}$, $\chi = 0.887$ corresponding to the wire structure with $d = 10$ mm, $r_0 = 0.2$ mm, $r_d = 0.3$ mm, $\varepsilon_1 = 5$; the mode numbers are indicated in figure.

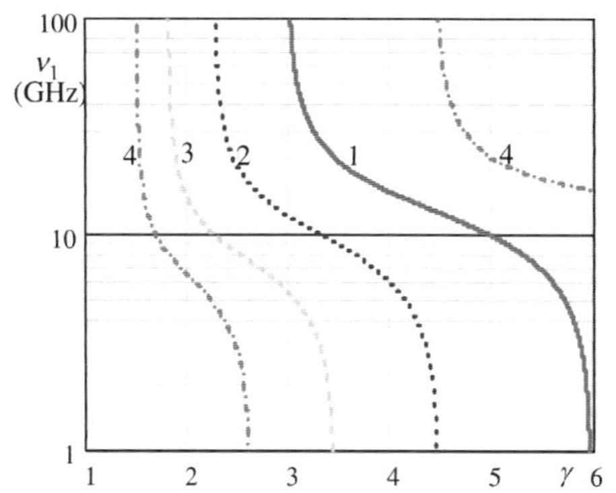

Figure 4. The 1st mode frequency depending on γ for $r_d = 0.3$ mm (curve 1), $r_d = 0.4$ mm (2), $r_d = 0.6$ mm(3), $r_d = 1$ mm (4); other parameters are the same as in figure 3.

determination of relatively small values of γ in some predetermined range. However it is not good for wide range of γ and for big values of γ.

4. The grid waveguide

Here we consider a structure that allows determining relatively big values of γ: a circular waveguide which has a grid wall with rectangular cells (figure 5). As usual, the z-axis coincides with the waveguide axis. The grid is characterized by period d_z for "z-wires" (which are parallel to the z-axis) and period d_ϕ for "ϕ-wires" (which are orthogonal to the z-axis).

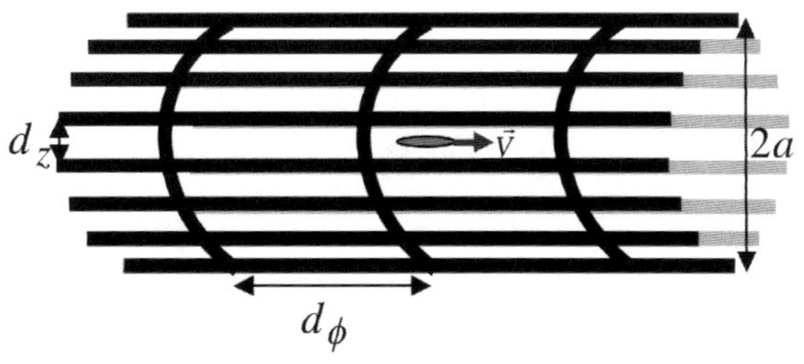

Figure 5. Geometry of the problem.

The grid wires are perfect, and the following conditions are assumed to be fulfilled:

$$r_0 << d_{z,\phi} << \Lambda \equiv \min(a, c/\omega)\,. \tag{10}$$

Owing to these inequalities we can use method of the averaged boundary conditions (ABC) [15]. In the case under consideration (where only TM-field is generated), the ABC has the form

$$E_{\omega z}\Big|_{r=\pm a} = -\frac{i\omega d_z}{2\pi c}\ln\left(\frac{d_z}{2\pi r_0}\right)\left(1+\frac{c^2}{\omega^2\delta}\frac{\partial^2}{\partial z^2}\right)\left(H_{\omega\phi}\Big|_{r=a+0} - H_{\omega\phi}\Big|_{r=a-0}\right), \tag{11}$$

where $\delta = (1 + d_\phi / d_z + \kappa)/(d_\phi / d_z + \kappa)$. Parameter κ is charged with contact between conductors in points of their intersections. In case of perfect contact $\kappa = 0$, and $\delta = (d_z + d_\phi)/d_\phi$, i.e. $\delta = 1$ for grid from z-wires only ($d_\phi = \infty$), $\delta = 2$ for grid with square cells ($d_z = d_\phi$), and $\delta \approx d_z / d_\phi >> 1$ in the case $d_\phi << d_z$.

Omitting derivation we write expressions for wave field ("wakefield") in the case of Gaussian bunch moving along the waveguide axis. The non-zero components of wakefield (existing only behind the charge) are

$$\begin{aligned} E_r^W &= \frac{4q\gamma}{a^2} \frac{k_0 a}{W(k_0 a)} \exp\left(-\frac{\omega^2\sigma^2}{2V^2}\right) \sin\left(\frac{\omega_0\zeta}{V}\right) \begin{Bmatrix} \mathrm{K}_0^2(k_0a)\mathrm{I}_1(k_0r) \text{ for } r<a \\ -\mathrm{K}_0(k_0a)\mathrm{I}_0(k_0a)\mathrm{K}_1(k_0r) \text{ for } r>a \end{Bmatrix}, \\ E_z^W &= \frac{4q}{a^2} \frac{k_0 a}{W(k_0 a)} \exp\left(-\frac{\omega^2\sigma^2}{2V^2}\right) \cos\left(\frac{\omega_0\zeta}{V}\right) \begin{Bmatrix} \mathrm{K}_0^2(k_0a)\mathrm{I}_0(k_0r) \text{ for } r<a \\ \mathrm{K}_0(k_0a)\mathrm{I}_0(k_0a)\mathrm{K}_0(k_0r) \text{ for } r>a \end{Bmatrix}, \\ B_\phi^W &= \beta E_r^W, \end{aligned} \quad (12)$$

where $W(x) = \mathrm{I}_1(x)\mathrm{K}_0(x) - \mathrm{I}_0(x)\mathrm{K}_1(x)$, $k_0 = \omega_0 V^{-1}\gamma^{-1}$, $\gamma = (1-\beta^2)^{-1/2}$, $\beta = V/c$, $\zeta = z - Vt$. Here k_0 is a real positive solution of the following dispersion equation:

$$\mathrm{I}_0(ka)\mathrm{K}_0(ka) = \chi_g, \quad (13)$$

where $\chi_g = \frac{\delta\beta^2 - 1}{\delta} \frac{d_z}{2\pi a} \ln\left(\frac{d_z}{2\pi r_0}\right)$. This equation can have only a single real root $k = k_0$. This root exists under condition $\chi_g > 0$, that is $\beta > \delta^{-1/2}$. Thus, propagating mode can be generated only in the case $\delta > 1$. It means that generation of radiation (wakefield) is possible only in the case of grid possessing both z-wires and ϕ-wires. This fact can be seemed wonderful because only current in z-wires are excited. However the distribution of the charge on the grid depends on the orthogonal ϕ-wires too. This circumstance is reflected in the ABC (11) owing to presence of the parameter $\delta \neq 1$.

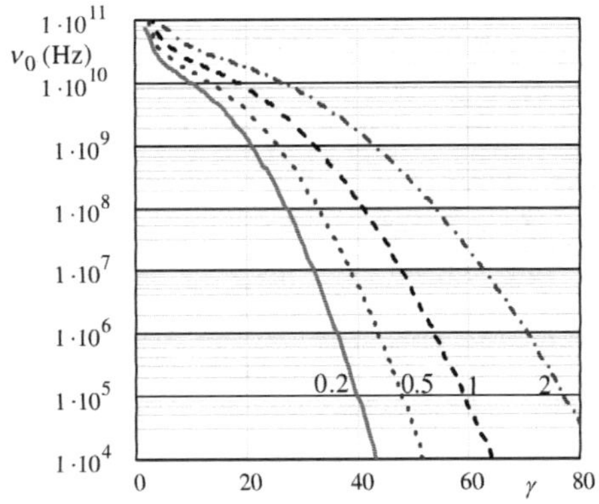

Figure 6: The mode frequency depending on γ; $a = 3$ cm, $r_0 = 0.5$ mm, $d_z = 2\pi a/38 \approx 5$ mm; magnitudes of d_ϕ(cm) are given close to the curves.

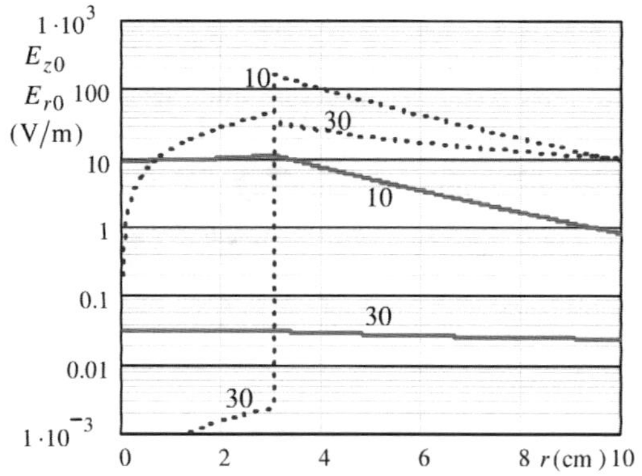

Figure 7: Amplitudes of components E_z^W (solid curves) and E_r^W (dotted ones) depending on r; $q = 1$pC, $\sigma = 3$mm, $a = 3$ cm, $r_0 = 0.5$ mm, $d_z = d_\phi = 2\pi a/38 \approx 5 =$ mm, magnitudes of γ are indicated near the curves.

One can obtain simple approximations for the propagating mode frequency in two particular cases:

$$\omega_0 \approx V\gamma/(2a\chi_g) \text{ for } \chi_g << 1 \text{ (not very large } \gamma\text{)}, \tag{14}$$

$$\omega_0 \approx \frac{2c}{a}\sqrt{\gamma^2-1}\,\exp(-C-\chi_g) \text{ for } \chi_g > 1\ (\gamma >> 1), \tag{15}$$

where $C = 0.577...$ is Euler constant. According to (15) the dependency of ω_0 on γ for large values of γ can be significant due to factor $\exp(-\chi_g)$ strongly decreasing with increase in γ. This fact is especially attractive for determination of particle energy.

Figures 6–9 show some results of computation of wave field. One can see that dependency of the mode frequency on Lorentz factor is essential for enough large range of magnitudes of γ (figure 6). For waveguide with radius of the order of several centimeters this range can be several decades. One can show that relative accuracy of determination of Lorentz factor $\Delta\gamma/\gamma$ has the same order as relative accuracy of measurement of frequency $\Delta\nu/\nu$ in some range of γ. The radial component of electric field on the outward surface of waveguide is the most convenient for measurement because it is much more than longitudinal component (figure 7). Dependency of the radial component on γ is shown in figure 8. One can see that its values for bunch with the charge of 1 pC is enough large for measurements. Figure 9 shows typical wakefield. Note that this computation was performed with account of finite conductivity of wires. Consequently, exponentially decrease of wakefield with increase of $|\zeta|$ occurs.

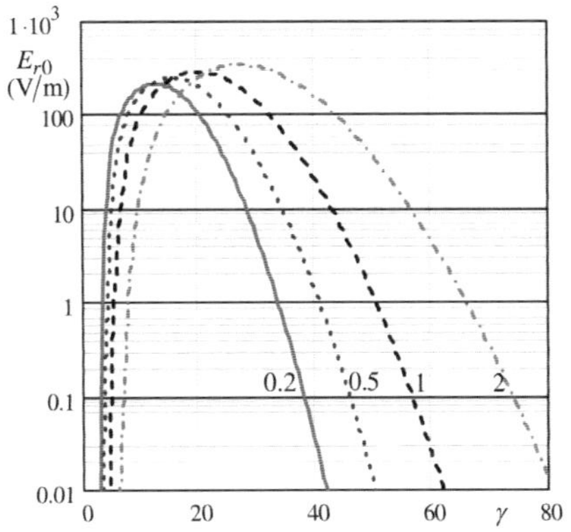

Figure 8. Amplitude of the component E_r^W on the outward surface of waveguide ($r = a + 0$) depending on γ; magnitudes of d_ϕ (cm) are indicated near the curves, other parameters are the same as in figure 7.

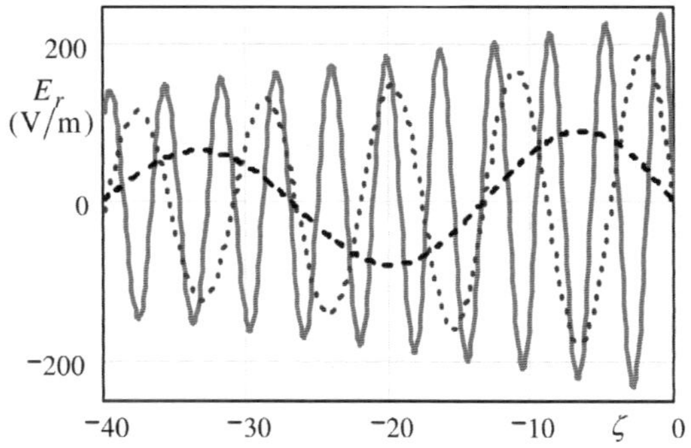

Figure 9. Component E_r^W on the outward surface of waveguide depending on the distance $\zeta = z - Vt$ for $\gamma = 15$ (solid curve), $\gamma = 20$ (dotted curve), $\gamma = 25$ (dashed curve). Conductivity of wires is $5 \cdot 10^7\,(\text{om m})^{-1}$ (copper), other parameters are the same as in figure 7.

Note that the structure under consideration can be examined as generator of monochromatic microwave radiation too. It is important that radiation has single propagating mode in our model where the averaged boundary conditions are used. Really, radiation has some spectrum of shorter waves with wavelengths $\leq d_\phi$. However these waves can be filtered out, and we will obtain almost monochromatic radiation with frequency ω_0 that can be varied with variation of the bunch velocity.

5. Conclusion

We have considered different variants of realization of new method of particle energy determination. This method consists in use of dependency of waveguide mode frequencies ω_m on Lorenz factor γ. For this method it is principal that the mode frequencies depend essentially on γ.

The first version consists in use of a thin dielectric layer. Decrease of the layer thickness results in increase of dependency of ω_m on γ and in some extension of the Lorenz-factor range which is suitable for measurement. The important advantage of this structure is that it is relatively simple for realization. But this method has essential limitations on γ (both above and below).

The second variant is based on use of a waveguide containing a system of parallel wires coated with a dielectric material. Periods of the structure are assumed to be small, therefore it can be considered as some "medium" (metamaterial). In the waveguide with such filling, the bunch can generate not only the ordinary modes but also additional ("anomalous") modes. These modes allow the energy determination with high accuracy. But this is possible only for some narrow range of γ, and the values of γ can not be large.

The third version consists in application of a circular waveguide having a grid wall with small rectangular cells. Analytical solution of the problem has been obtained with use of the averaged boundary conditions. It has been shown that such structure gives essential dependency of frequency on Lorentz factor in wide range of its values. As well this structure can be used for generation of monochromatic microwave radiation with tunable frequency depending on the particle energy.

Acknowledgments
This research was supported by the Saint Petersburg State University.

References

[1] Zrelov V P 1970 *Vavilov-Cherenkov Radiation in High-Energy Physics* (Jerusalem: Israel Program for Scientific Translations)
[2] Power J G, Conde M E, Gai W, Kanareykin A, Konecny R and Schoessow P 2000 *Phys. Rev. ST-AB* **3** 101302
[3] Kanareykin A 2010 *J. Phys.: Conf. Ser.* **236** 012032
[4] Poliektov V V, Vetrov A A, Trukhanov K A and Shvedunov V I 2008 *Instruments and Experimental Techniques* **51** 191
[5] Tyukhtin A V, Antipov S P, Kanareykin A and Schoessow P 2007 *Proc. Particle Accelerator Conf. PAC'07* (USA, Albuquerque, 2007) p 4156
[6] Tyukhtin A V 2008 *Tech. Phys. Lett.* **34** 884
[7] Tyukhtin A V 2009 *Tech. Phys. Lett.* **35** 263
[8] Tyukhtin A V, Schoessow P, Kanareykin A and Antipov S 2009 *AIP Conf. Proc.* **1086** 604
[9] Tyukhtin A V, Antipov S P, Kanareykin A and Schoessow P 2009 *Proc. Particle Accelerator Conf. PAC'09* (Canada, Vancouver, 2009) p 4033
[10] Tyukhtin A V, Doil'nitsina E G and Kanareykin A 2010 *Proc. Int. Particle Accelerator Conf. IPAC'10* (Japan, Kyoto, 2010) p 1071
[11] Tyukhtin A V 2011 *Proc. Int. Particle Accelerator Conf. IPAC'11* (Spain, San-Sebastian, 2011) p 1302
[12] Bolotovskiy B M 1961 *Physics – Uspekhi* **75** 295
[13] Demetriadou A and Pendry J B 2008 *J. Phys.: Condens.Matter* **20** 295222
[14] Tyukhtin A V and Doilnitsina E G 2011 *J. Phys. D: Applied Physics* **44** 265401
[15] Kontorovich M I, Astrakhan M I, Akimov V P and Fersman G A 1987 *Electrodynamics of grid structures* (Moscow: Radio i svyaz'; in Russian).

Self-amplified Cherenkov radiation from a relativistic electron in a waveguide partially filled with a laminated material

L Sh Grigoryan[1,3], **A R Mkrtchyan**[1], **H F Khachatryan**[1], **S R Arzumanyan**[1] **and W Wagner**[2]
[1]NAS Institute of Applied Problems in Physics, 25 Hr. Nersessian Str., 0014 Yerevan, Armenia
[2]Helmholz-Zentrum Dresden-Rossendorf, POB 510119, 01314 Dresden, Germany

E-mail: levonshg@mail.ru

Abstract. The radiation from a relativistic electron uniformly moving along the axis of cylindrical waveguide filled with laminated material of finite length is investigated. Expressions for the spectral distribution of radiation passing throw the transverse section of waveguide at large distances from the laminated material are derived with no limitations on the amplitude and variation profile of the layered medium permittivity and permeability. Numerical results for layered material consisting of dielectric plates alternated with vacuum gaps are given. It is shown that at a special choice of problem parameters, Cherenkov radiation generated by the relativistic electron inside the plates is self-amplified. The visual explanation of this effect is given and a possible application is discussed.

1. Introduction

The presence of matter essentially influences the radiation from a relativistic particle: Cherenkov radiation, transition radiation etc. [1–9]. The effects of interest arise in stratified and periodic media (see e.g. [2, 9]). The interfaces of media are widely used to control the radiation flow emitted by various systems. The conjuncture of three factors: the influence of material, the periodicity of medium and the presence of interfaces, provides a promising research area for development of high power controlled radiation sources.

An investigation of such a kind was carried out for a charged particle inside an infinite waveguide completely filled with a layered (spatially periodic) medium [10]. However, the case of Cherenkov radiation (CR) generation was not considered in [10] and has been partially investigated in [11, 12] as the emission of CR from relativistic particles inside an infinitely long waveguide filled with semi-infinite layered medium. It was shown that CR may self-amplify due to the presence of a waveguide and of a periodical medium. In the present work a more realistic problem of filling an infinitely long waveguide with periodic medium of finite length is solved, and visual explanation of obtained numerical results is given.

In [13] the authors reported the first direct observation of narrow-band terahertz coherent CR driven by subpicosecond electron bunches travelling along the axis of a hollow cylindrical dielectric-

[3] To whom any correspondence should be addressed.

lined waveguide (see also [14]). The results obtained both in [11, 12] and in the present work permit an increase in the power of coherent CR from subpicosecond electron bunches observed in [13, 14].

2. Formulation of the problem

Let us consider a relativistic particle of q charge uniformly moving with v velocity along the axis of an ideal and infinitely long cylindrical waveguide of R radius. We shall assume that the finite part of the waveguide is filled with a laminated material that weakly absorbs the radiation. The case of a waveguide filled with a stack of three plates is shown in Fig.1.

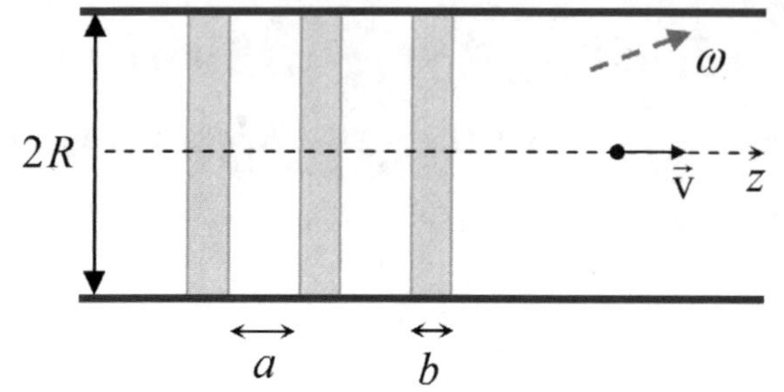

Figure 1. The radiation from a charged particle moving along the axis of cylindrical waveguide filled with a stack of three plates.

Now direct the Z axis of the cylindrical system of coordinates r, φ, z along the waveguide axis and assume that the charge travels from vacuum ($z < z_1$) to the laminated material ($z_1 < z < z_2$) and then from laminated material again to vacuum ($z > z_2$).

We shall assume that the permittivity ε_1 and permeability μ_1 of the layered medium are independent of the transverse coordinates r, φ, but alternate along the direction of particle motion according to an arbitrary law:

$$\varepsilon_1(z+l) = \varepsilon_1(z), \qquad \mu_1(z+l) = \mu_1(z) \tag{1}$$

(l is the period of layered medium). The permittivity ε and permeability μ of the system may be written in the form

$$\begin{aligned} &\varepsilon(z) = \varepsilon_1(z), \qquad \mu(z) = \mu_1(z) \qquad \text{when } z_1 < z < z_2 \qquad \text{(range 1)} \\ &\varepsilon(z) = 1, \qquad \mu(z) = 1 \qquad \text{when } z < z_1 \text{ or } z > z_2 \qquad \text{(range 2).} \end{aligned} \tag{2}$$

We have studied the energy of radiation

$$W = \sum_{n=1}^{\infty} \int_{\omega_n}^{\infty} I_n(\omega)\, d\omega = \sum_n W_n \tag{3}$$

emitted during the whole period of the particle motion and passing through the transverse section of waveguide in the $z = z_3$ point at large distances from the laminated material ($z_3 - z_2 >> l$). In (3) I_n and W_n are the spectral distribution and the energy of radiation at the n-th mode of waveguide respectively.

The aim of the paper is the identification, analysis and clear explanation of cases when the waveguide and the periodical structure of layered material jointly strongly influence the spectral distribution of CR from a particle.

3. The stages of analytical calculations

Taking into account the azimuthal symmetry of the problem and making the Fourier transform

$$f_\omega = \frac{1}{2\pi}\int f(t)\exp(i\omega t)\mathrm{d}t\,, \tag{4}$$

one may reduce the set of Maxwell equations to the solution of one equation [10]

$$[\varepsilon\frac{\mathrm{d}}{\mathrm{d}z}\left(\frac{1}{\varepsilon}\frac{\mathrm{d}}{\mathrm{d}z}\right)+\frac{\omega^2}{c^2}\varepsilon\mu-\frac{\alpha_n^2}{R^2}]A_n = \frac{\varepsilon}{\mathrm{v}}\frac{\mathrm{d}}{\mathrm{d}z}\left(\frac{1}{\varepsilon}\exp(i\omega z/\mathrm{v})\right)-i\frac{\omega}{c^2}\varepsilon\mu\exp(i\omega z/\mathrm{v})\,, \tag{5}$$

that determines the longitudinal component $E_{z\omega}$ of electric field strength by means of formula

$$\varepsilon E_{z\omega}(r,z)=\sum_{n=1}^{\infty}\frac{2q}{\pi R^2 J_1^2(\alpha_n)}J_0(\tfrac{\alpha_n}{R}r)A_n(z)\,. \tag{6}$$

Here α_n is the n-th root of the zeroth order Bessel function: $J_0(\alpha_n)=0$, and the coefficient $2q/\pi R^2 J_1^2(\alpha_n)$ is introduced for ease in treatment ($J_1(x)$ is the first order Bessel function). In these expressions $\varepsilon \equiv \varepsilon_\omega$ and $\mu \equiv \mu_\omega$ are Fourier transforms of the permittivity and permeability (with due regard for the frequency dispersion, the contribution of spatial dispersion being assumed negligibly small). We suppose that $\varepsilon_{1\omega}$ and $\mu_{1\omega}$ are complex valued functions for taking into account the absorption of radiation by the laminated material. Thus, (5) is the main equation of the problem under consideration.

In a hollow part of waveguide ($z > z_2$) the plane waves propagate:

$$A_n(z) = A_n^q(z)+\frac{i}{\omega}a_2\exp(ik_2 z)\,, \qquad k_2=\sqrt{\omega^2/c^2-\alpha_n^2/R^2}\,. \tag{7}$$

Here the first summand describes the known field of charge inside the hollow waveguide, and the 2-nd summand (a_2 is a dimensionless quantity) describes the free field (the radiation field), if the following condition is observed

$$\omega > \omega_n = \alpha_n c/R\,. \tag{8}$$

Far from the interface $z=z_2$ of the laminated material with vacuum the field of produced radiation is «separated» from the field of charged particle and freely propagates inside the waveguide along the positive sense of the axis Z. The spectral distribution of this radiation over the cross section of waveguide on its n-th bandwidth is determined by means of well-known formula

$$I_n(\omega)=\frac{4q^2}{\pi\omega}\frac{k_2|a_2|^2}{\alpha_n^2 J_1^2(\alpha_n)}\,. \tag{9}$$

For calculation of amplitude a_2 it is sufficient to compare (7) with the complete solution of (5) that is valid for all $-\infty < z < \infty$. Such solution may be obtained using the method of Green functions. It is convenient as it is based on the solution of homogeneous equation corresponding to (5). Inside the laminated material that equation transforms to

$$[\varepsilon_1\frac{\mathrm{d}}{\mathrm{d}z}\left(\frac{1}{\varepsilon_1}\frac{\mathrm{d}}{\mathrm{d}z}\right)+\frac{\omega^2}{c^2}\varepsilon_1\mu_1-\frac{\alpha_n^2}{R^2}]G(z)=0\,. \tag{10}$$

According to the Bloch theorem an arbitrary solution of (10) may be presented as a superposition

$$G(z)=c_1B_1(z)+c_2B_2(z) \qquad \text{(range 1)} \tag{11}$$

of Bloch waves

$$B_m(z)=w_m(z)\exp(i\mu_m k_1 z)\,, \quad \text{where } w_m(z+l)=w_m(z) \text{ and } \mu_m=\begin{cases}-1 & \text{when } m=1\\ +1 & \text{when } m=2\end{cases} \tag{12}$$

(travelling waves $\exp(\mp ik_1 z)$, modulated in the period of the layered medium). Here the quasiwave-number $k_1(\omega)$ is determined with an accuracy to the sign. To fix the sign we shall assume that

$$k_1 = k_1' + ik_1'', \qquad \text{where} \qquad \omega k_1''(\omega) \geq 0 \tag{13}$$

(the imaginary part $k_1'' \neq 0$, because the allowance for absorption of radiation by the material of layered medium is made).

It is known (see e.g., [15]) that one can represent the Bloch functions as sums of two arbitrary independent solutions $u_{1;2}(z)$ of equation (10):

$$B_m(z) = p_{1;m}u_1(z) - p_{2;m}u_2(z). \tag{14}$$

We shall need the following expressions

$$p_{1;m} = u_2(l+0) - \delta^{\mu_m}u_2(+0)$$
$$p_{2;m} = u_1(l+0) - \delta^{\mu_m}u_1(+0) \qquad \text{where} \qquad \delta = \exp(i\,k_1 l) \tag{15}$$

for expansion coefficients in (14). Using some arbitrariness in the choice of $u_{1;2}(z)$ we can simplify the calculations.

4. The general formula and its special cases

Omitting the intermediate calculations we now give the resultant expression

$$a_2 = \frac{\omega}{\mathrm{v}k_2\gamma_-}\left[\eta_2\xi_2 - \eta_1\xi_1 + (\gamma - \sigma)(\varepsilon_1^{-1} - 1) + \tfrac{1}{2}(1 - \tfrac{\mathrm{v}^2}{c^2})\left(\frac{\gamma_- - 2\sigma}{1 - k_2\mathrm{v}/\omega} + \frac{1 - \delta^{2N}}{1 + k_2\mathrm{v}/\omega}\right)\right] \tag{16}$$

for the case when the first range contains $N = 1;2;3...$ periods of layered medium: $z_2 - z_1 = N \cdot l$. In (16)

$$\xi_m = i\omega/\mathrm{v}\int_{+0}^{l+0}(\varepsilon_1^{-1} - \mu_1\mathrm{v}^2/c^2 + i\,\mathrm{v}\dot{\varepsilon}_1/\omega\varepsilon_1^2)\,u_m \exp(i\omega\, z/\mathrm{v})\,dz \tag{17}$$

and then

$$\eta_m = p_{m;2}B_{2+}^{-1}\frac{1 - \Delta_2^{-N}}{1 - \Delta_2} - p_{m;1}B_{1+}^{-1}\delta^{2N}\frac{1 - \Delta_1^{-N}}{1 - \Delta_1},$$
$$\gamma = B_1/B_{1+} - \delta^{2N}B_2/B_{2+}, \qquad \gamma_- = B_{1-}/B_{1+} - \delta^{2N}B_{2-}/B_{2+},$$
$$\sigma \cdot \Delta_1^N = B_1/B_{1+} - B_2/B_{2+}, \qquad \Delta_m = \delta^{\mu_m}\exp[i\,\omega\, l/\mathrm{v}] \qquad B_{m\pm} = B_m \pm \dot{B}_m/ik_2\varepsilon_1. \tag{18}$$

Functions $\varepsilon_1(z)$, $B_m(z)$ and $B_{m\pm}(z)$ in (16) and (18) shall be taken in $z = 0$ point, the point over the function implies the differentiation with respect to its argument. In (16) there are no limitations on the amplitude and variation profile of $\varepsilon_1(z)$ and $\mu_1(z)$.

Equations (16) and (18) are simplified when the waveguide is filled with weakly absorbing semi-infinite layered medium ($z_1 \to -\infty$), as $\delta^N \to 0$ when $N \to \infty$. An appropriate expression for a_2 (in other equivalent form) is given in [11]. In the particular case of $\varepsilon_1(z)$, $\mu_1(z) = const$ the equation (16) describes the radiation field of particle passing through the dielectric plate inside the waveguide [16].

5. Numerical results based on (16)

Below we shall consider a special, but highly advantageous case of a laminated material consists of N plates (each of b length) interleaved with vacuum gaps (the length of a gap is a, figure 1). The corresponding expression for a_2 is given in [17], but here the case of CR has not been investigated. Now investigate CR generated inside a waveguide filled with a stack of plate

Table 1. The parameters of the problem.

Curve	a/R	b/R
A	308.6	13.79
B	-	3x13.79
C	15.43	13.79

In figure 2 three curves of the spectral distribution $I_n(\omega)$ of CR from an electron on the 3rd mode of waveguide are shown. The thicknesses of plates and gaps in between are given in Table 1 and the permittivity and permeability of the material of plates are

$$\varepsilon_b = \varepsilon_b'' + i\varepsilon_b'' = 1.3 + 0.005i, \qquad \mu_b = 1 \tag{19}$$

respectively (we assume that ε_b is a weak function of the frequency in the wavelength range under consideration). The energy of electron is 1.2 MeV so that the Cherenkov condition is satisfied: $\mathrm{v} > \mathrm{c}/\varepsilon_b'^{1/2}$.

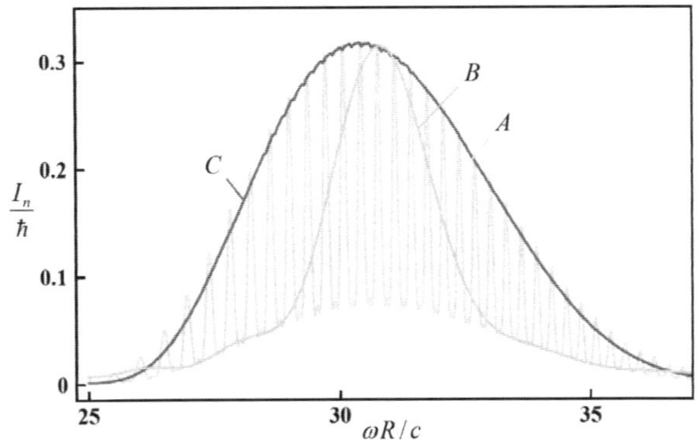

Figure 2. The spectral distribution of CR from an electron at the 3rd mode of waveguide.

In the case of curve A the waveguide is filled with a stack of $N = 3$ plates. The thicknesses of vacuum gaps and plates are selected as is given in table 1 (the row A). It is clear that the pulses of CR generated by the particle in different plates shall superimpose and, hence, the radiation spectrum has to be an oscillating curve. Now compare this curve with the curve B, when all the three plates are so close to each other that they merge and form one thick plate ($N = 1$, one plate filling of the waveguide, the row B in table 1). As was to be expected, the oscillation of spectrum disappears. And what is clear, the total energy of radiation for the curves A and B is the same as, in general, it does not change at interference:

$$W_3^A \approx W_3^B \approx 0.95c\hbar/R. \tag{20}$$

Add besides, that the location of the maximum in CR spectrum is described by the known simple formula:

$$\omega_n^{\max} \approx \frac{\alpha_n \mathrm{v}}{R\sqrt{\varepsilon_b' \mathrm{v}^2/c^2 - 1}}. \tag{21}$$

It is quite clear. It is unique that another situation is possible.

Now return to the case when the waveguide filling is a stack of three plates and select another special value of the thickness of vacuum gaps between the plates (table 1, the row C). The corresponding spectral distribution of CR is also depicted in figure 2 (the curve C). It is seen that at the transition $A \rightarrow C$ the oscillations practically disappear. Moreover, the total energy of radiation (the surface under the curve C) is almost twice larger than that for the curve A or B:

$$W_3^C = 1.76c\hbar/R. \tag{22}$$

In figure 3 it is shown the radiation spectra at the first five modes of waveguide (the number of mode is seen beside the curves B and C (see table 1). The curves B are given for comparison (waveguide filled with one thick plate).

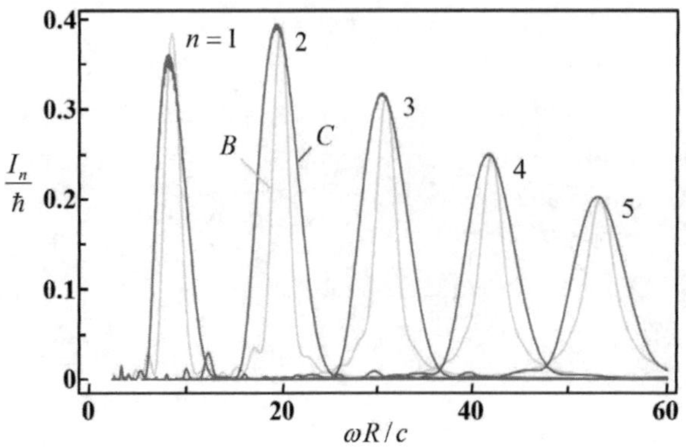

Figure 3. The radiation spectra at the first five modes of waveguide.

From figure 3 it follows that at the transition $B \to C$ the total energy of radiation increases at all five waveguide modes.

6. Visual explanation

In figure 4 two neighboring plates inside the waveguide are shown.

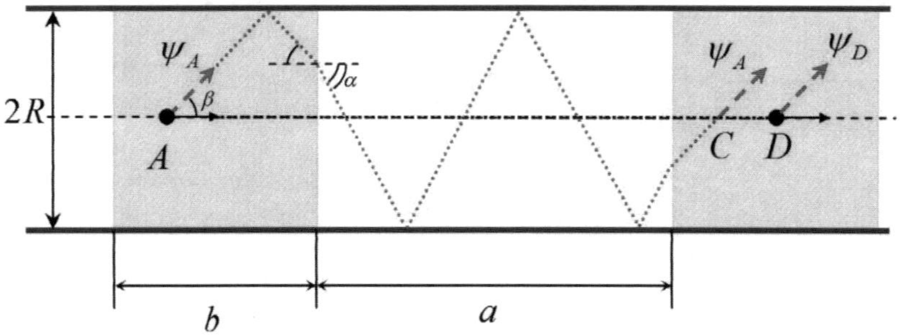

Figure 4. In case of $C \equiv D$ the pulse Ψ_D of CR generated by the particle in the vicinity of point D is «superimposed» on the pulse Ψ_A of CR generated by the particle in the vicinity of point A prior to that.

Now let us consider the instant when the relativistic particle was in the left plate and observe Ψ_A pulse of CR, generated by a particle in the immediate vicinity of point A, that is emitted at a specific angle β (angle of CR) with respect to the waveguide axis (dashed line). At propagation this pulse is found in the right plate and at some instant of time crosses the particle trajectory (the waveguide axis) in a certain point C at the same angle β. By this time the relativistic particle emits Ψ_D pulse of CR in the immediate vicinity of some (generally) other point D (see Fig.4). However, if the following two equations

$$\frac{a+b}{\mathrm{v}} = \frac{a}{\mathrm{c}\cos\alpha} + \frac{b\sqrt{\varepsilon_b'\mu_b'}}{\cos\beta}, \qquad \frac{a\,\mathrm{tg}\alpha + b\,\mathrm{tg}\beta}{2R} = 4s \tag{23}$$

are satisfied (α is an angle giving the direction of CR with respect to the waveguide axis in the vacuum between the plates, s is the integer), then the exceptional situation arises:

$$CD = 0 \quad \text{and} \quad AC = a + b, \tag{24}$$

when, first, the point D coincides with the point C and, second, this «joined» point C-D in the right-side plate is located where in the left-side plate the point A is located. For this reason Ψ_D pulse emitted at the same angle β in the vicinity of point D will "superimpose" on Ψ_A pulse emitted earlier in the 1st plate. As a result of interference the pulses may be either suppressed or amplified depending on the specific value of phase difference. The factor 4 (see e.g., [2]) in the right-hand side of the second equation of (23) insures the fulfillment of the condition of pulses constructive superposition.

Summarizing one concludes that if equations (23) are met, then the Cherenkov waves are generated inside the plate and at the same time constructively interfere with those emitted by the particle in the preceding plates. In other words, the in-phase superposition of CR pulses takes place in the zone of radiation formation. The in-phase superposition of electromagnetic oscillations at point C-D entails the total field increase in the zone of radiation formation. The force that brakes the motion of particle will also increase and the extra work of external force compelling the particle to move uniformly along the waveguide axis will be spent for generation of more powerful CR.

The thicknesses of vacuum gaps and plates for the curves C in figures.2, 3 (see table 1, the row C) have been determined from (23) at $s = 1$. This fact explains the reason of CR amplification (more precisely, self-amplification of CR) at all five waveguide modes at the transition from the waveguide filled with one thick plate (curves B in figures 2, 3) to the waveguide filled with a stack of shorter plates, obtained from this long plates by its equal sharing (curves C in figures 2, 3).

7. Discussion

In 2009 J.B. Rosenzweig with co-authors reported [13] the fist direct observation of narrow-band terahertz coherent CR driven by a subpicosecond electron bunch traveling along the axis of a hollow cylindrical dielectric-lined waveguide. The measurements indicate a peak power of 150 kW. In 2011 J.B. Rosenzweig with co-authors implement structure variations that gave improved result [14].

We propose to improve the experimental setup of [13, 14]. The hollow cylindrical dielectric inside the waveguide shall be cut in *N* identical parts (tubes) and then a periodic structure consisting of these tubes alternated with vacuum gaps shall be combined. If the length of each vacuum gap between the tubes is selected correctly, then (according to the preliminary calculations [12]) we expect nearly a *N*-fold increase in the peak power of coherent CR from electron bunches.

8. Conclusions

In this paper the radiation from a relativistic particle moving along the cylindrical waveguide axis has been investigated under the assumption that the waveguide is filled with a laminated material of finite length, the permittivity $\varepsilon_1(z)$ and permeability $\mu_1(z)$ of which are periodic functions of arbitrary shape.

1. Expression (16) for calculation of the spectral distribution (9) of radiation passing throw the transverse section of waveguide in vacuum (at large distances from the laminated material) have been derived with no limitations on the amplitude and variation profile of $\varepsilon_1(z)$ and $\mu_1(z)$.
2. Numerical results for CR from a relativistic electron uniformly moving along the axis of cylindrical waveguide filled with a stack of dielectric plates alternated with vacuum gaps have been investigated (Section 5).
3. The values of the problem parameters, for which the waveguide and the periodical structure of laminated material jointly strongly influence the spectral distribution of CR from an electron, have been investigated.
4. It has been shown that due to the existence of vacuum gaps between the plates the power of CR from an electron may be many times greater than that produced, when the plates in the stack are so close that one thick plate is formed (waveguide filled with a solid dielectric

without vacuum gaps). The visual explanation of this effect of self-amplification of CR have been given (Section 6).

5. It was proposed to use this effect for amplification of the power of coherent CR from subpicosecond electron bunches observed by J.B. Rosenzweig with co-authors (Section 7).

Acknowledgments
One of the authors (L.Sh.G.) is thankful to A.P. Potylitsyn for stimulating discussion and encouragement, N. F. Shulga, B. M. Bolotovskii for valuable comments and R. Chehab for the interest to these investigations.

References

[1] Jelley J V 1958 *Cherenkov Radiation and its Applications* (London: Pergamon Press)
[2] Bolotovskii B M 1961 Physics-Uspekhi 75 295
[3] Zrelov V I 1968 *Vavilov-Cherenkov Radiation and its Applications in High Energy Physics* (Moscow: Atomizdat) (in Russian)
[4] Garibian G M and Yang C 1983 *X-Ray Transition Radiation* (Yerevan: AN Arm. SSR Press) (in Russian)
[5] Ginzburg V L and Tsytovich V N 1990 *Transition Radiation and Transition Scattering* (Bristol: Adam Hilger)
[6] Frank I M 1988 *Vavilov-Cherenkov Radiation. Theoretical Aspects* (Moscow: Nauka) (in Russian)
[7] Berman B L and Datz S 1985 in: *Coherent Radiation Sources,* eds A W Saenz and H Uberall (Berlin: Springer-Verlag)
[8] Rullhusen P, Artru X and Dhez P 1998 *Novel Radiation Sources Using Relativistic Electrons* (Singapore: World Scientific)
[9] Potylitsyn A P 2009 *Radiation from Electrons in Periodic Structures* (Tomsk: NTL Press) (in Russian)
[10] Fainberg Y B and Khizhniak N A 1957 *Zh. Eksp. Teor. Fiz.* **32** 883 (1957 *Sov. Phys. JETP* **5** 720)
[11] Grigoryan L Sh, Khachatryan H F and Arzumanyan S R 2011 *IL Nuovo Cimento* **34C** 317
[12] Grigoryan L Sh, Khachatryan H F and Arzumanyan S R 2010 Amplification of coherent Cherenkov radiation from a one-dimensional electron bunch inside a layered dielectric-lined waveguide: Proc. of the Int. Conf. on "*The Technique of Microwave and THz Waves and its Application in Biomedical and Radar Thecnologies and in Remote Sensing*" (Ashtarak-Aghveran, Armenia, 23-25 September 2010) pp 32-37
[13] Cook A M, Tikhoplav R, Tochitsky S Y, Travish G, Williams O B and Rosenzweig J B 2009 *Phys. Rev. Lett.* **103** 095003
[14] Andonian G *et al.* 2011 *Appl. Phys. Lett.* **98** 202901
[15] Flugge S 1971 *Practical Quantum Mechanics 1* (Berlin: Springer-Verlag)
[16] Gazazian E D and Laziev E M 1963 *Izv. Nats. Akad. Nauk Arm. Fiz.* **16** 79 (in Russian)
[17] Begloian E A, Gazazian E D and Laziev E M 1976 *Radiotekhnika i Elektronika* **21** 164 (in Russian)

Direct Observation of a Semi-Bare Electron Coulomb Field Recover

G Naumenko, Yu Popov and M Shevelev
Tomsk Polytechnic University, Lenina str. 2, Tomsk, 634050, Russia

E-mail: naumenko@tpu.ru

Abstract. The problem of "semi-bare electron" was first considered in frame of quantum electrodynamics by E L Feinberg in 1980. In theory in frame of classical electrodynamics this problem was touched on in articles of N F Shul'ga and X Artru. In 2008 the experimental investigations of this phenomenon in millimeter wavelength region were started by the group of scientists, including authors of this article. Used technique allowed us to study this effect in macroscopic mode. In this paper we present the results of direct observation of a semi-bare electron coulomb field recovery. The semi-bare state was obtained by passing of electron beam through the hole in a conductive screen. Measured spatial distribution of electromagnetic field shows the process of recover of the electron coulomb field, which is followed by a forward radiation. The experiments were performed on the relativistic electron beam of the microtron of Tomsk Polytechnic University.

1. Introduction

The term "semi-bare electron" for the first time, probably was used in the article of E L Feinberg [1], where in the framework of quantum electrodynamics was considered the problem of scattering of a relativistic electron at a large angle. The author has noted that the electron state just after the scattering may be interpreted as the state of the electron, which partially lost its own Coulomb field (the "half-naked electron" or "semi-bare electron"). In the framework of classical electrodynamics, this problem was raised in articles of N F Shulga [2] and X Artru [3]. Most clearly, this problem may be represented in the view of the electron field as a field of pseudo-photons. The "pseudo-photon" method proposed by E Fermi [4] and developed by E J Williams [5] is widely used for the theoretical studies of electromagnetic processes (see for example [6] and [7]). According to this approach, the field of a charged particle may be replaced by a field of photons, which in this case are called the pseudo-photons (in [8] is used the term "virtual quanta" (this term should be differ from the term "virtual quanta" in quantum mechanics)). This approach provides the good accuracy for the ultra-relativistic particles when the particle velocity is close to the light velocity ($v \sim c$) (see [6]), and when the longitudinal electric field component of the particle is negligible. In this case the particle field has the properties like the properties of the field of real photons.

Let us consider the interaction of real photons with a thick conductive mirror of high reflectivity. It is clear that real photons are reflected almost completely, they do not penetrate the target, and they do not excite surface currents on the downstream surface of the target. Because the properties of pseudo-photons are close to the properties of real photons, we may expect that with the passage of electrons through a hole in a conducting screen the electron field particularly is reflected from the screen and downstream to the screen the electron will lose a part of its Coulomb field. It is clear that in the further evolution the Coulomb field of the electron will be restored, because far from the screen we always observe electrons with the usual Coulomb field.

We take an interest in the study of this field and its spatial distribution. In [2] the solution of Maxwell's equations for a similar problem (the passage of electrons through the conducting screen) is presented. In this work, the electrons, emitted from a screen, are represented by the superposition of the theta function $\theta(t - t_0)$ and the expression for the current density of a uniformly moving electron, where t_0 is the emitting time of an electron from the screen. This statement, however, may not be used for our problem, because it is equivalent to the case suddenly accelerated electron, while we are considering the dynamics of the field of a uniformly moving electron.

The problem of the field radiation of electrons emitted from the absorber in the form of "blackbody", was considered in [9], but the solution is done for the far zone (at infinity) and does not provide the information about the dynamics of the field within the zone of radiation formation, which is more interesting for us.

In [10] we studied the dynamics of the electron field in condition of the shadowing effect. The studies were based on the radiation characteristics of semi-bare electrons downstream to conductive and absorbing screens, or based on the evolution of the electron pseudo-photon field, diffracted on an electron dump, in which the electrons are stopped. These measurements are indirect, and they do not provide complete information about the electron field distribution during its evolution, although they give the general regularities in the field distribution.

In this article the direct measurement of the field characteristics of the electron bunch passing through a hole in conducting screen are presented. Such method allows a direct study of the electron field during the transit from the state of semi-bare electron, to the usual (stable) state.

2. Experimental set-up and procedure

The experiment was carried out in the extracted electron beam of the Tomsk Polytechnic University microtron with parameters presented in table 1.

Table 1. Electron beam parameters

Electron energy	6.1 MeV ($\gamma = 12$)	Bunch period	380 psec
Train duration	$\tau \approx 4$ μsec	Bunch population	$N_e = 6\times10^8$
Bunches in a train	$n_b \approx 1.6\cdot10^4$	Bunch length	$\sigma \approx 1.3 \sim 1.6$ mm

For the electron field measurement we use the probe (figure 1) based on the well-known technique, which is applied for the surface current measurement in strip-line beam position monitors [6].

Beam parameters of the microtron allow using the coherent properties in an interaction of the electron field with probe in the frequency region corresponding to the millimetre wavelength range. This possibility increases the probe magnitude response by the 8 orders (proportional to the bunch population) and makes this response achievable for the measurements using existing sensors.

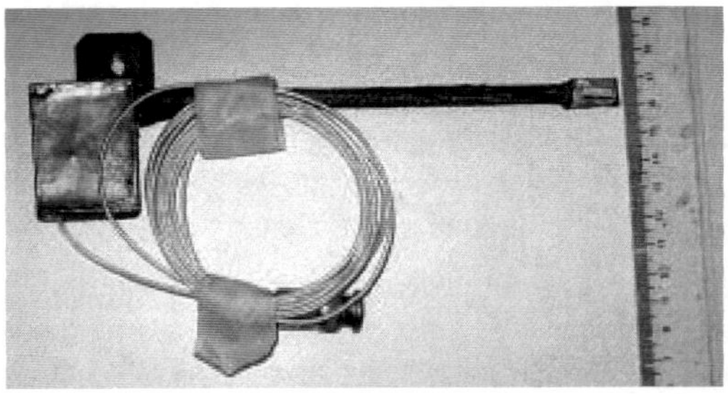

Figure 1. The photo of the probe for the electron field measurement.

The probe has a maximum sensitivity for the Fourier harmonics of the field corresponding to a wavelength equal to 16 mm. The characteristics of the probe were tested on the stand (figure 2) with a beam of real photons from the source with the wavelength equal to 11 mm.

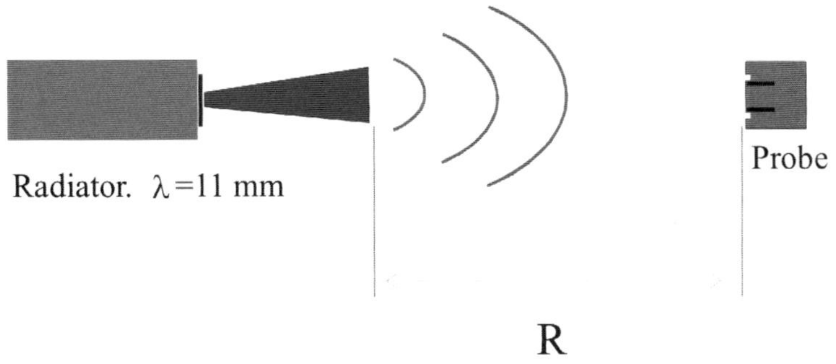

Figure 2. The scheme for testing of the probe properties.

We measured the probe response as a function of the distance R from the radiator (figure 3). To verify the nature of the probe response, the values of the measured dependence was approximated by the power function $y = a \times x^{-b}$ (figure 3).

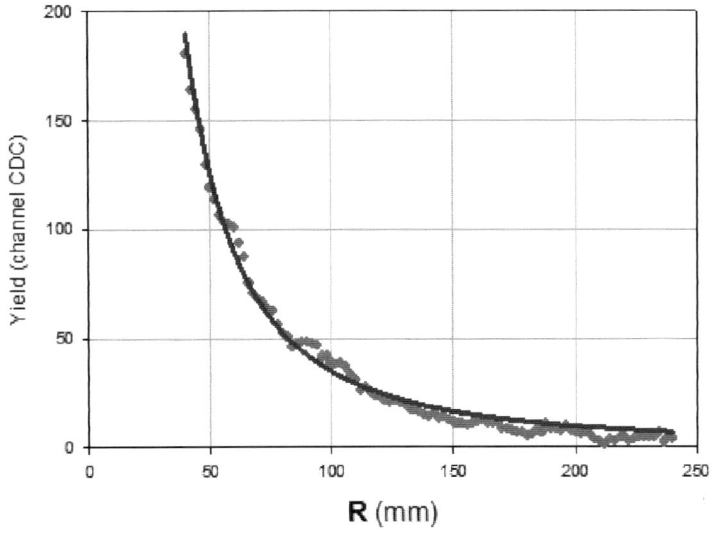

Figure 3. The probe response as a function of the distance R from the radiator. The points are the measured values and the sold line is the approximation $y = a \times x^{-b}$.

The value of power factor b was estimated as $b = 1.86 \pm 0.05$. Taking into account the finite radiation source size (6 mm × 12 mm), the experimental dependence is well approximated using the inverse quadratic dependence. Up to a scale factor, this corresponds to the measurement of squared field value.

3. Electron field distribution measurement

The measurements were performed using the scheme shown in figure 4. Relativistic electrons grouped in bunches of approximately 10^8 electrons, passing through a hole in a conducting screen.

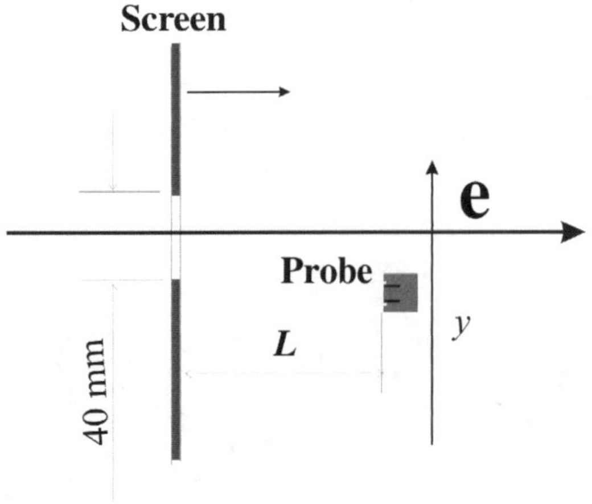

Figure 4. The scheme of the electron field distribution measurement.

Most part of the electron field of the pseudo-photons are reflected from the conducting screen. The remaining part of the Coulomb field of electrons passes through a hole in the screen. This forms an unstable state of the electrons, which partially lost its own Coulomb field. As this electron state must be restored for a stable state of normal electrons during the further evolution of the electrons, the distribution of electrons field must depend on the distance between the electron bunch and screen. This dependence was measured in an experiment using a probe. Keep in mind that the spatial resolution of the probe is ≈4 mm, so that within this interval the probe detects the average value of the field. Then, due to the axial symmetry of the electron bunch field, the average field in the center of the bunch should be zero. Ie, the transverse field distribution of the electron bunch must have a deep in the center of the beam. The field distribution was measured with the step of 2 mm in the direction perpendicular to the beam, and with step of 20 mm in variation of *L*. To exclude the beam divergence contribution, the probe was moved only in the transversal direction to the electron beam trajectory. The distance *L* was varied by moving the screen along the beam. The maximum value of *L* was limited because of the beam size in the hole of the screen due the beam divergence (≈ 0.1 radian in full). For measurement the background, the active surface of the probe was glued by the aluminium foil, using the conductive adhesive. The statistical error of the measurement was equal to 5 %. The photo of the experimental set-up is shown in figure 5.

Figure 5. The photo of the experimental set-up.

The background was measured for the same area and was subtracted from the measured dependence. The smoothed squared field magnitude dependence on the transverse coordinate y and on the distance L between the screen and the probe is shown in figure 6.

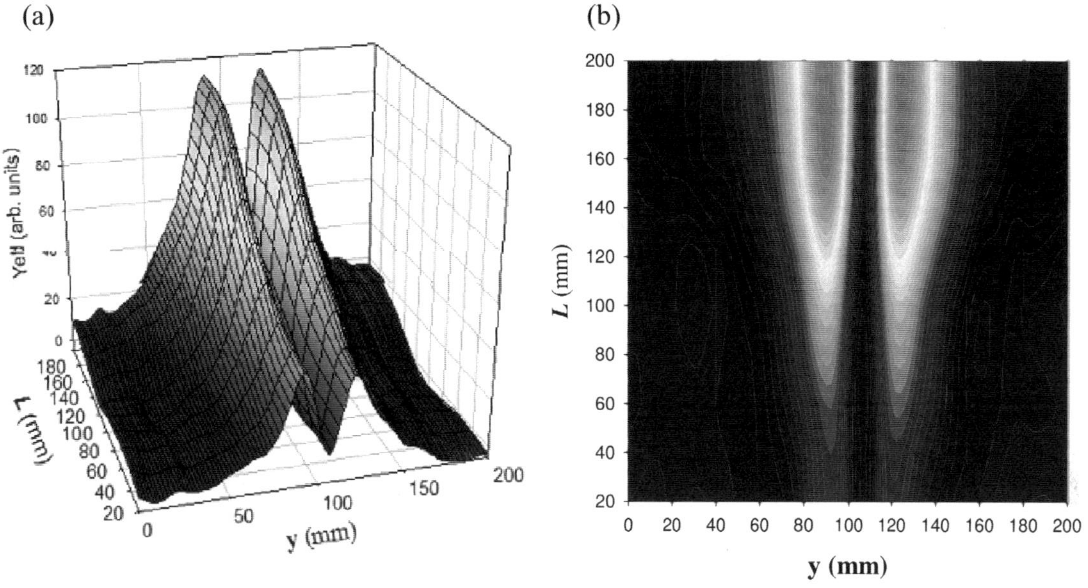

Figure 6. The smoothed squared field magnitude dependence on the transverse coordinate y and on the distance L between the screen and the probe. *a* – 3D view, *b* – the plane representation of this dependence.

4. Discussion

We see in figure 6 the build-up of the field with the increasing of the distance L from the screen. One may say that we observe the destructive interference of the forward transition radiation from the conductive screen with the field of electron (see [13]). However, according to our experimental result [12] the surface currents are not induced on a downstream surface of a thick conductive screen in the investigated spectral region. Therefore no radiation may be emitted from the screen in this spectral region. It is clear from Fourier representation of Maxwell's equations together with the continuously equation written as:

$$\begin{cases} \Delta \vec{E}_\omega + \omega^2 \vec{E}_\omega = 4\pi\left(\nabla\rho_\omega + i\omega\cdot\vec{j}_\omega\right) \\ div\ \vec{E}_\omega = 4\pi\rho_\omega \\ div\ \vec{j}_\omega + i\omega\cdot\rho_\omega = 0. \end{cases}$$

It is seen from this equation system, that if a surface current is absent in some frequency region, then the system became homogeneous and describes only plane waves without radiation sources in this frequency region. This result is in agreement with pseudo-photon viewpoint in presentation of the electromagnetic field of relativistic electrons.

For analysis the processes in the used geometry, let us consider the ideal case, when the relativistic electron moves along the axis z through a thick conductive infinite screen, which divides the space on the left and right half-spaces in plane $z = 0$. In this case we cannot use the model of suddenly accelerated electron, because in this case the electrons moves uniformly. If the electron moves in the left half-space, then no response may be observed in right half-space. Let now the electron crosses the screen at the point of time $t = 0$. After this point the electron current appears in the right half-space and no another currents are induced (see above). Therefore only the electron current may be considered as a source of radiation in the right half-space.

The electric field may be found from solution of Maxwell's equations in Fourier representation:

$$\vec{E} = \frac{i\omega}{k^2 - \omega^2}\left(\vec{j} - \frac{\vec{k}\cdot\vec{j}}{\omega^2}\vec{k}\right),$$

where $\vec{j}$ is the single electron current density.

For $0 < t < \infty$ $\vec{j}(\vec{k}) = \frac{e\vec{v}}{2\pi}\left(\frac{\delta(\omega + \vec{k}\cdot\vec{v})}{2} + i\frac{1}{2\pi(\omega + \vec{k}\cdot\vec{v})}\right)$, and for axial symmetry (as it is in our case) after inverse Fourier transform on z variable we obtain

$$E_\rho(z,k_\rho) = -i\frac{e}{8\pi^2}\frac{k_\rho}{k_\rho^2 + \frac{\omega^2}{\beta^2\gamma^2}}e^{\frac{i\omega z}{\beta}} + \frac{e\beta}{(2\pi)^3\omega}\int_{-\infty}^{\infty}\frac{e^{-ik_z z}}{k_z^2 + k_\rho^2 - \omega^2}\cdot\frac{k_z k_\rho}{\omega + k_z\cdot\beta}dk_z,$$

where ρ is the transversal coordinate in the axial system. Finally after full inverse transform we obtain

$$E_\rho(z,\rho) = \frac{1}{2\pi}\int_0^\infty E_\rho(z,k_\rho)\cdot J_0(k_\rho\cdot\rho)k_\rho dk_\rho, \quad (1)$$

where $J_0(k_\rho\cdot\rho)$ is the Bessel function.

The further analytical simplification of this expression is complicated, and we present in figure 7 (a) the numerical representation of $|E_\rho(z,\rho)|^2$ for the experimental conditions $\gamma = 12$ and $\lambda = 10\,\text{mm}$ (the grid for $\rho = 0$ is skipped due to the singularity in the expression under the integral in this point).

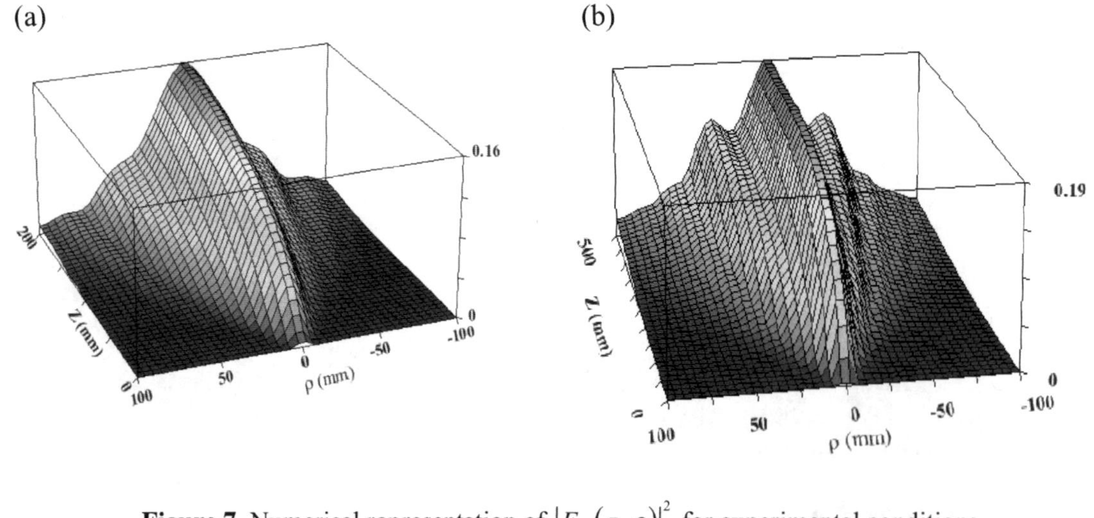

Figure 7. Numerical representation of $|E_\rho(z,\rho)|^2$ for experimental conditions.

We see in figure 7 (a) that Maxwell's equation solution also shows the recovery of the field of electron crossing the conductive screen. For the larger distance from the screen (see figure 7 (b)) we see the separation of the electron field and the radiation, which is named usually "transition radiation" and is usually supposed by mistake as a transition radiation from the screen.

In comparison with the experimental conditions, the theoretical calculations were performed for a single electron (not for a bunch of finite size) crossed the screen with the infinite small hole. Also the finite spatial resolution of the probe was not taken into account. Nevertheless both the experimental rasult and the theoretical calculation show the recovery of the electron field downstream to a conductive screen.

Finally we can conclude that we had observed in the experiment the semi-bare electron and the process off electron field recovery in macroscopic mode.

Acknowledgment
This work was partly supported by the warrant-order 1.226.08 of the Ministry of Education and Science of the Russian Federation.

References

[1] Feĭnberg E L 1979 *Sov. Phys. Usp.* **22** 6 479-479
[2] Shul'ga N F, and Syshchenko V V 2000 *Journal Physics of Atomic Nuclei* **11** 2018
[3] Artru X, Ray C Interferences between light emitted by charged particles crossing or skimming past optical fibers. *International Symposium RREPS'07*, Prague, 2007
[4] Fermi E 1924 *Zh. Phyz* **29** 315
[5] Williams E J 1935 *Kgl. Danske Videnskab. Selskab. Mat.-fys. Medd.* **13** 4
[6] Ter-Mikaelian N L 1972 High-Energy Electromagnetic Processes in Condensed Media *Wiley, New York*
[7] Ter-Mikaelian N L 1969 The Influence of the Medium on High Energy Processes at High Energies *Publishing House of Academy of Science of Armenia, Yerevan*
[8] Jackson J D 1998 Classical Electrodynamics, 3rd ed. *J. Willey&Sons, New-York*
[9] Bolotovskii B M, Serov A V 2009 *Phys. Usp.* **52** 487
[10] Naumenko G, Artru X, Potylitsyn A, et al. 2010 *Journal of Physics: Conference Series* **236** 012004.
[11] Sargsyan V 2004 Comparison of Stripline and Cavity Beam Position Monitors, *TESLA Report* **03**
[12] Naumenko G, Potylitsyn A, Popov Yu et al. 2010 *Journal of Physics: Conference Series* **236** 012024.
[13] BolotovskiI B N 1982 *Preprints of Lebedev Institute of Physics, Russian Academy of Sciences* **140** 95

Journal of Physics: Conference Series **357** (2012) 012006 doi:10.1088/1742-6596/357/1/012006

Radiation of a charge moving in wire metamaterial perpendicularly to the main axis

V V Vorobev, A V Tyukhtin
Radiophysics Dep., St Petersburg State University
1 Ul'yanovskaya Str., St Petersburg, 198504, Russia

E-mail: vorobjovvictor@gmail.com, tyukhtin@bk.ru

Abstract. We consider electromagnetic radiation of a charge which moves in a volume structure consisting of thin parallel wires and having small period. The structure is described by the permittivity tensor. The charge velocity is directed perpendicularly to the wires. Analytical and numerical investigations are performed for the case of point charge. Some unusual properties of radiation are noted. In particular, it is shown that radiation propagates along the wires and concentrates nearby certain "rays" behind the charge. The wave field doesn't decrease with increase in distance along these "rays". The cases of Gaussian and rectangular bunches with negligible thickness are considered as well. Prospects of use of structure under consideration for diagnostics of bunches are noted.

1. Introduction

It is well known that Cherenkov radiation is widely used for detection of charged particles and in beam diagnostics [1]. Prospects of development of this technique depend on materials which can be used for generation of radiation. Note that properties of radiation in the presence of media with dispersion and anisotropy can be very unusual [1–8]. At the present time, special attention is given to so-called "metamaterials" which are artificial periodic structures with relatively small spacing. They can be considered as some "media" and characterized by effective permittivity and permeability. Metamaterials allow obtaining different properties which cannot be achieved with help of traditional media. For example, they can be similar to left-handed medium (LHM). Radiation of moving charges in the presence of LHM was investigated intensively [2–7]. Different metamaterials are known as well, and they attract large attention. Meanwhile radiation processes in them were analyzed rarely.

Here we consider the case when the charge moves through a 3-D system of parallel wires. Such structure was intensively analyzed in recent years [9–12] but radiation of moving charge in the presence of it was not considered. It is assumed that $d \ll \Lambda \equiv \min(\lambda, \Delta)$, where d is a period of structure, λ is a typical wavelength, and Δ is the distance of variation of the "incident" field (i.e. the field in the case when the structure is absent). Different approaches to derivation of macroscopic characteristics of this structure can be found in papers [9–12]. If the x-axis is parallel to the wires the expression of permittivity tensor has the diagonal form:

$$\hat{\varepsilon} = \begin{pmatrix} \varepsilon_{\parallel} & 0 & 0 \\ 0 & 1 & 0 \\ 0 & 0 & 1 \end{pmatrix}, \tag{1}$$

where

$$\varepsilon_{\parallel}(\omega,k_x)=1-\frac{\omega_{\mathrm{p}}^2}{\omega^2-k_x^2c^2+2\mathrm{i}\omega_{\mathrm{d}}\omega},\qquad \omega_{\mathrm{p}}^2=\frac{2\pi c^2}{d^2\left[\ln(d/r_0)-\mathrm{C}\right]}. \tag{2}$$

Here r_0 is a wire radius, ω_{d} is a small constant ($\omega_{\mathrm{d}}\ll\omega_{\mathrm{p}}$) which is responsible for losses, and C is some constant which has the order of 1 and depends on the ratio r_0/d [9, 11, 12].

In this paper we will analyze the case when the charge velocity is perpendicular to the main axis of the "metamedium" (which is parallel to wires). Note that in another case when the charge moves parallel to wires, the radiation can be generated only for wires possessing non-conducting coating. For such structure the component $\varepsilon_{\parallel}$ has more complex form [12, 13].

2. Electromagnetic field of a point charge

In this section we perform analytical investigation of the point charge moving with a constant velocity $\vec{V}=V\vec{e}_z=c\beta\vec{e}_z$ perpendicularly to the main axis of the medium which is characterized by the permittivity tensor (1), (2). For this source, the current density is $\vec{j}=\vec{e}_z c\beta q\cdot\delta(x,y,\zeta)$, where $\zeta=z-c\beta t$, and its Fourier transform is $\vec{j}_{\omega,\vec{k}}=\vec{e}_z c\beta q\cdot\delta(\omega-k_z c\beta)/(2\pi)^3$.

The electromagnetic field can be found out from Maxwell equations with help of the Fourier transformation technique. Note that the initial fourfold Fourier integrals are reduced to the threefold ones due to the Dirac delta-function $\delta(\omega-c\beta k_z)$. Omitting some simple transformations we write here the exact result for the electric force:

$$\vec{E}=-\frac{q\mathrm{i}}{2\pi^2c^3\beta}\int_{\mathbb{R}}\mathrm{d}k_x\int_{\mathbb{R}}\mathrm{d}k_y\int_{\mathbb{R}}\vec{u}\frac{\exp\left(\mathrm{i}k_xx+\mathrm{i}k_yy+\mathrm{i}\omega\zeta/(c\beta)\right)}{\left(k_x^2-k_{x0}^2\right)\left(k_x^2-k_{x1}^2\right)\left(k_x^2-k_{x2}^2\right)}\mathrm{d}\omega, \tag{3}$$

where

$$u_x=k_xc^2\left(k_x^2-k_{x0}^2\right)\left(k_x^2-k_{x2}^2\right), \tag{4}$$

$$u_y=k_y\left[\frac{\omega^2}{c^2}\left(\omega^2-c^2k_x^2-\omega_{\mathrm{p}}^2\right)-\left(k_x^2+k_y^2+\frac{\omega^2}{c^2\beta^2}\right)\left(\omega^2-c^2k_x^2\right)\right], \tag{5}$$

$$u_z=c\beta\omega\left(k_x^2-k_{x0}^2\right)\left[\left(k_x^2-\frac{\omega^2}{c^2}\right)\frac{1-\beta^2}{\beta^2}-\frac{\omega_p^2}{c^2}\right]-c\beta\omega k_y^2\frac{\omega_{\mathrm{p}}^2}{c^2}; \tag{6}$$

$$k_{x0}^2=-\frac{\omega^2}{c^2}\frac{1-\beta^2}{\beta^2}-k_y^2,\qquad k_{x1}^2=-\frac{\omega^2}{c^2}\frac{1-\beta^2}{\beta^2}-\frac{\omega_{\mathrm{p}}^2}{c^2}-k_y^2,\qquad k_{x2}^2=\frac{\omega^2}{c^2}. \tag{7}$$

Here we tend dissipation to zero ($\omega_{\mathrm{d}}\to0$) but we should take into account that $k_{x2}=\omega/c+i0$ (the pole $k_x=k_{x2}$ is situated above the integration path and the pole $k_x=-k_{x2}$ is situated below it). Note that the formulae (7) give solutions of the dispersion equation.

Writing the integrals over the variable k_x in (3) as the sum of residues we obtain double integrals. It is interesting that the integral for E_x can be expressed in the explicit form [13]:

$$E_x=\frac{1}{\sqrt{2\pi}}\frac{2qx\omega_{\mathrm{p}}^{3/2}\beta^{5/2}}{c^3\left(1-\beta^2\right)}K_{3/2}\left[\left(\frac{\omega_{\mathrm{p}}^2\beta^2}{1-\beta^2}\right)\left(\frac{\zeta^2}{c^2\beta^2}+\frac{\rho^2\beta^2}{c^2\left(1-\beta^2\right)}\right)\right]^{1/2}\left[\frac{\zeta^2\left(1-\beta^2\right)}{c^2\beta^2}+\frac{\rho^2\beta^2}{c^2}\right]^{-3/4}, \tag{8}$$

where $K_\nu(\xi)$ is modified Hankel function of the second kind and variable ρ is radial distance from the charge trajectory ($\rho=\sqrt{x^2+y^2}$). Due to the presence of modified Hankel function the E_x component exponentially decreases with increase in ρ and ζ. Thus this component is a part of quasi-static field, and the wave field (field of radiation) does not contain it.

Behavior of E_y and E_z is principally different. They both are cumbersome expressions that consist of contributions of poles $\pm k_{x0}$, $\pm k_{x1}$, $\pm k_{x2}$. Two of these poles are fully imaginary and produce contributions characterized by exponential decrease (so-called quasi-static field). Only the contribution of poles $\pm k_{x0}$ gives the wave field (further it is signed by index W):

$$\begin{Bmatrix} E_y^{\mathrm{W}} \\ E_z^{\mathrm{W}} \end{Bmatrix} = \frac{-q\omega_{\mathrm{p}}^2}{2\pi c^5\beta^2}\int_{\mathbb{R}} \mathrm{d}k_y \int_{\mathbb{R}} \mathrm{d}\omega \begin{Bmatrix} k_y c\beta \\ \omega \end{Bmatrix} \omega \exp\left(\mathrm{i}\frac{\omega}{c}|x| + \mathrm{i}\frac{\omega}{c\beta}\zeta + \mathrm{i}k_y y\right)\left(\frac{\omega^2}{c^2\beta^2} + k_y^2\right)^{-1}\left(\frac{\omega^2}{c^2\beta^2} + k_y^2 + \frac{\omega_{\mathrm{p}}^2}{c^2}\right)^{-1}. \quad (9)$$

These expressions can be analytically calculated using tabulated integrals [13]:

$$E_y^{\mathrm{W}} = -2\beta q \frac{y(\zeta + \beta|x|)}{\xi^4}\left[1 - \frac{\omega_{\mathrm{p}}^2}{2c^2}\xi^2 K_2\left(\frac{\omega_{\mathrm{p}}}{c}\xi\right)\right], \quad (10)$$

$$E_z^{\mathrm{W}} = -\frac{\beta q}{c^2}\left[\frac{c^2\left(y^2 - (\zeta + \beta|x|)^2\right)}{\xi^4} + \frac{(\zeta + \beta|x|)^2 \omega_{\mathrm{p}}^2}{\xi^2} K_2\left(\frac{\omega_{\mathrm{p}}}{c}\xi\right) - \frac{\omega_{\mathrm{p}} c}{\xi} K_1\left(\frac{\omega_{\mathrm{p}}}{c}\xi\right)\right], \quad (11)$$

where $\xi^2 = y^2 + (\zeta + \beta|x|)^2$.

One can make the same transformations for the components of the magnetic field as well. The result is that H_x exponentially decreases with ρ and ζ the same way as E_x, and only components H_y and H_z have the wave part which are equal to the components of the electrical field up to sign:

$$H_z^{\mathrm{W}} = E_y^{\mathrm{W}} \operatorname{sgn} x, \qquad H_y^{\mathrm{W}} = -E_z^{\mathrm{W}} \operatorname{sgn} x. \quad (12)$$

Using expansions of the modified Hankel functions one can find the following approximate expressions for field components in the vicinity of lines $\zeta + \beta|x| = 0$ in the plane $y = 0$ (i.e. $\xi \to 0$):

$$E_y^{\mathrm{W}} \approx -\frac{\beta q \omega_{\mathrm{p}}^2}{c^2}\frac{y(\zeta + \beta|x|)}{\xi^2}, \quad (13)$$

$$E_z^{\mathrm{W}} \approx \frac{\beta q}{c^2}\left[\frac{\omega_{\mathrm{p}}^2(\zeta + \beta|x|)^2}{2\xi^2} + \frac{\omega_{\mathrm{p}}^2}{2}\ln\frac{\omega_{\mathrm{p}}\xi}{c} + \frac{\omega_{\mathrm{p}}^2}{4}(2\mathrm{C_E} - 1)\right], \quad (14)$$

where $\mathrm{C_E} \approx 0.577$ is Euler constant. Formulae (12) – (14) show that components E_y^{W} and H_z^{W} tend to zero on the lines $\zeta + \beta|x| = 0$ in the plane $y = 0$, components E_z^{W} and H_y^{W} have logarithmic singularity on them. When the viewpoint is shifted along these lines (i.e. conditions $y = \mathrm{const}$ and $\zeta + \beta|x| = \mathrm{const}$ are fulfilled) components E_y^{W}, E_z^{W}, H_y^{W} and H_z^{W} do not change. The Pointing vector $\vec{S} = c(4\pi)^{-1}\left[\vec{E} \times \vec{H}\right]$ has the only component along wires and in fact it is proportional to the square of the full electrical wave field. The energy flow $\int_\Sigma \vec{S}\mathrm{d}\vec{\Sigma}$ through some square Σ is limited and does not depend on the distance from the charge trajectory.

Thus, the charge moving perpendicularly to the wires emits Vavilov-Cherenkov radiation that is concentrated in some small vicinity of lines $\zeta + \beta|x| = 0$ in the plane $y = 0$ behind the charge. These waves propagate along the wires and exist for arbitrary velocity of the charge. Computation of electrical field components confirms behavior described above (figure 1).

Explanation of this behavior can be given using the dispersion equation which follows from Maxwell equations for plane wave $\sim \exp\left(\mathrm{i}k_x x + \mathrm{i}k_y y + \mathrm{i}k_z z - \mathrm{i}\omega t\right)$:

$$-\varepsilon_{\parallel} k_x^2 - k_y^2 - k_z^2 + \varepsilon_{\parallel}\omega^2/c^2 = 0. \quad (15)$$

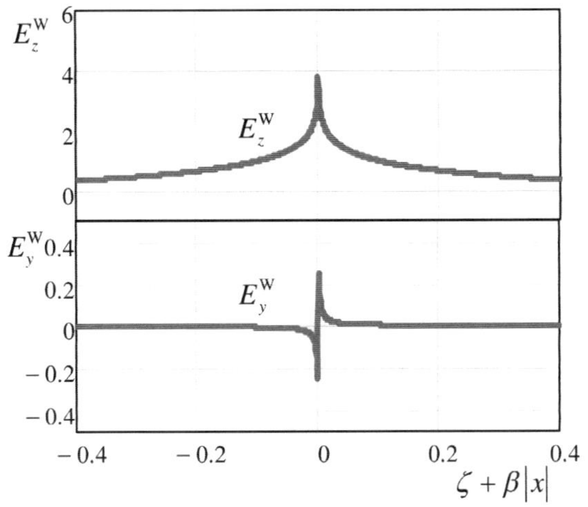

Figure 1. Electrical field components (in units $q\omega_\mathrm{p}^2/c^2$) depending on distance (in units c/ω_p). E_z^W is given for $y=0$, and E_y^W is shown for $y=0.01c/\omega_\mathrm{p}$.

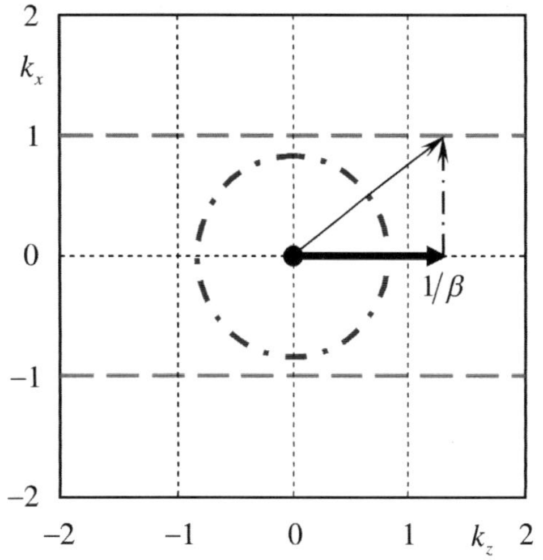

Figure 2. The first (dashed line) and the second (dashed-and-dotted line) solutions of the dispersion equation. Arrows are the wave-vector and its projections (in units ω/c).

Since $\varepsilon_\parallel$ depends on k_x in concordance with (2), one can obtain two following roots of (15):

$$k_{x\alpha}^2=\frac{\omega^2}{c^2},\qquad k_{x\beta}^2=\frac{\omega^2}{c^2}-k_z^2-k_y^2-\frac{\omega_\mathrm{p}^2}{c^2}. \tag{16}$$

In the wave vector space we have two parallel planes $k_x=\pm\omega/c$ (the first solution) and a sphere with radius less than ω/c for frequencies greater than ω_p (the second solution). Since the field of moving charge includes plane wave with $k_z=\omega/(c\beta)>\omega/c$, the waves of the second type are not generated. However the waves of the first type are exited for any velocity (figure 2). Note that the group velocity for these waves is always directed along x-axis and equal to the speed of light in the vacuum.

3. Electromagnetic fields of the bunches

Further we will analyze fields of bunches which have negligible thickness but some finite length. Two types of longitudinal distribution of the charge are considered: Gaussian and rectangular. They are described by the following formulas accordingly:

$$\rho_\mathrm{G}(\zeta)=\exp\left(-\zeta^2/2\sigma^2\right)/\left(\sqrt{2\pi}\sigma\right),\quad \rho_\mathrm{R}(\zeta)=\left[\Theta(\zeta+\sigma)-\Theta(\zeta-\sigma)\right]/(2\sigma), \tag{17}$$

where parameter σ determines the length of the bunch, and $\Theta(z)$ is Heaviside function.

The electromagnetic field of the bunch moving perpendicular to the wires has the form of convolution product:

$$\vec{E}(\vec{r},t)=\frac{1}{2\pi}\int_{-\infty}^{+\infty}\rho(\zeta-\zeta')\vec{E}_\delta(x,y,\zeta',t)\mathrm{d}\zeta', \tag{18}$$

where $\vec{E}_\delta(x,y,\zeta,t)$ is field of point charge obtained above. Further we give some results of computations of these integrals.

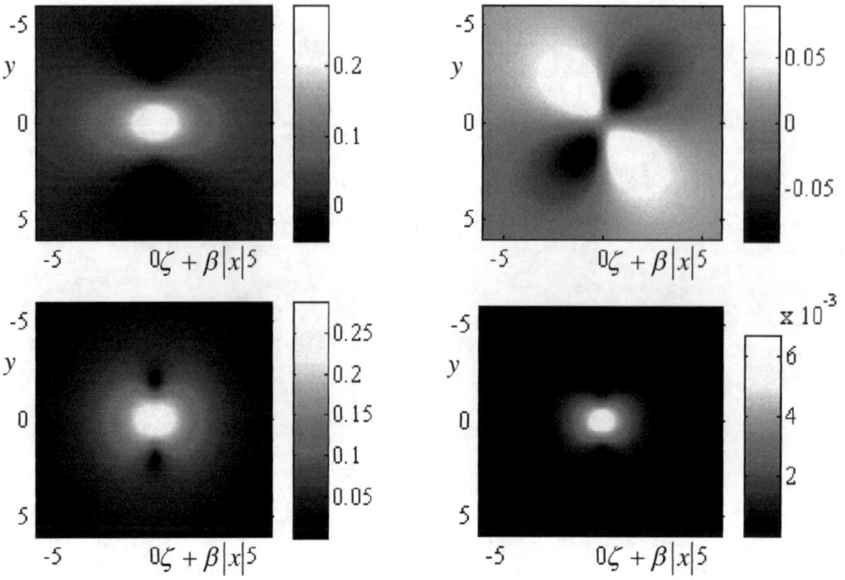

Figure 3. «Snapshots» for E_z^{W} (left top), E_y^{W} (right top), modulus of $\vec{E}^{\mathrm{W}}$ (left bottom) and absolute value of Pointing vector $\vec{S}^{\mathrm{W}}$ (right bottom) in the case of Gaussian bunch at $x = \mathrm{const}$; $\sigma = c/\omega_{\mathrm{p}}$, distances are given in units c/ω_{p} , field strength in units $q\omega_{\mathrm{p}}^2/c^2$, energy flow density in units $q^2\omega_p^4/c^3$.

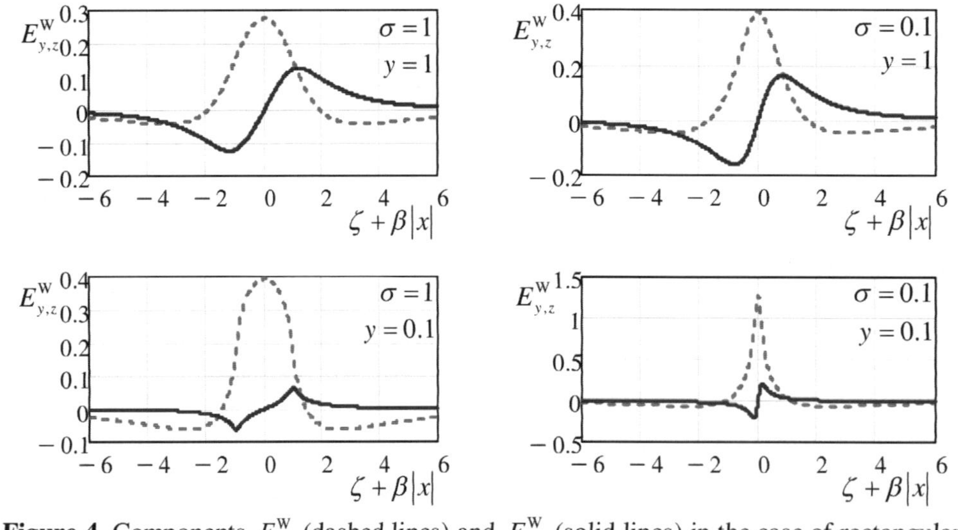

Figure 4. Components E_z^{W} (dashed lines) and E_y^{W} (solid lines) in the case of rectangular bunch. Distances and parameter σ are given in units c/ω_{p} , intensities in units $q\omega_{\mathrm{p}}^2/c^2$.

Figure 3 represents the intensities of the electrical field components and magnitude of the Pointing vector on the plane $x = \mathrm{const}$ for the case of Gaussian bunch. One can see that the components E_z^{W} and H_y^{W} have a maximum at $\zeta = -\beta|x|$, $y = 0$. They decrease with increase in $\zeta + \beta|x|$ more slowly than with increase in y. Components E_y^{W} and H_z^{W} are alternating-sign. They are equal to zero in the point $\zeta = -\beta|x|$, $y = 0$ and have extremes in all four quadrants. Behavior of absolute value of $\vec{E}^{\mathrm{W}}$ is similar to behavior of E_z^{W} because this component is usually more than E_y^{W}. Note that the peak of the Pointing vector's absolute value is especially clear.

Analogous graphs for the rectangular bunch have a very similar appearance; therefore such images are not presented here. Instead them we give usual graphics of the electric force at $y = \mathrm{const}$ (figure 4). As one can see, component E_y^{W} has sharp peaks in the plane $y = \mathrm{const} << c/\omega_{\mathrm{p}}$ but $y \neq 0$ (figure 4, bottom graphs). We can use measurements of positions of these peaks for determination of the bunch length.

4. Conclusion

Electromagnetic field of the charge moving perpendicularly to wires in the wire metamaterial was considered. Analytical and numerical investigations were shown that Vavilov-Cherenkov radiation has unusual properties in this situation. It has been shown that radiation propagates along the wires. The radiation field has two components of electric force: one is parallel to the charge velocity and another is perpendicular to it and to the wires. The magnetic force behaves itself the similar way. The radiation field does not decrease with increase in distance along some "rays" behind the charge (if the losses are negligible). In the case of point charge, radiation concentrates nearby these "rays". One of electric field component has the logarithmic singularity on these "rays" while the other is equal to zero on them. Computations of the wave fields of bunches with finite length show that the structure under consideration can be used for diagnostics of bunches.

Acknowledgments

This research was supported by the Saint Petersburg State University.

References

[1] Zrelov V P 1970 *Vavilov-Cherenkov Radiation in High-Energy Physics* (Jerusalem: Israel Program for Scientific Translations)
[2] Tyukhtin A V and Galyamin S N 2008 *Phys. Rev. E* **77** 066606
[3] Averkov Yu O, Kats A V and Yakovenko V M 2009 *Phys. Rev. B* **79** 193402
[4] Galyamin S N, Tyukhtin A V, Kanareykin A and Shoessow P 2009 *Phys. Rev. Lett.* **103** 194802
[5] Duan Z Y, Wu B-I, Xi S, Chen H S and Chen M 2009 *Progress in Electromag. Research* **90** 75
[6] Galyamin S N and Tyukhtin A V 2010 *Journal of Physics: Conference Series* **236** 012003
[7] Galyamin S N and Tyukhtin A V 2010 *Physical Review B* **81** 235134
[8] Belonogaya E S, Galyamin S N and Tyukhtin A V 2011 *J. Opt. Soc. of America. B* **28** 2871
[9] Belov P A, Tretyakov S A and Viitanen A J 2002 *J. Electromagn. Waves Appl.* **16** 1153
[10] Demetriadou A and Pendry J B 2008 *J. Phys.: Condens.Matter* **20** 295222
[11] Maslovski S I and Silveirinha M G 2009 *Phys. Rev. B* **80** 245101
[12] Tyukhtin A V and Doilnitsina E G 2011 *Journal Of Physics D: Applied Physics* **44** 265401
[13] Tyukhtin A V, Doilnitsina E G and Kanareykin A 2010 *Proc.1st Int. Particle Accelerator Conf. IPAC'10* (Kyoto, Japan, 23–28 May 2010) 1071
[14] Prudnikov A P, Brychkov Yu A and Marichev O I 1983 *Integrals and Series. Special Functions.* (Moscow: Nauka)

Journal of Physics: Conference Series **357** (2012) 012007 doi:10.1088/1742-6596/357/1/012007

Spin light of a neutron in classical and quantum theories

V A Bordovitsyn, O A Konstantinova

Department of theoretical physics, National Research Tomsk State University, Lenin avenue 36, Tomsk, 634050, Russia

E-mail: olgakonst87@mail.ru, vabord@sibmail.com

Abstract. In this paper, we try to give a systematic exposition of the classical and quantum description of relativistic neutrons radiation. Here we show that the correspondence principle in the spin precession entails the complete agreement of the classical and quantum properties of radiation (total power, angular distribution, linear and circular polarization in a full accordance with the results of well-known quantum theory of A. Ternov, V. Bagrov, A. Khapaev. Finally, we give estimates for the time and length of radiation self-polarization of the neutrons.

1 Introduction

Spin light is a radiation associated with the spin precession of the proper magnetic moment of elementary particles [1].

In the classical theory the spin precession of the neutron with a magnetic moment $\mu_N = -|\mu| = -1,93\mu_{nucl.} < 0$ in the uniform magnetic field $\vec{H} = (0,0,H)$ is described by solution of Bargmann-Michel-Telegdi (BMT) equation [2]

$$\Pi^{\mu\nu} = \Pi_1^{\mu\nu} - \frac{1}{\Omega} q^{[\mu\alpha}\Pi_{\alpha 1}^{\nu]} \sin\Omega\tau - (2\Pi_1^{\mu\nu} + \frac{1}{\Omega^2} q^{[\mu\alpha} q_{\alpha\beta}\Pi_1^{\beta\nu]})(1-\cos\Omega\tau)\alpha\alpha.$$

Here $\Pi^{\mu\nu}$ is a dimensionless spin tensor with the invariant

$$\frac{1}{2}\Pi_{\mu\nu}\Pi^{\mu\nu} = \frac{1}{2}\Pi_{\mu\nu 1}\Pi_1^{\mu\nu} = 1.$$

Index "1" corresponds to the initial conditions, a brackets denote an antisymmetrization $a^{[\mu}b^{\nu]} = a^\mu b^\nu - a^\nu b^\mu$, Ω is a neutron spin precession frequency in the units of the proper time τ:

$$\Omega = \frac{2|\mu|H}{\hbar}\sqrt{1-\beta^2 cos^2\alpha} = \omega\gamma, \qquad (1)$$

when it moves in accordance with $\vec{\beta} = \beta(sin\alpha, 0, cos\alpha)$, the tensor $q^{\mu\nu}$ is space-like part of the tensor $h^{\mu\nu} = -2(|\mu|/\hbar)H^{\mu\nu}$.

Description of the spin precession in quantum theory is given by a nonstationary wave function $\Psi(\vec{r},t)$ of Dirac-Pauli equation, and the initial conditions are given by the formula (see [3])

$$\hat{\vec{\Pi}}\psi(\vec{r},0) = \vec{\lambda}\psi(\vec{r},0). \qquad (2)$$

In the following it is showed that for the same initial conditions the identical solutions for the spin precession as in classical theory can be obtained.

2 Classical theory of neutron radiation

The classical theory of neutron spin light was built by V.A. Bordovitsyn et al [2]. Here we give a comparison of this theory with the quantum theory of radiation of the neutron.

2.1 Total power of radiation

At the heart of the classical theory of radiation is the field stress tensor in the wave zone [4]

$$\tilde{H}^{\mu\nu} = -|\mu| c \frac{\ddot{\Pi}^{[\mu\lambda}\tilde{r}_\lambda \tilde{r}^{\nu]}}{(\tilde{r}_\rho v^\rho)^3} \sim \frac{1}{\tilde{r}}.$$

Here $\tilde{r}^\mu = (\tilde{r}, \vec{\tilde{r}})$ is the light-like vector drawn from the world point of neutron to the observer at time $\tilde{t}$, the dot denotes the derivative with respect to the proper time τ. In addition, there are $\bar{H}$ and $\bar{\bar{H}}$ tensors of the mixed and convective fields correspondingly.

$$\bar{H} \sim \frac{1}{\tilde{r}^3}; \bar{\bar{H}} \sim \frac{1}{\tilde{r}^2}.$$

Knowing the stress tensor fields, we can construct a tensor density of energy in the wave zone

$$\tilde{\mathscr{E}}^{\mu\nu} = -\frac{1}{4\pi}(\tilde{H}^{\mu\rho}\tilde{H}^{\nu}_{\rho} - \frac{1}{4}g^{\mu\nu}\tilde{H}_{\alpha\beta}\tilde{H}^{\alpha\beta})$$

or

$$\tilde{\mathscr{E}}^{\mu\nu} = -\frac{c^2\mu^2}{4\pi}\frac{\tilde{r}_\alpha \ddot{\Pi}^{\alpha\rho}\ddot{\Pi}_{\rho\beta}\tilde{r}^\beta}{(\tilde{r}_\rho v^\rho)^6}\tilde{r}^\mu \tilde{r}^\nu.$$

To obtain the radiation power in the most general form, we will proceed from the expression

$$\tilde{P}^\mu = \frac{1}{c}\oint \tilde{\xi}^{\mu\nu} d\sigma_\nu.$$

Here $d\sigma_\nu$ is the element of the spatial hypersurface: $d\sigma_\nu = \varepsilon^2 e_\nu d\Omega_0 c d\tau$, $\varepsilon = \tilde{r}\gamma[1-(\vec{n}\vec{\beta})] = -\tilde{r}_\rho v^\rho / c$; $\vec{n}$ is the unit vector along the direction of wave motion; e_ν is a unit space-like vector of Rohrlich; $e_\nu e^\nu = 1$; $e_\nu v^\nu = 0$; $\tilde{r}^\mu = \varepsilon(e^\mu + v^\mu/c)$; $d\Omega_0$ is the element of solid angle in a rest frame; $d\Omega_0 = c^2/(n_\rho v^\rho)^2 d\Omega$; in addition $\Pi^{\alpha\rho}v_\alpha = 0$ and $\ddot{\Pi}^{\alpha\rho}v_\alpha = 0$ due to the property of $\Pi^{\alpha\rho}$ tensor.

Further we provide the calculation using the technique of covariant integration [2] and guiding by the subsidiary equations above. Thus we obtain a general expression for the power of radiation

$$\frac{d\tilde{P}^\mu}{d\tau} = \frac{\mu^2 v^\mu}{3c^5}\ddot{\Pi}_{\alpha\beta}\ddot{\Pi}^{\alpha\beta}. \tag{3}$$

Next we consider the particular type of spin precession. Expression of $\ddot{\Pi}_{\alpha\beta}$ is obtained from the classical solution of the BMT

$$\begin{cases} \Pi_x = -|\zeta_\perp|\sin\omega t/\sqrt{1-\beta^2\cos^2\alpha}, \\ \Pi_y = |\zeta_\perp|\gamma\cos\omega t, \\ \Pi_z = 0. \end{cases}$$

Here the initial condition for a spin vector $\vec{\zeta}|_{t=0} = (0, \zeta_\perp, 0)$ It should be noted, that this case is equal to the initial spin direction parallel to the Y axis as in the quantum theory (see the condition (2)). After averaging over the time from (3) the following formula for the total power of radiation is obtained

$$\bar{W} = \frac{2^5}{3}\frac{\mu^6 H^4}{c^3\hbar^4}\frac{(1-\beta^2 cos^2\alpha)^2}{(1-\beta^2)^2}, \tag{4}$$

which is in a full accordance with the result of quantum theory of Ternov-Bagrov-Khapaev [5].

2.2 Angular distribution and polarization

To study the polarization properties, we use the following considerations. The angular distribution $dW/d\Omega$ can be obtained with respect to formula [2]

$$dW = \frac{\tilde{r}^2 c}{4\pi} \int \bar{E}^2 (1 - (\vec{n}\vec{\beta})) d\Omega,$$

here $\bar{E} = \sqrt{E_\theta{}^2 + E_\varphi{}^2}$ is the electromagnetic field strength tensor, obtained from the tensor of the radiation field by the rule $E_s = (\vec{E}\vec{n}_s)$, where index $s = \theta$, φ and $\vec{n}_s$ are the vectors perpendicular to the normal vector $\vec{n} = [\vec{n}_\theta \vec{n}_\varphi]$ (these vectors define the polarization). In the explicit form

$$\vec{n}_\theta = (\cos\theta\cos\varphi, \cos\theta\sin\varphi, -\sin\theta), \quad \vec{n}_\varphi = (-\sin\varphi, \cos\theta, 0), \quad \vec{n} = (\sin\theta\cos\varphi, \sin\theta\sin\varphi, \cos\theta).$$

The relation between $\vec{n}$ and E_s

$$E_s = -|\mu| \frac{[\vec{n}[\vec{n}[(\vec{n} - \vec{\beta})\ddot{\Pi}]]]}{c^2 \gamma^3 \tilde{r} (1 - \vec{n}\vec{\beta})^3}.$$

Finally, to obtain the angular distribution in a more convenient form the transformation of rotation of the coordinate axes should be done to let the $\vec{\beta}$ axis coincides with the X' axis (right-hand turn around the Y axis)

$$\vec{n}' = (\sin\theta' \cos\varphi', \sin\theta' \sin\varphi', \cos\theta'),$$

$$\vec{n} = (\cos\alpha \sin\theta' \cos\varphi' + \sin\alpha \cos\theta', \sin\theta' \sin\varphi', -\sin\alpha \sin\theta' \cos\varphi' + \cos\alpha \cos\theta').$$

Thus, one can obtain the angular distribution of the radiation with respect to the certain polarization

$$\frac{dW_s}{d\Omega} = \frac{8\mu^6 H^4}{\pi c^3 \hbar^4} \frac{(1 - \beta^2 \cos^2\alpha)^2}{(1 - \beta\cos\theta')^5} E_s{}^2.$$

For the components of linear polarization in the case of an axially symmetric spin-light ($\alpha = 0$), we have

$$E_\theta = -\frac{\mu\zeta_\perp \Omega^2}{\tilde{r}\gamma^2 c^2} \frac{(\cos\theta - \beta)\sin(\omega t - \varphi)}{(1 - \beta\cos\theta)^3},$$

$$E_\varphi = -\frac{\mu\zeta_\perp \Omega^2}{\tilde{r}\gamma^2 c^2} \frac{\cos(\omega t - \varphi)}{(1 - \beta\cos\theta)^2},$$

where Ω is determined in (1); $\zeta = \pm 1$. In the case of an axially nonsymmetric spin-light ($\alpha = \pi/2$) easy to obtain

$$E_\theta = -\frac{\mu\zeta_\perp \Omega^2}{\tilde{r}\gamma^3 c^2} \frac{\cos\theta(\cos\varphi \sin\omega t - \gamma^2 \sin\varphi \cos\omega t)}{(1 - \beta\sin\theta\cos\varphi)^3},$$

$$E_\varphi = -\frac{\mu\zeta_\perp \Omega^2}{\tilde{r}\gamma^3 c^2} \frac{\sin\varphi\sin\omega t + \gamma\cos\omega t(\cos\varphi - \beta\sin\varphi)}{(1 - \beta\cos\theta)^3}.$$

Then, for the components of the right ($s = 1$) and the left ($s = -1$) circular polarization we have correspondingly

$$E_s = \frac{1}{\sqrt{2}} \frac{\mu\zeta_\perp \Omega^2}{\tilde{r}\gamma^2 c^2} \frac{(1 + s\cos\theta)(1 + s\beta)}{(1 - \beta\cos\theta)^3}.$$

It should be noted that the components of circular polarization can be obtained from the equations of motion automatically. One can see that the angular distribution ($\alpha = 0$) has an azimuthal symmetry in contrast to the angular distribution ($\alpha = \pi/2$)

After integrating over angles and after averaging over time we obtain correspondingly the following expressions for the total radiation power

$$\bar{W} = \frac{2^5}{3}\frac{\mu^6 H^4}{c^3 \hbar^4},$$

$$\bar{W}_\theta = \frac{4 + 2\beta^2 - 2\beta^4}{16}\bar{W},$$

$$\bar{W}_\varphi = \frac{12 - 2\beta^2 + \beta^4}{16}\bar{W}.$$

For the circular polarization we have

$$\bar{W}_{\pm 1} = \frac{1}{2}(1 \pm \frac{1}{2}\beta)\bar{W}.$$

From this in the nonrelativistic case one can obtain

$$\bar{W}_\theta = \frac{1}{4}\bar{W}, \quad \bar{W}_\varphi = \frac{3}{4}\bar{W},$$

and correspondingly in the ultrarelativistic case:

$$\bar{W}_\theta = \frac{5}{16}\bar{W}, \quad \bar{W}_\varphi = \frac{11}{16}\bar{W}.$$

3 Comparison with the quantum theory of Ternov-Bagrov-Khapaev

Now we show that absolutely all the properties of neutron radiation in classical and quantum theory are the same.

In the classical theory of radiation frequency can be found from the condition of invariance of 4-dimensional scalar product

$$k_\mu v^\mu|_{\beta \to 0} = k_{\mu 0} v^{\mu 0} = inv.$$

Here k^μ is the 4-dimensional light-like wave vector, $k^\mu = \tilde{\omega}/c(1, \vec{n})$, $k_\mu k^\mu = 0$, $v^\mu = (v^0, \vec{v})$ is a 4-dimensional velocity; $\tilde{\omega}$ is the radiation frequency,

$$\tilde{\omega} = \frac{\tilde{\omega}_0}{\gamma(1 - (\vec{n}\vec{\beta}))}. \tag{5}$$

Thus, in the rest frame radiates the fundamental frequency is radiated. It coincides with the mechanical precession frequency. The relation between radiation frequency in the rest frame $\tilde{\omega}_0$ and precession frequency ω is the following

$$\tilde{\omega}_0 = \omega\gamma = \Omega.$$

Let us turn to the laboratory frame. On the basis of (5) we get

$$\tilde{\omega} = \frac{\omega}{1 - (\vec{n}\vec{\beta})} = \frac{2|\mu|H}{\hbar}\frac{\sqrt{1 - \gamma^2\cos^2\alpha}}{1 - \beta\cos\theta'}. \tag{6}$$

Now we show that the same formula can be obtained in the relativistic quantum theory. Thereby, in the quantum theory radiation frequency can be obtained on the basis of conservation laws of energy and momentum (in the dimensionless form), taking into account the spin states

$$\begin{cases} \gamma_\zeta - \gamma'_{\zeta'} = \hbar\tilde{\omega}/m_0c^2, \\ \vec{b} - \vec{b}' = \hbar\tilde{\omega}\vec{n}/m_0c^2, \end{cases}$$

here γ_ζ is obtained from the solution of Dirac-Pauli equation (with respect to spin). It is easy to see that

$$\tilde{\omega} = \frac{2\zeta|\mu|H\sqrt{1-\beta^2\cos^2\alpha}}{\hbar(1-\beta\cos\theta')},$$

which is exactly the same with (6), when ζ=1. This corresponds to the established in [5] selection rule: only a neutron spin of which was oriented in the direction of the magnetic field emits. In this case the radiation necessarily accompanied by a change of spin direction.

All the essential features of the neutron radiation follows the intermediate relation, derived from the relation

$$\frac{|\mu|H}{m_0c^2}(\zeta\sqrt{1-\beta^2\cos^2\alpha} - \zeta'\sqrt{1+\beta^2\cos^2\alpha}) = \frac{\hbar\omega\gamma_\zeta}{m_0c^2}(1-\beta\cos\theta') > 0. \tag{7}$$

Let's turn to the rest frame ($\beta \to 0$). Formula (7) becomes the following (this condition is satisfied only if $\zeta = -\zeta' = 1$)

$$\frac{|\mu|H}{m_0c^2}(\zeta - \zeta') = \frac{\hbar\omega\gamma_\zeta}{m_0c^2} > 0; \quad \zeta = -\zeta' = 1.$$

In the rest frame the relation between the power of radiation W_0 and the total probability of the radiation w_0 is

$$W_0 = \hbar\tilde{\omega}_0 w_0.$$

where

$$w_0 = \frac{32}{3}\frac{|\mu|^5 {H_0}^3}{c^3\hbar^4}| < \zeta'|\vec{\sigma}|\zeta > |^2,$$

and the matrix element built on Pauli matrices $\vec{\sigma}$ is easily to obtained by suitable spinors

$$< \zeta'|\vec{\sigma}|\zeta > = \frac{1-\zeta\zeta'}{2}(\vec{\imath} + i\zeta\vec{\jmath}) + \frac{1+\zeta\zeta'}{2}(\zeta\vec{k}). \tag{8}$$

The first member of the right part of (8) corresponds to the spin-flip transitions ($\zeta = -\zeta'$), and the second one corresponds to the transitions without flip ($\zeta = \zeta'$). Then we take into account the known fact that here all the transitions are spin-flip transitions [5]. It leads to

$$| < \zeta'|\vec{\sigma}|\zeta > |^2 = 2\frac{1-\zeta\zeta'}{2} + \frac{1+\zeta\zeta'}{2}|_{\zeta=-\zeta'} = 2,$$

and

$$W_0 = \frac{2^7}{3}\frac{|\mu|^5 {H_0}^3}{c^3\hbar^4}.$$

Such result we can see in works [5] and [6]. However, comparing with (4) one can find a two orders difference in the coefficient. This is due to the fact that in classical theory, we performed a time averaging. While the authors of articles below didn't performed it. Accordingly there is a need to average over initial states and summation over final

$$\sum_{\zeta\zeta'}\overline{\frac{1-\zeta\zeta'}{2}} = \frac{1}{2}\sum_{\zeta\zeta'}\frac{1+\zeta\zeta'}{2} = \frac{1}{4}.$$

It is easy to see that after this procedure all of the differences in the coefficients are vanished. Other features are exactly the same.

4 Radiative self-polarization of neutrons

This effect was first discovered for relativistic electrons in the quantum theory of synchrotron radiation, and later confirmed experimentally (see, for example, [7]). In the case of neutron spin-light radiation self-polarization time in quantum theory is determined by the reciprocal of the total probability of radiation [6].

$$T_0 = \frac{1}{w_0} = \frac{3}{32}\frac{c^3\hbar^4}{|\mu|^5 {H_0}^3}\frac{1}{|<\zeta'|\vec{\sigma}|\zeta>|^2} = \frac{3}{64}\frac{c^3\hbar^4}{|\mu|^5 H^3}. \tag{9}$$

To get the same effect in the classical theory let us deal with the rest frame where the spin precession leads to the minimum potential energy:

$$U = -(\vec{\mu}\vec{H})|_{\beta=0} = (\vec{\mu}_0\vec{H}_0) = \mu_0 H_0 \cos\lambda|_{\lambda\to\pi} \to min,$$

when the neutron spin is oriented against the direction of the magnetic field; λ is the angle between the direction of magnetic field and magnetic moment. Thus, the magnetic moment and also spin are oriented against the direction of magnetic field.

Considering the process of radiation as a non stationary in the rest frame where $\vec{\mu}|_{\beta=0} = -\mu_0\vec{\zeta} = -\mu_0(\vec{\zeta}_\parallel + \vec{\zeta}_\perp)$ we obtained the following equation for the radiative self-polarization [8]

$$\frac{d\zeta_\parallel}{d\tau} = \frac{1}{2T_0}(1 - {\zeta_\parallel}^2),$$

The solutions of this equation for an initially unpolarized neutron leads to

$$\zeta_\parallel(\tau) = \frac{exp(-\tau/T_0) - 1}{exp(-\tau/T_0) + 1}|_{\tau\to\infty} \to 1,$$

where T_0 is given by (9). Taking into account the time of radiative self-polarization for the neutron with the velocity $\vec{\beta} = \beta(\sin\alpha, 0, \cos\alpha)$ we find the expression

$$T = \frac{3}{64}\frac{c^3\hbar^4}{|\mu|^5 H^3}\frac{1}{\gamma^2(1 - \beta^2 cos^2\alpha)^{3/2}}.$$

To sum up, for a relativistic 10^{13}GeV neutron and $H = 10^6$Gs, $\alpha = \pi/2$ we find the time $T \approx 1,3 \cdot 10^{-7}$sec and length of the self-polarization $L \approx 4 \cdot 10^3$sm. However, as shown by this calculation, for the neutron to compared with the electron this effect is small and becomes noticeable only at high magnetic fields and at very high energies. The classical and the quantum theories of the spin light and radiative self-polarization of the neutron fully adequate to each other.

Acknowledgments

And now in closing, the authors wish to take this opportunity to thank Ms. L.V. Puzyrevich (Research Institute of Yad TPU) who gave the important advises in publishing of this paper. The work was partially supported by the Federal Targeted Program "Scientific and scientific - pedagogical personnel of innovative Russia", contract No 02.740.11.0238; No P789.

References

[1] Bordovitsyn V A, Ternov I M, Bagrov V G 1995 *Usp. Fiz. Nauk* **165** 1083
[2] Bordovitsyn V A 1999 *Synchrotron Radiation Theory and it's Development. In Memory of Ternov* ed V A Bordovitsyn (Singapore: World Scientific)
[3] Bordovitsyn V A, Konstantinova O A 2010 *Rus. Phys. J.* **53** 1210
[4] Bordovitsyn V A, Konstantinova O A, Nemchenko E A 2010 *J. Phys.: Conf. Series* **236** 012002
[5] Ternov I M, Bagrov V G, Khapaev A M 1965 *JETF* **21** 613
[6] Lyuboshits V L 1966 *Yad.Fiz.* **4** 2693
[7] Sokolov A A, Ternov I M 1986 *Radiation from Relativistic Electrons*, (New York: AIP)
[8] Ternov I M, Bordovitsyn V A, Sorokin S V 1983 *Izv. VUZov Fiz.* **8** 122

Journal of Physics: Conference Series **357** (2012) 012008 doi:10.1088/1742-6596/357/1/012008

Radiation from a Charged Particle Flying Through a Dielectric Ball

S R Arzumanyan
NAS Institute of Applied Problems in Physics, 25 Hr. Nersessian Str., 0014 Yerevan, Armenia

E-mail: arzumsr@mail.ru

Abstract. The radiation from a relativistic charged particle uniformly moving through the centre of a dielectric ball is investigated. Analytical expressions for spectral and spectral-angular distributions of the radiated energy are derived and numerical results for the spectrum of radiation are given. It is shown, that the spectral distribution of Cherenkov radiation generated by the relativistic particle inside a dielectric ball at specified frequencies is strongly influenced by the ball-vacuum boundary.

1. Introduction

The presence of matter may essentially influence the characteristics of high energy electromagnetic processes, generating new types of phenomena [1]. The synchrotron radiation from a charged particle rotating in a homogeneous medium is considered in [2] (see also [3, 4]). It has been shown that the interference between synchrotron radiation and Cherenkov radiation (CR) leads to interesting consequences. New interesting phenomena occur in the case of inhomogeneous media.

The investigations of radiation from a charged particle rotating along the equatorial orbit about/inside a dielectric ball showed [5-7] that when the Cherenkov condition for the ball material and particle speed is satisfied, there appear high narrow peaks in the spectral distribution of the number of electromagnetic field quanta emitted to the outer space for some specific values of the ratio of ball-to-particle orbit radii. It is worthwhile to point that the similar phenomenon takes place in the case of cylindrical symmetry for, e.g., the radiation emitted from a charged particle rotating about/inside a dielectric cylinder [8] (for a more general case of a particle moving along a helical orbit see [9]). The radiation from a relativistic particle moving along a rectilinear trajectory passing through the centre of a dielectric or metallic ball was investigated in [10] (see also [11]). However, in [10,11] the phenomena of the absorption and dispersion of electromagnetic waves inside the ball material were not considered. In the present paper the dielectric losses of electromagnetic waves inside the ball material are taken into account, and the features of CR from a relativistic particle conditioned by the influence of the ball surface are analyzed.

2. Formulation of problem and basic formulas

In [12, 13] a method is proposed for the evaluation of the Green function of the electromagnetic field in a medium consisting of $N \geq 1$ spherically-symmetric layers having a common center and different permittivities. Based on this method, a formula is derived for the energy of radiation from a charged particle moving along an arbitrary trajectory in such medium. The special cases of a charged particle rotating about or inside a dielectric ball have been investigated in [5–7]. In the present paper another example is investigated when a charged particle flies rectilinearly and uniformly in a medium with the permittivity ε_1 and passes through a centre of a dielectric ball (with the radius r_1 and the dielectric permittivity ε_0) immersed in such medium (see figure 1, the magnetic permeability of matter is taken to be 1). The trajectory of the particle is described by the equation $z = \mathrm{v}t$.

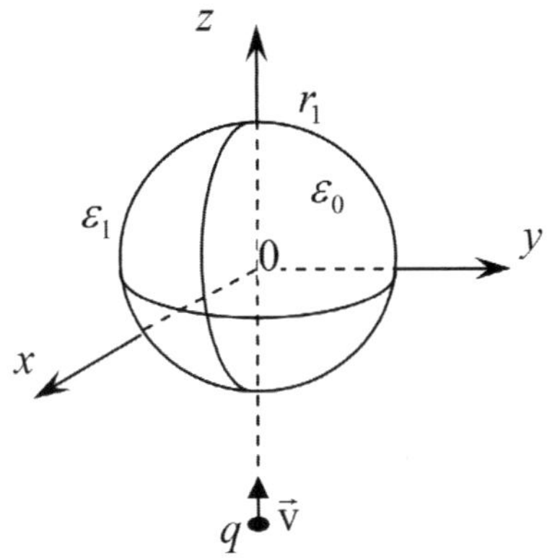

Figure 1. A particle flying from a medium into a ball.

In the case under consideration the current density can be written as

$$\vec{j}(\vec{r},t) = \frac{q\vec{\mathrm{v}}}{\pi r^2 \sin\theta}\begin{cases}\delta(r-\mathrm{v}t)\delta(\theta) & \text{for} \quad t>0,\\ \delta(r+\mathrm{v}t)\delta(\theta-\pi) & \text{for} \quad t<0,\end{cases} \tag{1}$$

where q and v are the charge and velocity of the particle, and θ is the polar angle of corresponding spherical system of coordinates r,θ,φ. For the corresponding Fourier transform one has

$$\vec{j}(\vec{r},\omega) = \frac{1}{2\pi}\int \vec{j}(\vec{r},t)e^{i\omega t}dt = \frac{q\vec{\mathrm{v}}}{2\pi^2 \mathrm{v} r^2 \sin\theta}[\delta(\theta)e^{i\frac{\omega}{\mathrm{v}}r} + \delta(\theta-\pi)e^{-i\frac{\omega}{\mathrm{v}}r}]. \tag{2}$$

Expanding $\vec{j}(\vec{r},\omega)$ in spherical vectors $\vec{X}^{(\mu)}_{lm}$ [14], we can evaluate the corresponding expansion coefficients:

$$j^{lm}_{\mu}(r) \equiv \int \vec{j}(\vec{r},\omega)\vec{X}^{(\mu)*}_{lm} d\Omega. \tag{3}$$

As a result we find

$$j^{lm}_1 = \frac{q\delta_{m0}}{2\pi r^2}\sqrt{\frac{2l+1}{4\pi}}[e^{i\frac{\omega}{\mathrm{v}}r} - (-1)^l e^{-i\frac{\omega}{\mathrm{v}}r}], \qquad j^{lm}_2 = j^{lm}_3 = 0. \tag{4}$$

Similarly, one can expand the vector potential $\vec{A}(\vec{r},\omega)$ of the electromagnetic field in spherical vectors $\vec{X}^{(\mu)}_{lm}$. In the Lorentz gauge the expansion coefficients A^{lm}_{μ} are given by the expressions

$$\frac{(2l+1)cr_1^2}{4\pi}\begin{bmatrix}A^{lm}_1(r)\\ A^{lm}_2(r)\end{bmatrix} = \begin{bmatrix} lu_{\underline{l-1}} + (l+1)u_{\underline{l+1}} \\ \sqrt{l(l+1)}(u_{\underline{l-1}} - u_{\underline{l+1}})\end{bmatrix}^{lm}_{r} + \gamma_l[lB^{lm}_{\underline{l-1}} - (l+1)B^{lm}_{\underline{l+1}}]\begin{bmatrix} lP_{\underline{l-1}} + (l+1)P_{\underline{l+1}} \\ \sqrt{l(l+1)}(P_{\underline{l-1}} - P_{\underline{l+1}})\end{bmatrix}_{(r,r_1)} \tag{5}$$

(for details see [13]). The magnetic type multipoles ($\mu = 3$) are not generated: $A^{lm}_3(r) = 0$. Here

$$u^{lm}_{\underline{l_1}}(r) \equiv \int_0^{\infty} P_{\underline{l_1}}(r,x)j^{lm}_1(x)x^2 dx,$$

$$B_{\underline{l_1}}^{lm} = \frac{q\delta_{m0}}{2\pi}\sqrt{\frac{2l+1}{4\pi}}\{\lambda_0 j_l(\lambda_0 r_1)\int_{r_1}^{\infty} h_{\underline{l_1}}^{(1)}(\lambda_1 x)[e^{i\frac{\omega}{v}x} - (-1)^l e^{-i\frac{\omega}{v}x}]dx +$$
$$+\lambda_1 h_l^{(1)}(\lambda_1 r_1)\int_0^{r_1} j_{\underline{l_1}}(\lambda_0 x)[e^{i\frac{\omega}{v}x} - (-1)^l e^{-i\frac{\omega}{v}x}]dx\}, \tag{6}$$

with the notations

$$\gamma_l = \frac{1/\varepsilon_0 - 1/\varepsilon_1}{l z_{l-1}^l + (l+1)z_{l+1}^l},\quad \lambda_i = \frac{\omega}{c}\sqrt{\varepsilon_i},\quad z_\upsilon^l = \frac{\lambda_1 j_\upsilon(\lambda_0 r_1)h_l^{(1)}(\lambda_1 r_1)/\varepsilon_1 - \lambda_0 j_l(\lambda_0 r_1)h_\upsilon^{(1)}(\lambda_1 r_1)/\varepsilon_0}{\lambda_1 j_\upsilon(\lambda_0 r_1)h_l^{(1)}(\lambda_1 r_1) - \lambda_0 j_l(\lambda_0 r_1)h_\upsilon^{(1)}(\lambda_1 r_1)} \tag{7}$$

and $h_l^{(1)}(y) = j_l(y) + i n_l(y)$ ($j_l(y)$ and $n_l(y)$ are the spherical Bessel and Neumann functions respectively). In (6) the following notations are used

$$f_{\underline{l}}(\tau) \equiv f_l(\tau)/[\lambda_0 j_{l-1}(\lambda_0 r_1)h_l^{(1)}(\lambda_1 r_1) - \lambda_1 j_l(\lambda_0 r_1)h_{l-1}^{(1)}(\lambda_1 r_1)], \tag{8}$$

and the definition of the function $P_l(r,x)$ is given in [13]. For the case under consideration ($r > r_1$) one obtains

$$P_l(r,r_1) = h_l^{(1)}(\lambda_1 r)j_l(\lambda_0 r_1). \tag{9}$$

From the relations

$$h_l^{(1)}(x) \approx (-i)^{l+1}\frac{e^{ix}}{x},\quad j_l(x) \approx \frac{1}{x}\sin(x - l\pi/2),\qquad x >> l \tag{10}$$

for the spherical Bessel and Hankel functions [15, 16] it follows that at large distances from the ball one has

$$A_\mu^{lm}(r) \approx \delta_{m0} a_\mu^l \frac{e^{i\lambda_1 r}}{\lambda_1 r} + \ldots,\qquad r \to \infty. \tag{11}$$

Here

$$\begin{bmatrix} a_1^l \\ a_2^l \end{bmatrix} = \frac{(-i)^l q}{c r_1^2\sqrt{\pi(2l+1)}}\begin{bmatrix} l\mathrm{w}_{\underline{l-1}} - (l+1)\mathrm{w}_{\underline{l+1}} \\ \sqrt{l(l+1)}(\mathrm{w}_{\underline{l-1}} + \mathrm{w}_{\underline{l+1}}) \end{bmatrix},\qquad a_3^l = 0 \tag{12}$$

and

$$\mathrm{w}_{\underline{l_1}} \equiv J_{\underline{l_1}}(<) + i\lambda_1 r_1^2\left[a_{\underline{l_1}}^{(12)} J_{\underline{l_1}}(>) + a_{\underline{l_1}}^{(22)} H_{\underline{l_1}}\right] + \gamma_l j_{\underline{l_1}}(\lambda_0 r_1)\left[lQ_{\underline{l-1}} - (l+1)Q_{\underline{l+1}}\right]. \tag{13}$$

In (13) the following notations are introduced

$$J_{\underline{l_1}}(<) \equiv \int_0^{r_1} j_{\underline{l_1}}(\lambda_0 x)[e^{i\frac{\omega}{v}x} - (-1)^l e^{-i\frac{\omega}{v}x}]dx,\quad J_{\underline{l_1}}(>) \equiv \int_{r_1}^{\infty} j_{\underline{l_1}}(\lambda_1 x)[e^{i\frac{\omega}{v}x} - (-1)^l e^{-i\frac{\omega}{v}x}]dx,$$
$$H_{\underline{l_1}} \equiv \int_{r_1}^{\infty} h_{\underline{l_1}}^{(1)}(\lambda_1 x)[e^{i\frac{\omega}{v}x} - (-1)^l e^{-i\frac{\omega}{v}x}]dx,\quad Q_{\underline{l_1}} = \lambda_0 j_l(\lambda_0 r_1)H_{\underline{l_1}} + \lambda_1 h_l^{(1)}(\lambda_1 r_1)J_{\underline{l_1}}(<). \tag{14}$$

According to [12] the spectral-angular and spectral distributions of the radiation energy (during the all time of charge motion) are determined by the expressions

$$\frac{dI}{d\omega\, d\Omega} = \frac{c}{\sqrt{\varepsilon_1}}\left|\sum_l a_2^l \vec{X}_{l0}^{(2)}\right|^2 \tag{15}$$

and

$$\frac{dI}{d\omega} = \frac{c}{\sqrt{\varepsilon_1}}\sum_l |a_2^l|^2 \tag{16}$$

respectively.

Journal of Physics: Conference Series **357** (2012) 012008 doi:10.1088/1742-6596/357/1/012008

3. Numerical results

The dependence of dimensionless function $dI/\hbar d\omega$ on the cyclic frequency ω of the radiation is given in figure 2 for an electron flying through a dielectric ball surrounded by the vacuum ($\varepsilon_1 = 1$). The radius of the ball made of melted quartz and its permittivity are equal $r_1 = 4\,\text{cm}$ and $\varepsilon_0 = \varepsilon_0' + i\varepsilon_0'' = 3.78(1+0.0001i)$ [17] respectively (dispersion of electromagnetic waves inside the ball made of melted quartz is not taken into account). We consider the radiation with the wavelength of the order of ball size ($10 \le \omega \le 70\,\text{GHz}$) and the electron with energy 2MeV, so that the Cherenkov condition

$$\mathrm{v} > c/\sqrt{\varepsilon_0'} \tag{17}$$

for the velocity of electron and the substance of ball is satisfied. The dashed line in figure 2 corresponds to CR from an electron with the same energy moving in an infinite and transparent medium with $\varepsilon_0 = \varepsilon_0'$ under the condition that the radiation is collected from the path having the length $2r_1$ (diameter of the ball).

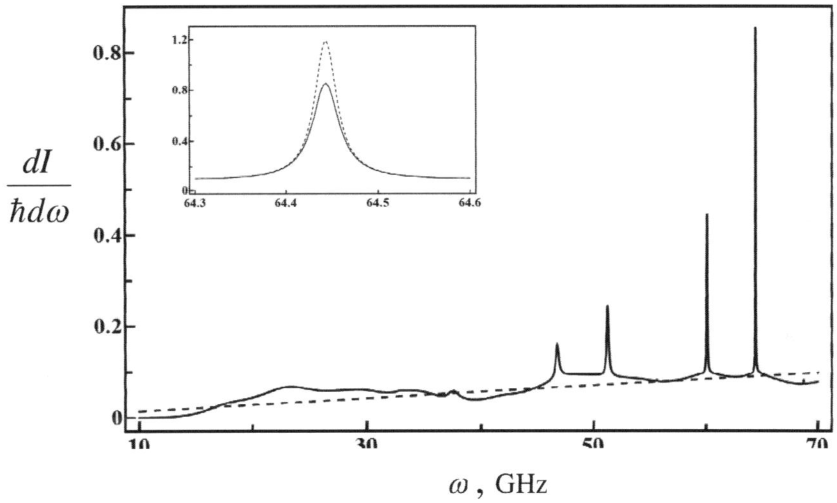

Figure 2. Radiation spectrum from an electron passing through the centre of a ball (continuous curve). The Cherenkov condition (17) is satisfied. For explanation see the text.

The radiation intensities (full and dashed lines) considerably differ only in the vicinity of the dedicated "resonant" frequencies (for example, $\omega \approx 64.45\,\text{GHz}$) with the wavelength of the order of r_1. At these frequencies the values of $dI/\hbar d\omega$ are almost ten times larger than those for neighbouring frequencies. The presence of the peaks indicates that the spectrum of CR generated by the particle inside the substance of the ball is strongly modified due to the influence of the ball-vacuum boundary. Moreover, if the dielectric losses of the radiated energy inside the substance of the ball are decreased, then the quantity $dI/\hbar d\omega$ practically is not influenced outside the regions near the peaks and it considerably increases at "resonant" frequencies $\omega \approx \omega_r$. This conclusion follows from the comparison of the curves in the top left corner of figure 2, where the dotted and continuous curves correspond to the ball made of loss-free dielectric and of melted quartz respectively. One can explain the abovementioned increase of the radiation intensity as result of a constructive superposition of the electromagnetic waves numerously reflected from the internal surface of the ball. Note that the situation here is qualitatively different compared to the case of dielectric plate. CR from the electron with energy 2 MeV is completely reflected from the boundaries of plate with dielectric permittivity $\varepsilon_0 = 3.78(1+0.0001i)$ immersed in the vacuum and it is confined inside the plate. Only the transition radiation penetrates outside the plate. But the intensity of transition radiation is considerably smaller than CR generated inside the ball.

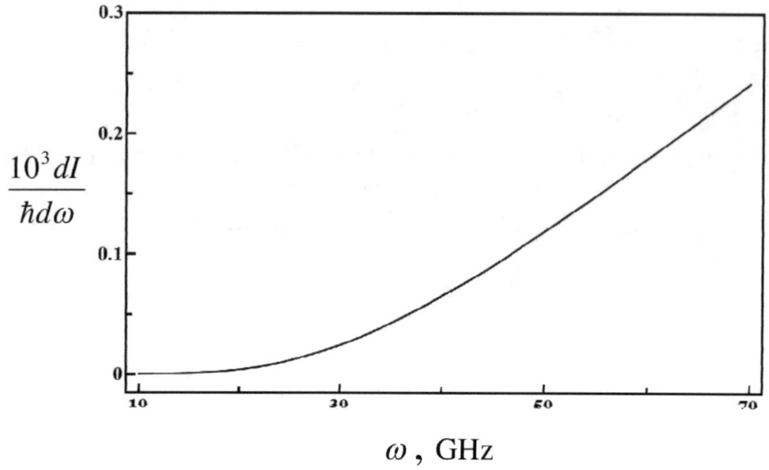

Figure 3. Spectral distribution of the radiation energy from an electron passing through the centre of the ball when the Cherenkov condition (17) is not satisfied.

The spectral distribution of radiated energy is plotted in figure 3 for the case when the Cherenkov condition (17) is not satisfied ($\varepsilon_0 = 1.05$, the values of the other parameters are the same as those in figure 2). As it is seen from figure 3, the spectral density of the radiation (transition radiation) is a monotonic function of the frequency and it is considerably smaller compared with the spectral density of CR in figure 2.

4. Conclusions

In this paper we have derived analytical expressions for the spectral and spectral-angular distributions of the radiation energy from a charged particle flying through the centre of a dielectric ball. Numerical results for the spectrum of the radiation energy from an electron with the energy 2MeV are given for two cases when the Cherenkov condition (17) is satisfied (figure 2) and when it is not satisfied (figure 3). The dispersion of the electromagnetic waves inside the ball material is not taken into account in numerical calculations. According to the data given in figures 2, 3, the spectral density of the radiated energy in the first case is almost ten times larger than the corresponding quantity in the second one. One can explain this difference by the dominance of CR over the transition radiation in the frequency range under consideration. In addition to the overall increase of the radiation intensity, narrow peaks appear in figure 2 at certain "resonant" frequencies. These peaks are related with the influence of the ball-vacuum interface on the CR generated by the particle inside the ball. It is noteworthy, that the total energy of the radiation (in the range of frequencies under consideration) is approximately equal to the radiation energy from an electron moving rectilinearly in an infinite transparent medium with $\varepsilon_0 = \varepsilon_0'$ when the radiation is accumulated from the path with the length equal to the diameter of the ball. This fact indicates that the narrow peaks at "resonant" frequencies in figure 2 are due to the spectral redistribution of the radiation influenced by the multiple reflection of CR from the ball-vacuum interface.

Acknowledgements
The author is grateful to L.SH. Grigoryan, H.F. Khachatryan for stimulating discussions and continued interest to this work, A.A. Saharian for assistance.

References
[1] Rullhusen P, Artru X and Dhez P 1998 *Novel Radiation Sources Using Relativistic Electrons* (Singapore: World Scientific)
[2] Tsytovich V N 1951 *Westnik MGU* **11** 27 (in Russian)
[3] Zrelov V P 1968 *Vavilov-Cherenkov Radiation and its Applications in High Energy Physics* (Moscow: Atomizdat) (in Russian)
[4] Kitao K 1960 *Prog. Theor. Phys.* **23** 759
[5] Grigoryan L Sh, Khachatryan H F and Arzumanyan S R 1998 *Izv. Nats. Akad. Nauk Arm. Fiz.* (Engl. Transl.: *J. Contemp. Phys.*) **33** 267 (arXiv:cond-mat/0001322)
[6] Grigoryan L Sh, Khachatryan H F and Arzumanyan S R 2002 *Izv. Nats. Akad. Nauk Arm. Fiz* (Engl. Transl.: *J. Contemp. Phys.*) **37** 327
[7] Grigoryan L Sh, Khachatryan H F, Arzumanyan S R and Grigoryan M L 2006 *Nucl. Instr. and Meth.* **B 252** 50
[8] Kotanjyan A S, Khachatryan H F, Petrosyan A V and Saharian A A 2000 *Izv. Nats. Akad. Nauk Arm. Fiz.* (Engl. Transl.: *J. Contemp. Phys.*) **35** 115; Kotanjyan A S and Saharian A A 2001 *Izv. Nats. Akad. Nauk Arm. Fiz.* (Engl. Transl.: *J. Contemp. Phys.*) **36** 310; Kotanjyan A S and Saharian A A 2002 *Mod. Phys. Lett.* **A 17** 1323
[9] Kotanjyan A S and Saharian A A 2005 *J. Phys.* **A 38** 4275; Saharian A A, Kotanjyan A S and Grigoryan M L 2007 *J. Phys.* **A 40** 1405; Saharian A A and Kotanjyan A S 2007 *J. Phys.* **A 40** 10641; Saharian A A and Kotanjyan A S 2009 *J. Phys.* **A 42** 135402
[10] Afanasiev G N, Kartavenko V G and Stepanovsky Yu P 2003 *J. Math. Phys.* **44** 4026
[11] Pogorzelski R and Yeh C 1973 *Phys. Rev.* **A 8** 137
[12] Arzumanyan S R, Grigoryan L Sh and Saharian A A 1995 *Izv. Nats. Akad. Nauk Arm. Fiz* (Engl. Transl.: *J. Contemp. Phys.*) **30** 99
[13] Arzumanyan S R, Grigoryan L Sh, Saharian A A and Kotanjyan Kh V 1995 *Izv. Nats. Akad Nauk Arm. Fiz.* (Engl. Transl.: *J. Contemp. Phys.*) **30** 106
[14] Berestetskii V B, Lifshitz E M and Pitaevskii L P 1980 *Quantum Electrodynamics* (Moscow: Nauka) (in Russian)
[15] Jackson J D 1962 *Classical Electrodynamics* (New York-London: John Willey & Sons)
[16] Korn G A and Korn T M 1968 *Mathematical Handbook* (New York: McGraw Book Company)
[17] Voronkova E M, Grechushnikov B N, Distler G I and Petrov I P 1965 *Optical Materials for Infrared Technology* (Moscow: Nauka) (in Russian)

Journal of Physics: Conference Series **357** (2012) 012009 doi:10.1088/1742-6596/357/1/012009

Synchrotron radiation inside a dielectric waveguide

A S Kotanjyan[1] and A A Saharian[1,2]

[1] Department of Physics, Yerevan State University, 1 Alex Manoogian Street, 0025 Yerevan, Armenia

[2] Institute of Applied Problems in Physics, 25 Nersisyan Street, 0014 Yerevan, Armenia

E-mail: anna.kotanjyan@gmail.com, saharian@ysu.am

Abstract. We investigate the electromagnetic fields generated by a charged particle rotating inside a dielectric cylinder immersed into a homogeneous medium. The radiation intensity for the modes propagating inside the cylinder is evaluated by using two different ways: by evaluating the work done by the radiation field on the charge and by evaluating the energy flux through the cross-section of the cylinder. The relation between these two quantities is discussed. We investigate the relative contributions of the bound modes and the modes propagating at large distances from the cylinder to the total radiation intensity. The presence of the cylinder can lead to the considerable increase of the synchrotron radiation intensity.

1. Introduction

Unique characteristics of the synchrotron radiation have resulted in its extensive applications. Related to this, the investigation of mechanisms for the control of the radiation parameters is of great interest. Particularly important is the study of the influence of a medium on the spectral and angular characteristics. It is well-known that the medium can essentially change the characteristics of the high-energy electromagnetic processes. Moreover, new types of phenomena arise, such as the Cherenkov and transition radiations.

In [1]-[3] we have investigated the synchrotron radiation from a charge rotating around/inside a dielectric cylinder enclosed by a homogeneous medium. It has been shown that under the Cherenkov condition for the material of the cylinder and the particle velocity, strong narrow peaks appear in the angular distribution of the radiation intensity in the exterior medium. At these peaks the radiated energy exceeds the corresponding quantity in the case of a homogeneous medium by several orders of magnitude. Similar features for the radiation generated by a charge moving along a helical orbit have been discussed in [4, 5].

In the previous investigations the radiation intensity is considered in the exterior region at large distances from the cylinder. In addition to this part there is radiation which propagates inside the cylinder and it is of interest to investigate the relative contribution of these modes to the total intensity. Here we discuss the radiation intensity inside a dielectric cylinder emitted by a charge rotating inside the cylinder.

2. Radiation intensity inside a dielectric cylinder

Consider a dielectric cylinder of radius ρ_1 and with dielectric permittivity ε_0 and a point charge q rotating inside a cylinder. The radius of the rotation orbit and the velocity of the charge will be denoted by ρ_0, $\rho_0 < \rho_1$, and v respectively. We assume that the system is immersed in a homogeneous medium with permittivity ε_1. In the cylindrical coordinate system (ρ, ϕ, z) with the z-axis directed along the cylinder axis, the only nonzero component of the current density is given by the expression

$$j_\phi = \frac{q}{\rho} v\delta(\rho - \rho_0)\delta(\phi - \omega_0 t)\delta(z), \tag{1}$$

where $\omega_0 = v/\rho_0$ is the angular velocity of the charge.

Inside the cylinder the radiation fields can be expanded over the eigenmodes of the dielectric cylinder. The latter are the solutions of the equation

$$U_m \equiv V_m \left(\varepsilon_0 |\lambda_1| \rho_1 \frac{J'_m}{J_m} + \varepsilon_1 \lambda_0 \rho_1 \frac{K'_m}{K_m}\right) - m^2 \frac{\lambda_0^2 + |\lambda_1|^2}{\lambda_0^2 |\lambda_1|^2} \left(\varepsilon_1 \lambda_0^2 + \varepsilon_0 |\lambda_1|^2\right) = 0, \tag{2}$$

where $J_m = J_m(\lambda_0 \rho_1)$ and $K_m = K_m(|\lambda_1|\rho_1)$ are the Bessel and MacDonald functions, the prime means the derivative with respect to the argument of the function, and

$$V_m = |\lambda_1| \rho_1 \frac{J'_m}{J_m} + \lambda_0 \rho_1 \frac{K'_m}{K_m}. \tag{3}$$

In (2), we use the notations

$$\lambda_j^2 = m^2 \omega_0^2 \varepsilon_j / c^2 - k_z^2, \quad j = 0, 1, \tag{4}$$

with k_z being the projection of the wave vector on the z-axis. For the modes propagating inside the cylinder one has $\lambda_1^2 < 0$. We denote by $k_z = \pm k_{m,s}$, $k_{m,s} > 0$, $s = 1, 2, \ldots, s_m$, the solutions to the equation (2) and define

$$\lambda_{m,s} = \rho_1 \sqrt{m^2 \omega_0^2 \varepsilon_0 / c^2 - k_{m,s}^2}, \; \lambda_{m,s}^{(1)} = \rho_1 \sqrt{k_{m,s}^2 - \varepsilon_1 m^2 \omega_0^2 / c^2}. \tag{5}$$

The radiation intensity propagating inside the dielectric cylinder can be computed in two different ways. In the first one we evaluate the work done by the radiation field on the charged particle:

$$I = -\int d\rho d\varphi dz \rho \, j_\phi E_\phi, \tag{6}$$

with E_ϕ being the azimuthal component of electric field. This leads to the result (details can be found in [6])

$$\begin{aligned} I &= \sum_{m=1}^{\infty} I_m = \frac{q^2 v^2}{2\varepsilon_0 \omega_0} \sum_{m=1}^{\infty} \sum_{s=1}^{s_m} \frac{\lambda_{m,s}^{(1)} J_m^{-2}}{m \alpha'_m(k_{m,s}) K_m} \sum_{l=\pm 1} \frac{J_{m+l}(\lambda_{m,s}\rho_0/\rho_1)}{V_m - lmu} \\ &\times \sum_{p=\pm 1} J_{m+p}(\lambda_{m,s}\rho_0/\rho_1) \left[\left(\frac{m^2\omega_0^2}{c^2}\varepsilon_0 + k_{m,s}^2\right) \frac{K_{m+p}}{V_m - pmu} - \frac{\lambda_{m,s}^2}{\rho_1^2} \frac{K_{m-p}}{V_m + pmu}\right], \end{aligned} \tag{7}$$

where

$$\alpha_m = \frac{U_m}{(\varepsilon_1 - \varepsilon_0)(V_m^2 - m^2 u^2)}, \; u = \frac{\lambda_0}{|\lambda_1|} + \frac{|\lambda_1|}{\lambda_0}, \tag{8}$$

with $\lambda_0 = \lambda_{m,s}$ and $|\lambda_1| = \lambda_{m,s}^{(1)}$. In the limit $\rho_1 \to \infty$, from (7) we find the intensity for the synchrotron radiation in a homogeneous medium with dielectric permittivity ε_0.

Alternatively, the radiation intensity inside the dielectric cylinder can be obtained by evaluating the energy flux through the cross-section of the dielectric cylinder:

$$I_f^{(\text{in})} = \int_0^{\rho_1} d\rho \int_0^{2\pi} d\phi \, \rho \mathbf{n}_z \cdot \mathbf{S} \tag{9}$$

where $\mathbf{n}_z$ is the unit vector along the z-axis and $\mathbf{S}$ is the Poynting vector. This leads to the formula:

$$
\begin{aligned}
I_f^{(\mathrm{in})} &= \sum_{m=1}^{\infty} I_{f,m}^{(\mathrm{in})} = \frac{q^2 v^2 \rho_1^2}{8\varepsilon_0 \omega_0} \sum_{m=1}^{\infty} \sum_s \frac{k_{m,s} \lambda_{m,s}^{(1)2} J_m^{-4}}{m \alpha_m'^2(k_{m,s}) K_m^2} \left[\sum_{l=\pm 1} \frac{J_{m+l}(\lambda_{m,s}\rho_0/\rho_1)}{V_m - lmu} \right]^2 \\
&\times \sum_{p=\pm 1} \frac{K_{m+p}}{V_m - pmu} \left[\left(\frac{m^2 \omega_0^2}{c^2} \varepsilon_0 + k_{m,s}^2 \right) \frac{K_{m+p}}{V_m - pmu} - \frac{\lambda_{m,s}^2}{\rho_1^2} \frac{K_{m-p}}{V_m + pmu} \right] \\
&\times \left[J_{m+p}'^2 + \left(1 - \frac{(m+p)^2}{\lambda_{m,s}^2} \right) J_{m+p}^2 \right]. \qquad (10)
\end{aligned}
$$

This formula gives the flux in the region $z > 0$. By the symmetry of the problem we have the same flux in the region $z < 0$ and the total flux will be $2I_f^{(\mathrm{in})}$. Again, it can be seen that in the limit $\rho_1 \to \infty$, from (10) the expression for the radiation intensity in a homogeneous medium is obtained.

In figure 1, by the black points we present the number of the radiated quanta at a given harmonic m per period of the charge rotation,

$$N_m = \sum_s N_{m,s} = \frac{T I_m}{\hbar m \omega_0}, \; T = \frac{2\pi}{\omega_0}, \qquad (11)$$

for the electron energy $E_e = 2$ MeV and for the values of the parameters $\varepsilon_1 = 1$, $\varepsilon_0 = 3$, $\rho_1/\rho_0 = 1.2$. The red points correspond to the number of the radiated quanta evaluated by the flux through the cylinder cross-section:

$$N_{f,m}^{(\mathrm{in})} = \frac{2T I_{f,m}^{(\mathrm{in})}}{\hbar m \omega_0}. \qquad (12)$$

The factor 2 in the last expression corresponds to that we have the same amount of the energy flux in the region $z < 0$. The number of the radiated quanta increases with the appearance of new modes. The numerical data in figure 1 show that $N_m \neq N_{f,m}^{(\mathrm{in})}$, i.e., the number of the radiated quanta evaluated by the energy flux through the cross-section of the cylinder and by the work done by the radiation field differ. As it will be shown in the next section, the reason for this difference is that for the eigenmodes of the dielectric cylinder there is also energy flux in the exterior region located near the surface of the cylinder.

3. Radiation intensity outside a cylinder

For a charge rotating inside a dielectric cylinder, in Ref. [2] we have evaluated the average energy flux per unit time through the cylindrical surface of large radius ρ ($\rho \gg \rho_1$) coaxial with the cylinder. The expression for the angular density of the radiation intensity on a given harmonic m with the frequency $m\omega_0$ is given by the expression

$$\frac{dI_m^{(\mathrm{ex})}}{d\theta} = \frac{q^2 v^2 m^2 \omega_0^2 \sqrt{\varepsilon_1}}{\pi^2 c^3 \rho_1^2} \sin\theta \left[|C_m^{(1)} - C_m^{(-1)}|^2 + |C_m^{(1)} + C_m^{(-1)}|^2 \cos^2\theta \right], \qquad (13)$$

where we have defined

$$C_m^{(p)} = \frac{J_{m+p}(\lambda_0^{(\mathrm{ex})}\rho_0)}{W_{m+p}^J} + p\lambda_1^{(\mathrm{ex})} H_m(\lambda_1^{(\mathrm{ex})}\rho_1) \frac{J_{m+p}(\lambda_0^{(\mathrm{ex})}\rho_1)}{2\alpha_m W_{m+p}^J} \sum_{l=\pm 1} \frac{J_{m+l}(\lambda_0^{(\mathrm{ex})}\rho_0)}{W_{m+l}^J}, \qquad (14)$$

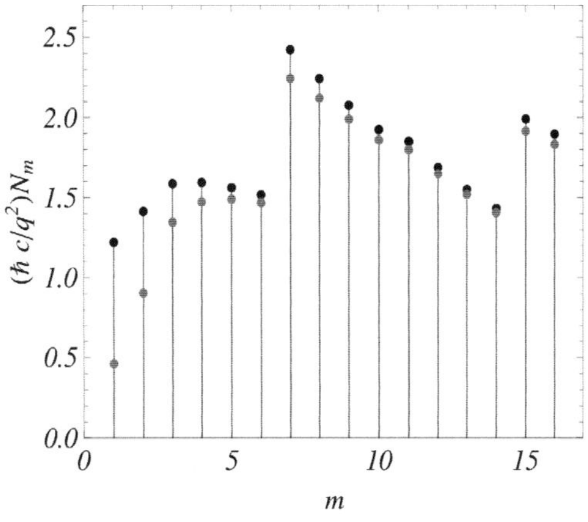

Figure 1. The number of the radiated quanta, per period of the rotation, on a given mode m for the electron energy $E_e = 2$ MeV and for the values of the parameters $\varepsilon_1 = 1$, $\varepsilon_0 = 3$, $\rho_1/\rho_0 = 1.2$. The black/red points correspond to the number of the quanta evaluated by formula (11)/(12).

and

$$\lambda_0^{(\mathrm{ex})} = \frac{m\omega_0}{c}\sqrt{\varepsilon_0 - \varepsilon_1 \cos^2\theta},\ \lambda_1^{(\mathrm{ex})} = \frac{m\omega_0}{c}\sqrt{\varepsilon_1}\sin\theta, \tag{15}$$

with θ being the angle between the radiation direction and the axis of the cylinder. In (14),

$$W_m^J = \lambda_1^{(\mathrm{ex})} J_m(\lambda_0^{(\mathrm{ex})}\rho_1) H'_m(\lambda_1^{(\mathrm{ex})}\rho_1) - \lambda_0^{(\mathrm{ex})} H_m(\lambda_1^{(\mathrm{ex})}\rho_1) J'_m(\lambda_0^{(\mathrm{ex})}\rho_1), \tag{16}$$

and $H_m(x) = H_m^{(1)}(x)$ is the Hankel function of the first kind.

The properties of the radiation corresponding to (13) are described in detail in [2] (see also [4] for a more general case of helical motion inside a dielectric cylinder). In particular, it was shown that under the Cherenkov condition for dielectric permittivity of the cylinder and the velocity of the particle image on the cylinder surface, strong narrow peaks are present in the angular distribution for the number of radiated quanta. At these peaks the radiated energy exceeds the corresponding quantity for a homogeneous medium by several orders of magnitude. For the illustration, in figure 2 we plot the angular density of the number of the quanta radiated in the exterior region per period of the charge rotation,

$$\frac{dN_m^{(\mathrm{ex})}}{d\theta} = \frac{T}{\hbar m\omega_0}\frac{dI_m^{(\mathrm{ex})}}{d\theta}, \tag{17}$$

with the radiation intensity given by (13). The corresponding parameters are as follows: $E_e = 2$ MeV, $\varepsilon_1 = 1$, $\varepsilon_0 = 3$, $m = 16$, $\rho_1/\rho_0 = 1.2$. The dashed curve corresponds to the synchrotron radiation in the vacuum in the absence of the dielectric cylinder ($\varepsilon_0 = 1$). We see the presence of the strong narrow peak at $\theta \approx 42.63$ (in degrees). The height of this peak is ≈ 463.4 and the width ≈ 0.05. At large distances from the cylinder for the total number of the radiated quanta one has $N_m^{(\mathrm{ex})} = 1.748$. Note that for the number of quanta radiated on the eigenmodes of the cylinder we have $N_m = 1.897$ (see figure 1). For the synchrotron radiation from an electron in the vacuum ($\varepsilon_0 = 1$) one has $N_m^{(0)} = 0.352$. As it is seen, the presence of the cylinder considerably increases the radiation intensity.

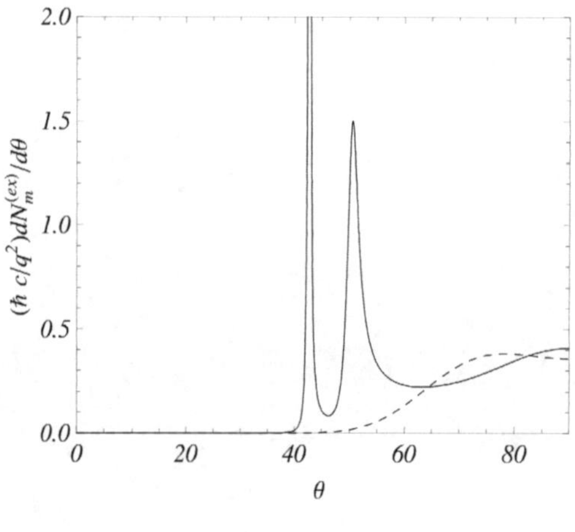

Figure 2. The angular density of the number of the quanta radiated on the harmonic $m = 16$ in the exterior region per period of the charge rotation as a function of θ (in degrees). The full curve is plotted for the values of parameters $E_e = 2$ MeV, $\varepsilon_1 = 1$, $\varepsilon_0 = 3$, $m = 16$, $\rho_1/\rho_0 = 1.2$. The dashed curve is for the radiation in the vacuum ($\varepsilon_0 = 1$).

In addition to the radiation corresponding to (13), in the exterior region there are also radiation fields localized near the surface of the dielectric cylinder. These fields are the tails of the eigenmodes for the dielectric cylinder and exponentially decay in the exterior region for $\rho/\rho_1 \gg 1/\lambda^{(1)}_{m,s}$. For a fixed value of $\rho > \rho_1$ and in the limit $z \to \infty$ the corresponding energy flux through a plane perpendicular to the axis of the cylinder is given by the expression

$$\begin{aligned} I_f^{(\mathrm{ex})} &= \sum_{m=1}^{\infty} I_{f,m}^{(\mathrm{ex})} = \frac{q^2 v^2 \rho_1^2}{8\varepsilon_1 \omega_0} \sum_{m=1}^{\infty} \sum_s \frac{k_{m,s} \lambda_{m,s}^{(1)2} J_m^{-4}}{m \alpha_m'^2 (k_{m,s}) K_m^2} \left[\sum_{l=\pm 1} \frac{J_{m+l}(\lambda_{m,s}\rho_0/\rho_1)}{V_m - lmu} \right]^2 \\ &\times \sum_{p=\pm 1} \frac{J_{m+p}}{V_m - pmu} \left[\left(\frac{m^2 \omega_0^2}{c^2} \varepsilon_1 + k_{m,s}^2 \right) \frac{J_{m+p}}{V_m - pmu} - \frac{\lambda_{m,s}^{(1)2}}{\rho_1^2} \frac{J_{m-p}}{V_m + pmu} \right] \\ &\times \left[K_{m+p}'^2 - \left(1 + \frac{(m+p)^2}{\lambda_{m,s}^{(1)2}} \right) K_{m+p}^2 \right]. \end{aligned} \tag{18}$$

Now it can be explicitly checked that

$$I_m = 2\left(I_{f,m}^{(\mathrm{in})} + I_{f,m}^{(\mathrm{ex})} \right).$$

In figure 3 we have plotted the number of the radiated quanta per period of the rotation on a given mode m for different values of m. The values of the parameters are as follows: $E_e = 2$ MeV, $\varepsilon_1 = 1$, $\varepsilon_0 = 3$, $\rho_1/\rho_0 = 1.2$. The black points correspond to the radiation in the exterior region at large distances from the cylinder, $N_m^{(\mathrm{ex})} = \int_0^{\pi/2} d\theta (dN_m^{(\mathrm{ex})}/d\theta)$ (see (17)), and the red points are for the radiation on the eigenmodes of the dielectric cylinder, $N_m^{(\mathrm{in})} = \sum_s N_{m,s}$ (see (11)). For the comparison with the synchrotron radiation in the vacuum, in figure 4, for the same values of the parameters, we plot the total number of the radiated quanta, $N_m = N_m^{(\mathrm{in})} + N_m^{(\mathrm{ex})}$, in the presence (black points) and in the absence (red points) of the dielectric cylinder. As it is seen from these graphs, the presence of the cylinder essentially increases the radiation intensity.

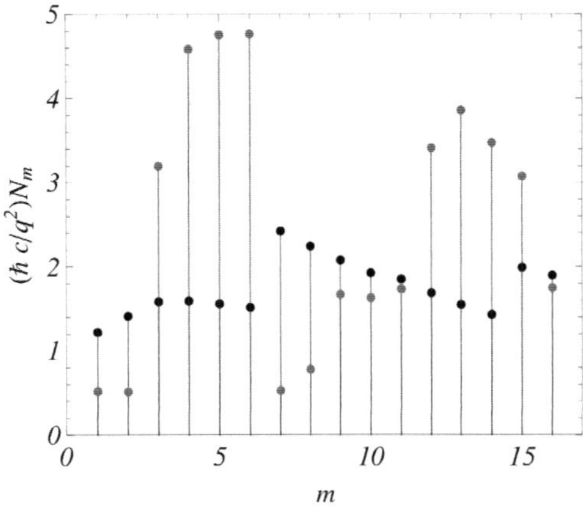

Figure 3. The number of the radiated quanta per period of the rotation on a given mode m for different values of m and for the values of the parameters $E_e = 2$ MeV, $\varepsilon_1 = 1$, $\varepsilon_0 = 3$, $\rho_1/\rho_0 = 1.2$. The black points correspond to the radiation in the exterior region and the red points are for the radiation emitted on the eigenmodes of the dielectric cylinder.

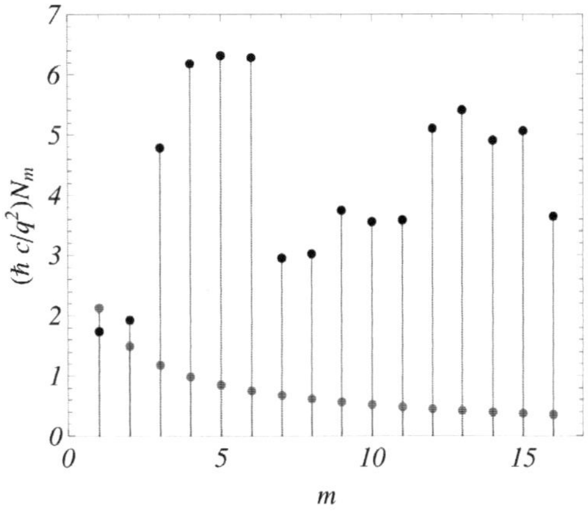

Figure 4. The total number of the radiated quanta, $N_m = N_m^{(\mathrm{in})} + N_m^{(\mathrm{ex})}$, in the presence (black points) and in the absence (red points) of the dielectric cylinder. The values of the parameters are the same as those for figure 3.

4. Conclusion

We have investigated the synchrotron radiation from a charged particle rotating inside a dielectric cylinder surrounded by a homogeneous medium. Firstly we have evaluated the work done by the radiation field on the charged particle. Further, we have computed the energy flux through the cross-section of the cylinder and it is shown that these quantities differ.

The radiation field in the exterior region consists of two parts. The first one corresponds to the radiation propagating at large distances from the cylinder. The angular density of the corresponding radiation intensity on a given harmonic m is given by formula (13). For large values of m and under the Cherenkov condition for dielectric permittivity of the cylinder and the velocity of the particle image on the cylinder surface, strong narrow peaks appear in the angular distribution for the radiation intensity. The second part of the radiation field in the exterior region corresponds to the modes of the dielectric cylinder. The corresponding fields are located near the surface of the cylinder These fields exponentially decay at distances from the cylinder surface larger than the radiation wavelength. The energy flux in the exterior region corresponding to the eigenmodes of the cylinder is given by the expression (18). We have explicitly checked that the radiation intensity evaluated by the work done by the radiation field on the charged particle is equal to the radiation intensity evaluated by the energy flux through the plane perpendicular to the cylinder axis if we take into account the contribution of the radiation field in the exterior region.

References

[1] Grigoryan L Sh, Kotanjyan A S and Saharian A A 1995 *J. Contemp. Phys.* **30** 1
Kotanjyan A S, Khachatryan H F, Petrosyan A V and Saharian A A 2000 *J. Contemp. Phys.* **35** 1
Kotanjyan A S and Saharian A A 2001 *J. Contemp. Phys.* **36** 7
[2] Kotanjyan A S and Saharian A A 2002 *J. Contemp. Phys.* **37** 263
[3] Kotanjyan A S and Saharian A A 2002 *Mod. Phys. Lett. A* **17** 1323
Kotanjyan A S 2003 *Nucl. Instrum. Methods B* **201** 3
[4] Saharian A A and Kotanjyan A S 2005 *J. Phys. A* **38** 4275
[5] Saharian A A and Kotanjyan A S and Grigoryan M L 2007 *J. Phys. A* **40** 1405
Saharian A A and Kotanjyan A S 2007 *J. Phys. A* **40** 10641
Arzumanyan S R, Grigoryan L Sh, Khachatryan H F, Kotanjyan A S and Saharian A A 2008 *Nucl. Instrum. Methods B* **266** 3703
Saharian A A and Kotanjyan A S 2009 *J. Phys. A* **42** 135402
[6] Saharian A A and Kotanjyan A S 2011 Synchrotron radiation inside a dielectric cylinder *Preprint* arXiv:1105.5245.

Radiation of a Charge in a Waveguide with a Boundary between Two Dielectrics

T Yu Alekhina and A V Tyukhtin
Radiophysics Department, Physical Faculty of St. Petersburg State University,
1 Ulyanovskaya str., St. Petersburg, 198504, Russia

E-mail: tanya@niirf.spbu.ru, tyukhtin@bk.ru

Abstract. We analyze the electromagnetic field of a charged particle moving uniformly and crossing a boundary between two homogenous dielectrics in a metal cylindrical waveguide. We make an analysis of exact expressions for components of waveguide modes with the methods of the complex variable function theory. An original algorithm for computation of the field components is presented as well. Two instances are investigated in detail: the particle is flying from vacuum into dielectric and from dielectric into vacuum. In the first case, Cherenkov radiation (so-called 'wakefield') is partially compensated with the free field. We demonstrate areas where the waveguide modes practically coincide with ones in the regular waveguide and areas where the boundary influence is principal. For the second case, we demonstrate the possibility of generation of large quasi monochromatic radiation (Cherenkov-transition radiation) in the vacuum region at certain conditions.

1. Introduction

Investigation of a field of a particle bunch in a waveguide loaded with dielectric is important for the wakefield acceleration technique and for other problems in the accelerator physics. One of them consists in analysis of effect of the boundary on the wave field when the bunch flies into (or from) the dielectric structure. It should be noticed that a charge field structure in a regular waveguide loaded with dielectric was analyzed frequently (see, for example, [1, 2]). However, the case of an irregular waveguide with two different media has not been examined sufficiently, although such investigation is important as well. This especially concerns the study of the electromagnetic field (EMF) structure in contrast to the energetic characteristics analyzed in papers [3, 4]. Note that the case of a semi-infinite waveguide with the metal end was analyzed in papers [5, 6]. But this problem differs essentially from the one considered here because matching of fields in different parts of the waveguide is of great importance. Note as well that analogous problem with a boundary between vacuum and cold plasma was investigated in [7]. But the Cherenkov radiation (CR) is not generated in such situation; therefore it differs radically from the case under consideration.

In present paper we consider the electromagnetic field (EMF) generated by a point charge particle q moving in the metal circular waveguide of radius a along its axis (z-axis) through the interface ($z=0$) between homogeneous isotropic dielectrics with permittivity ε_1 ($z<0$) and ε_2 ($z>0$) (figure 1). The media have no dispersion but small losses are taken into account. The charge moves uniformly with a velocity $\vec{V}=c\beta\vec{e}_z$ and intersects the boundary at the moment $t=0$.

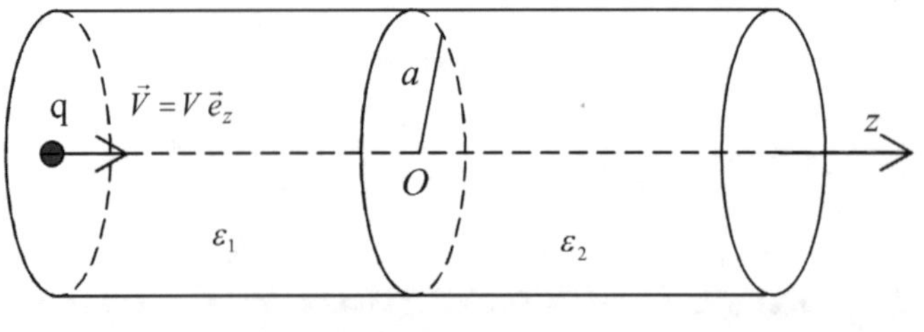

Figure 1. Geometry of the problem

The analytical solution for this problem is found traditionally as an expansion into a series of eigenfunctions of the transversal operator [3]. This expression has two summands: $\vec{E}_{1,2} = \vec{E}^q_{1,2} + \vec{E}^b_{1,2}$. The first one ($\vec{E}^q_{1,2}$) gives the field in the regular waveguide with homogeneous filling. V.L. Ginzburg [8] called it the 'forced' field. It contains CR if the charge velocity exceeds the Cherenkov threshold. The second summand ($\vec{E}^b_{1,2}$) is connected with the influence of the boundary and gives so-called 'free' field. It includes transition radiation (TR).

We investigate the exact solution with analytical and computational methods. Analytical research is an asymptotic investigation using the steepest descent technique. Computations are based on original algorithm using certain transformation of the initial integration path. Such approaches were applied as well in some papers concerning both boundless homogenous media [9] and problems with interface between two media [10, 11] including the case of waveguide partially filled with a cold plasma [7].

Further we study two cases in detail: (I) flying from vacuum ($\varepsilon_1 = 1$) into dielectric ($\varepsilon_2 > 1$), (II) flying from dielectric ($\varepsilon_1 > 1$) into vacuum ($\varepsilon_2 = 1$).

2. The case of flying from vacuum into dielectric

The exact expressions for the forced and free fields can be easily derived from the ones for the general case of the boundary between arbitrary homogeneous isotropic media [6] if we take into account that $\mu_1 = \mu_2 = 1$ and dispersion is negligible. The study of the forced field (which can contain CR) does not cause any difficulties [1]. So, we pay the main attention to research of the free field. Here we give expressions for longitudinal components of the free field:

$$E^b_{z1,2} = \frac{2iqc^2\beta}{\pi a^4 \varepsilon_{1,2}} \sum_{n=1}^{\infty} \frac{\chi_{0n}^2 J_0(\chi_{0n} r/a)}{J_1^2(\chi_{0n})} \int_{-\infty}^{+\infty} d\omega\, \widetilde{B}_{n1,2} \exp\left[-i\left(\omega t - k_{z1,2}|z|\right)\right], \tag{1}$$

$$\widetilde{B}_{n1,2} = \frac{\left(\varepsilon_{1,2} - \varepsilon_{2,1}\right)\left[\omega\left(\pm\varepsilon_{1,2}\beta^2 \mp 1\right) - \beta\sqrt{\omega^2\varepsilon_{2,1} - \omega_n^2}\right]}{\widetilde{g}(\omega)\left[\omega^2\left(1 - \varepsilon_{1,2}\beta^2\right) + \beta^2\omega_n^2\right]\left(\omega \pm \beta\sqrt{\omega^2\varepsilon_{2,1} - \omega_n^2}\right)}, \tag{2}$$

$$k_{z1,2} = \sqrt{\omega^2\varepsilon_{1,2} - \omega_n^2}\Big/c, \quad \widetilde{g}(\omega) = \varepsilon_1\sqrt{\omega^2\varepsilon_2 - \omega_n^2} + \varepsilon_2\sqrt{\omega^2\varepsilon_1 - \omega_n^2}, \tag{3}$$

where $\omega_n = \chi_{0n}c/a$, χ_{0n} is the nth zero of the Bessel function ($J_0(\chi_{0n}) = 0$), ($\operatorname{Im} k_{z1,2} > 0$).

In the analytical method, asymptotic expressions for the free field components of each mode can be obtained with the steepest descend technique [12]. We perform this analysis for the case of negligible losses taken into account for the determination of the location of the integrand singularities only. The first step in such research is to study the singularities of integrands in (1) in a complex plane (ω). One can show that integrands have the following singularities:

- four branch points $\pm\widetilde{\omega}_n^{(1)} = \pm\omega_n - i\delta_1$ and $\pm\widetilde{\omega}_n^{(2)} = \pm\omega_n\big/\sqrt{\varepsilon_2} - i\delta_2$;
- two poles on the imaginary axis (ω): $\pm\omega_{0n}^{(1)} = \pm i\beta\omega_n\left(1 - \beta^2\right)^{-1/2}$;

– two poles on the imaginary axis $\pm\omega_{0n}^{(2)} = \pm i\beta\omega_n\left(1-\varepsilon_2\beta^2\right)^{-1/2}$ if $\beta < \varepsilon_2^{-1/2}$ or on the real axis $\pm\omega_{0n}^{(2)} = \pm\beta\omega_n\left(\varepsilon_2\beta^2 - 1\right)^{-1/2} - i\delta_3$ if $\beta > \varepsilon_2^{-1/2}$.
Here $\delta_1, \delta_2, \delta_3$ are positive infinitesimal quantities. Therefore, all of the singularities take place below a real axis (figure 2 (a)).

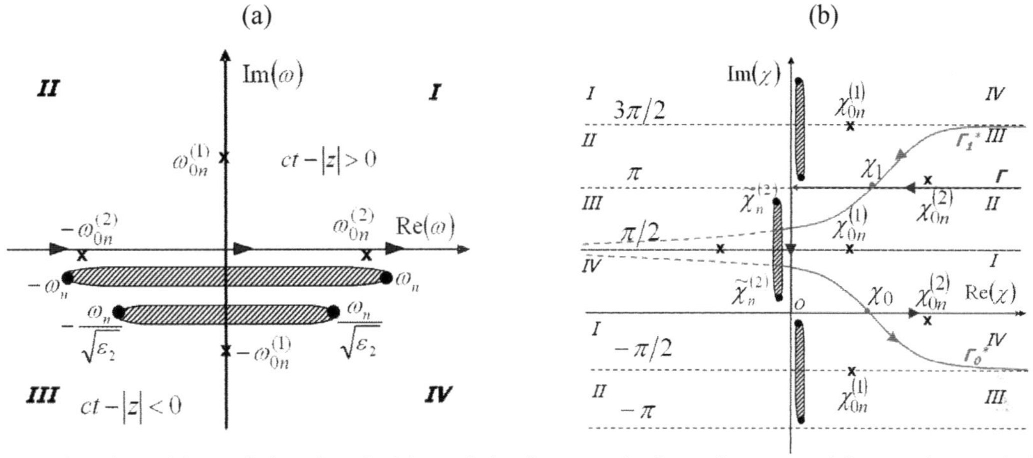

Figure 2. Disposition of the singularities of the integrands, branch cuts and integration path in a complex plane of ω (a) and χ (b) for the free field components in vacuum if $\beta > \varepsilon_2^{-1/2}$. The dashed parts of contour Γ^* are situated in the 'lower' sheet of the Riemann surface.

Since the medium is passive it is necessary to guarantee exponential decrease of the free field with increase of $|z|$. It means that $\mathrm{Re}\sqrt{\omega^2-\omega_n^2} > 0$ and $\mathrm{Re}\sqrt{\omega^2\varepsilon_2-\omega_n^2} > 0$ for real values of ω. It is convenient to use these definitions of radicals for all complex values of ω as well. Thus we define the branch cuts by the equations

$$\mathrm{Re}\sqrt{\omega^2-\omega_n^2} = 0, \qquad \mathrm{Re}\sqrt{\omega^2\varepsilon_2-\omega_n^2} = 0. \tag{4}$$

The integration path goes along the upper edge of the cuts (figure 2 (a)).

For obtaining asymptotic expressions, we use the steepest descend technique. At first, we make the following change of variables for the vacuum region:

$$\omega = \omega_n\, ch\chi, \qquad \sqrt{\omega^2-\omega_n^2} = \omega_n\, sh\chi. \tag{5}$$

This replacement removes the pair of branch points $\pm\tilde{\omega}_n^{(1)} = \pm\omega_n - i\delta_1$. The correspondence between quadrants of the complex plane (ω) and areas in the plane (χ) is presented in figure 2 (with Roman numerals). The initial integration path in the plane (χ) is the contour Γ. The branch point $\tilde{\omega}_n^{(2)}$ and poles $\omega_{0n}^{(1)}$, $\omega_{0n}^{(2)}$ turn into $\tilde{\chi}_n^{(2)}$, $\chi_{0n}^{(1)}$ and $\chi_{0n}^{(2)}$ respectively (figure 2). There are two saddle points on contour Γ: $\chi_0 = \mathrm{arch}(ct/R)$ and $\chi_1 = \chi_0 + i\pi$, where $R = \sqrt{(ct)^2 - z^2}$. The steepest descending paths (SDP) Γ^* consist of two branches (Γ_0^* and Γ_1^*).

The poles and the branch points can be crossed in the transformation of contour Γ into the SDP Γ^* passing through the saddle points $\chi_{0,1}$, and the contributions from the corresponding singularities should be included in asymptotic expressions. So, if $\mathrm{Re}\,\chi_{0,1} < \mathrm{Re}\,\chi_{0n}^{(2)}$ the poles $\chi_{0n}^{(2)}$ can be crossed (figure 2 (b)). The contributions from the poles are evaluated by the residue theorem. They give so-called Cherenkov-transition radiation (CTR). But in this case there is no CTR in vacuum because of peculiarities of integrands.

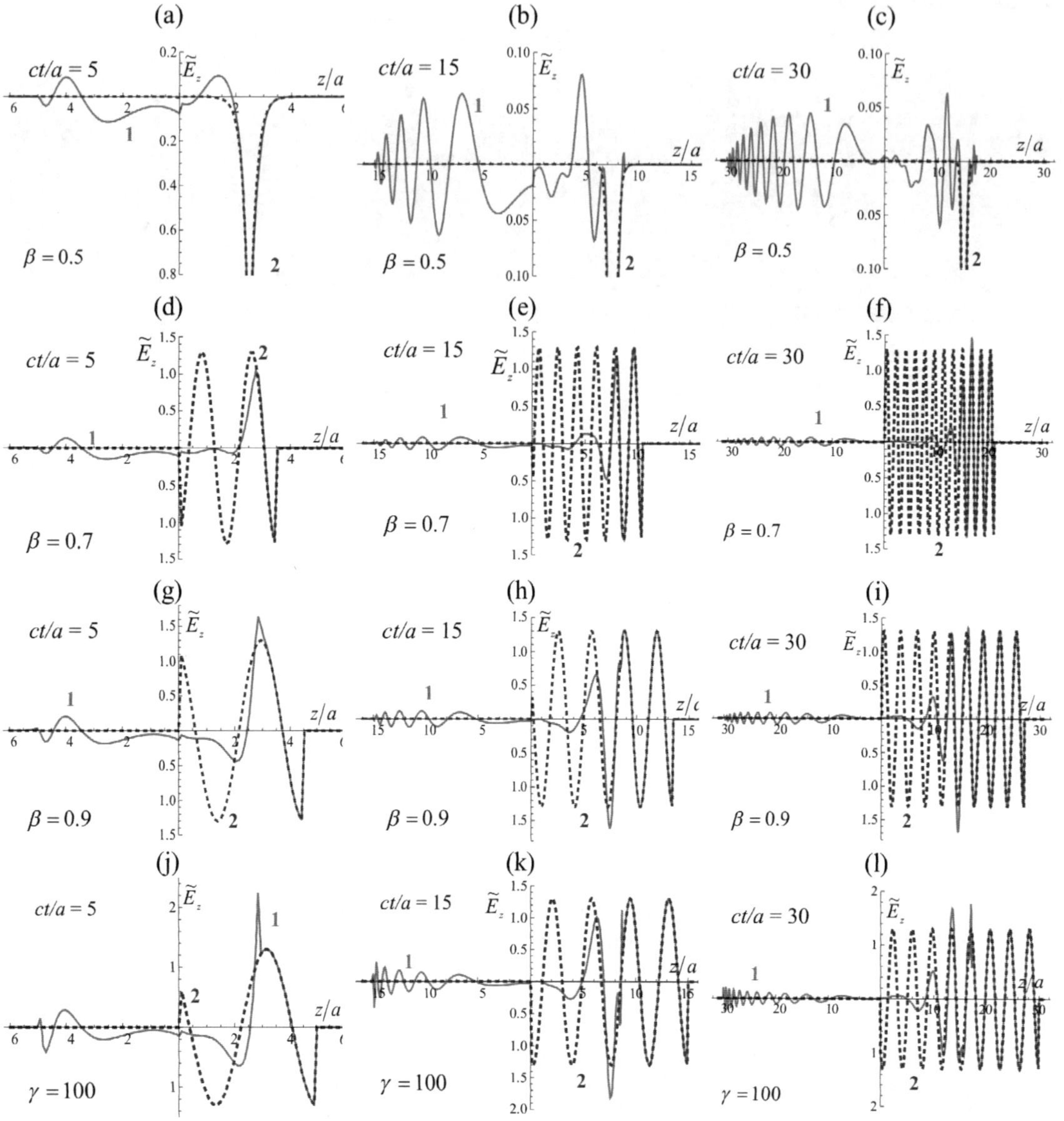

Figure 3. The case of flying from vacuum into dielectric. Dependence of normalized longitudinal component $\widetilde{E}_z = E_z \pi c^2 / 2q\omega_0^2$ of the first mode of the whole field (continuous line 1) and the forced one (dashed line 2) on distance z/a for different dimensionless times *ct/a* and velocities β (or $\gamma = 1/\sqrt{1-\beta^2}$); $\omega_0 = 2\pi \cdot 10\text{GHz}$, $\varepsilon_1 = 1$, $\varepsilon_2 = 3$, $a = 5$ mm. Sighting point is on the waveguide axis $r = 0$.

A similar investigation can be made for the free field in the dielectric area. Here it is convenient to make another change of variables:

$$\omega=\omega_n ch\chi\,\varepsilon_2^{-1/2}, \qquad \sqrt{\omega^2\varepsilon_2-\omega_n^2}=\omega_n sh\chi\,\varepsilon_2^{-1/2}. \tag{6}$$

In this case not only the poles but also the branch points can be crossed in transformation of Γ into Γ^*. For the pole contributions one can obtain the following expression:

$$E_{z2}^{\rm CTR}=\frac{4q}{a^2\varepsilon_2}\sum_{n=1}^{\infty}\frac{J_0(\chi_{0n}r/a)}{J_1^2(\chi_{0n})}\cos\left[\chi_{0n}(ct\beta-z)a^{-1}(\varepsilon_2\beta^2-1)^{-1/2}\right]\Theta(z_2-z), \quad z_2=ct/\beta\varepsilon_2. \tag{7}$$

$E_{z2}^{\rm CTR}$ is the exact formula for the forced field taken with an opposite sign. So, there is a compensation for the forced field with a part of free one in some domain from the boundary at $z<z_2$.

For numerical calculations, the exact integral representations (1) are used. The efficient algorithm developed is based on a certain transformation of the initial integration path in the complex plane (ω). Earlier, such an algorithm was used for the computation of the field in different dispersive unbounded or semi-bounded media [9, 10] and in a waveguide with semi-bounded cold plasma [6]. As it is shown in figure 2, the poles are located near the integration path Γ. This leads to the rather abrupt behaviour of the integrands in (1). The numerical algorithm is adapted for overcoming this difficulty. We transform contour Γ into some new contour in the upper-half plane (ω) for $ct-\sqrt{\varepsilon_{1,2}}\,|z|<0$ (before the 'wave front' $z_{f1,2}=ct/\sqrt{\varepsilon_{1,2}}$) and into another contour in the lower-half plane (ω) for $ct-\sqrt{\varepsilon_{1,2}}\,|z|>0$ (behind the 'wave front'). These new contours should bypass all of the singularities (the poles and the branch points) and then go parallel to SDP. Note that we can choose optimal parameters of contours for each computation.

Further we present the behaviour of the first mode of the longitudinal component E_z of the total field and the forced one in vacuum and dielectric for different velocities of β and at different moments ct/a (figure 3). If $\beta<\beta_{C2}=1/\sqrt{\varepsilon_2}\approx 0.56$ the total field is TR everywhere except some small neighbourhood of the charge (figure 3, a, b, c). That is also true for all velocities in the vacuum area. If $\beta>\beta_{C2}$ Cherenkov radiation is generated in dielectric, and the total field here is a combination of CR and TR. In the domain $0<z<z_2=ct/(\beta\varepsilon_2)$ the forced wave field (so-called 'wakefield') is compensated by some part of the free field which is equal to the wakefield taken with an opposite sign. Note that the point z_2 is determined with the group velocity $v_{g2}=c/\beta\varepsilon_2$ in a regular waveguide. The compensation domain is large for the velocities closed to the Cherenkov threshold (figure 3, d,e, f), and it is reduced with increase in β (figure 3, g, h, i). As a result, there is the area where the wave field practically coincides with the wakefield in the regular waveguide ($z_2<z<c\beta t$). As well, there is the area where the boundary influence is principal ($z<z_2$). Note that the first area is large if ε_2 takes on a large value. This conclusion is important for the wakefield acceleration technique. It is explained the fact that limited dielectric structure can be applied for this technique, and the length of structure can be easily estimated.

3. The case of flying from dielectric into vacuum

Similar investigation was made for the case of flying from dielectric into vacuum. In this situation the contribution of poles can exist: this is so-called Cherenkov-transition radiation (CTR) as in the case of two semi-infinite media without metal wall [9–11]. One can show that this radiation in vacuum takes place at $\beta_{C1}=\varepsilon_1^{-1/2}<\beta<\beta_{CT1}=(\varepsilon_1-1)^{-1/2}$. The lower threshold β_{C1} is connected with condition of the Cherenkov radiation generation. Presence of the upper threshold β_{CT1} is explained by total internal reflection of Cherenkov waves from the boundary (that takes place at $\beta>\beta_{CT1}$). So, there is the reflected wave of CR in dielectric in the domain $|z|<z_1=ct/(\beta\varepsilon_1)$:

$$E_{z1}^{\mathrm{CTR}} = \frac{4q}{a^2\varepsilon_1}\left(\varepsilon_1\beta^2 - 1 - \sqrt{1-\beta^2(\varepsilon_1-1)}\right)\left(1+\sqrt{1-\beta^2(\varepsilon_1-1)}\right)^{-1}\left(1+\varepsilon_1\sqrt{1-\beta^2(\varepsilon_1-1)}\right)^{-1}$$
$$\times\sum_{n=1}^{\infty} J_0(\chi_{0n}r/a)J_1^{-2}(\chi_{0n})\cos\left[\chi_{0n}\left(ct\beta-|z|\right)\left(a\sqrt{\varepsilon_1\beta^2-1}\right)^{-1}\right]\Theta(z_1-|z|), \quad (8)$$

In vacuum in the area $z < z_3 = ct\sqrt{\left|1-\beta^2(\varepsilon_1-1)\right|}\Big/\beta$ there is the transmitted wave of CR:

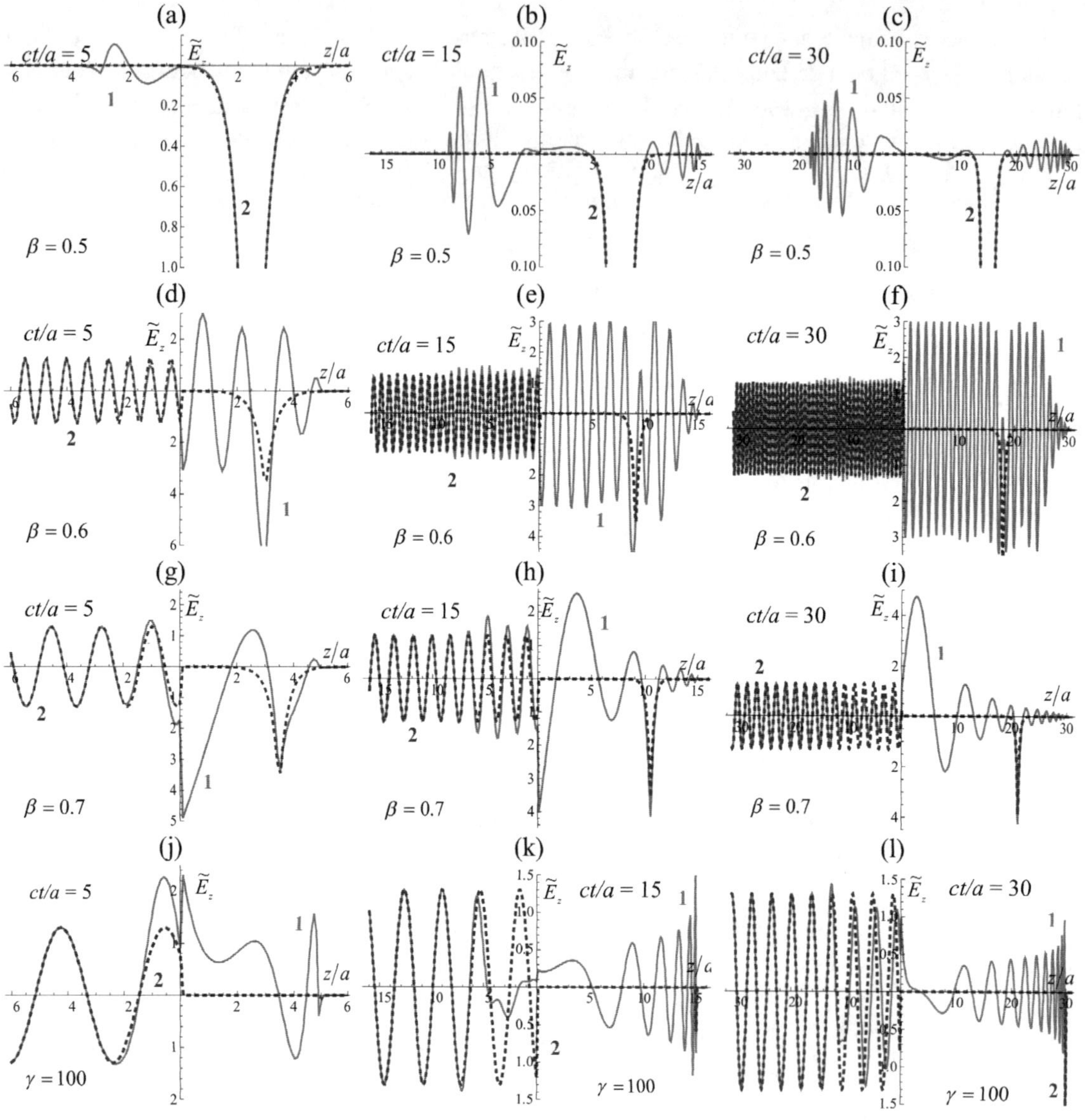

Figure 4. The case of flying from dielectric into vacuum. The same as for figure 3 at $\varepsilon_1 = 3$, $\varepsilon_2 = 1$, $\beta_{C1} \approx 0.56$, $\beta_{CT1} \approx 0.71$.

$$E_{z2}^{CTR} = \frac{8q}{a^2\varepsilon_2}\frac{1}{\left(1+\varepsilon_1\sqrt{1-\beta^2(\varepsilon_1-1)}\right)}\sum_{n=1}^{\infty}\frac{J_0(\chi_{0n}r/a)}{J_1^2(\chi_{0n})}\cos\left[\frac{\chi_{0n}\left(ct\beta - z\sqrt{1-\beta^2(\varepsilon_1-1)}\right)}{a\sqrt{\varepsilon_1\beta^2-1}}\right]\Theta(z_3-z), \quad (9)$$

So, CR can penetrate from dielectric into vacuum at velocities $\varepsilon_1^{-1/2} < \beta < (\varepsilon_1-1)^{-1/2}$.

The case of flying from dielectric into vacuum is presented in figure 4. If $\beta < \beta_{C1}$, the total field is TR in dielectric and in vacuum everywhere except the charge vicinity (figure 4, a, b, c). If $\beta > \beta_{C1}$ CR emerges in dielectric. The total wave field here consists of CR, reflected wave of CR (in the area $|z| < z_1$) and TR. One can find that reflected wave of CR differs only a little from CR at $\beta_{C1} < \beta < \beta_{CT1}$ (figures 4 (d-i)) and significantly at $\beta > \beta_{CT1}$ (figure 4, j, k, l).

In vacuum, the total wave field consists of TR and transmitted wave of CR (CTR). CTR exists at $\beta_{C1} < \beta < \beta_{CT1}$ in the area $z < z_3$, $z_3 = ct\sqrt{|1-\beta^2(\varepsilon_1-1)|}/\beta$ (figure 4, d,e,f). At $\beta > \beta_{CT1}$ this field decreases exponentially with the distance from the boundary. So, CR penetrates from dielectric into vacuum at some conditions. It should be noted that similar effect takes place at the boundary between vacuum and 'left-handed medium' [10] where the CTR has 'reversed' directivity. Note as well that there is some increase and concentration of TR near the "wave front" $z_f = ct$ in vacuum for the case of the ultra-relativistic particle (figure 4, j, k, l).

3. Conclusion

Underline the main conclusions of the work. If the charge flies from vacuum into dielectric there is area where the wave field practically coincides with the wakefield in the regular waveguide and area where the boundary influence is principal. The first area is large if the dielectric permittivity takes on a large value. This conclusion is important for the wakefield acceleration technique. If the charge flies from dielectric into vacuum with certain velocity a large quasi monochromatic radiation can be generated in the vacuum region. This conclusion is of interest for development of new methods of generation of electromagnetic radiation.

Acknowledgments
This research was supported by Saint Petersburg State University.

References

[1] Bolotovskii B M 1962 *Sov. Phys. Usp.* **4** 781
[2] Altmark A M, Kanareykin A D and Sheinman I L 2005 *Thech. Physics* **50** 87
[3] Barsukov K A 1959 *Zh. Exps. Teor. Fiz.* **37** 1106–09
[4] Barsukov K A 1960 *Zh. Tekh. Fiz.* **30** 1337
[5] Burshtein Ye L and Voskresenskii G V 1963 *Zh. Tekh. Fiz* **33** 34
[6] Balakirev V A, Onishchenko I N, Sidorenko D Yu and Sotnikov G V 2001 *Zh. Exps. Teor. Fiz.* **120** 41
[7] Alekhina T Yu and Tyukhtin A V 2011 *Physical Review E* **83** 066401
[8] Ginzburg V L and Tsytovich V N 1990 *Transition Radiation and Transition Scattering* (London: Hilger) p. 445
[9] Tyukhtin A V and Galyamin S N 2008 *Phys. Rev. E* **77** 066606
[10] Galyamin S N and Tyukhtin A V 2010 *Phys. Rev. B* **81** 35134
[11] Galyamin S N and Tyukhtin A V 2011 *Technical Physics Letters* **37** 317
[12] Felsen B and Marcuvitz N 2003 *Radiation and Scaterring of Waves* (New Jersey: Wiley Interscience) p. 888

Diffracted transition radiation of a relativistic electron in the artificial periodic multilayer medium

S Blazhevich[1], I Kolosova[2], A Noskov[2]
[1] National Research University Belgorod State University, Belgorod, Russia
[2] Belgorod University of Cooperation, Economics and Low, Belgorod, Russia

E-mail: blazh@bsu.edu.ru

Abstract. A theory of coherent X-ray radiation of a relativistic electron crossing the artificial periodic medium in the Laue scattering geometry is constructed. The expressions describing the spectral and angular characteristics of radiation in the direction of Bragg scattering are obtained and investigated. By analogy with the radiation emission in a crystalline medium this radiation is considered as the result of coherent summation of the contributions of two radiation mechanisms: parametric (PXR) and diffracted transition (DTR). It is shown that the yield of DTR from layered target can be more than one order higher than the yield in single crystal radiator, under similar conditions. The manifestations of the Borrmann effect for DTR in the artificial multilayer environment are demonstrated for a Laue scattering geometry.

1. Introduction

When a charged particle crosses the entrance surface of the crystal plate the transition radiation arises (TR) [1], which then is diffracted by a system of parallel atomic planes of the crystal, forming the diffracted transition radiation DTR [2–4]. At the same time a charged particle Coulomb field is scattered by a system of parallel atomic planes of the crystal, creating a parametric X-ray radiation (PXR) [5–7]. In the scheme of the symmetric reflection when the system of diffracting atomic planes is perpendicular (in the case of Laue scattering geometry) or parallel (in the case of Bragg scattering) to the surface of the crystal plate, the radiation mechanisms in the two-wave approximation of dynamic diffraction theory were considered in [8–11].

In the general case of asymmetric reflection of the radiation from the plate when the diffracted atomic planes make an arbitrary angle with the surface of the plate, the dynamic effects of PXR and DTR are considered in [12–15], where it was shown that by changing the asymmetry of reflection, we can significantly increase the radiation yield. Traditionally, the radiation of a relativistic particle in a periodically layered structure was considered in the Bragg scattering geometry for the case where the reflecting layers are parallel to the entrance surface, i.e. for the case of symmetric reflection. The radiation in a periodic layered structure is usually viewed as resonant transition radiation [5, 16]. In the works [17], the radiation from an artificial periodic structure was represented as the sum of diffracted transition radiation (DTR) and parametric X-ray radiation (PXR).

In the cited works the radiation of relativistic particles in an artificial periodic structure was considered only in the Bragg scattering geometry for the special case of symmetric reflection of the particle field with respect to the target surface, when the diffracted layers are parallel to the target surface.

In the present paper we consider the coherent X-ray radiation scattering in the Bragg direction generated by relativistic electron crossing the artificial periodic structure in the Laue scattering geometry. By analogy with the crystalline environment the coherent radiation is considered as the sum of PXR and DTR contributions. On the basis of two-wave approximation of dynamic diffraction theory [18] the expressions describing the spectral and angular characteristics of radiation are derived.

2. Amplitude of the radiation

We analyze the radiation emitted by a relativistic electron passing through a multilayer structure (Figure 1) consisting of periodically arranged amorphous layer with thickness a and b respectively ($T = a + b$ is the structure period) with the dielectric susceptibility χ_a and χ_b respectively.

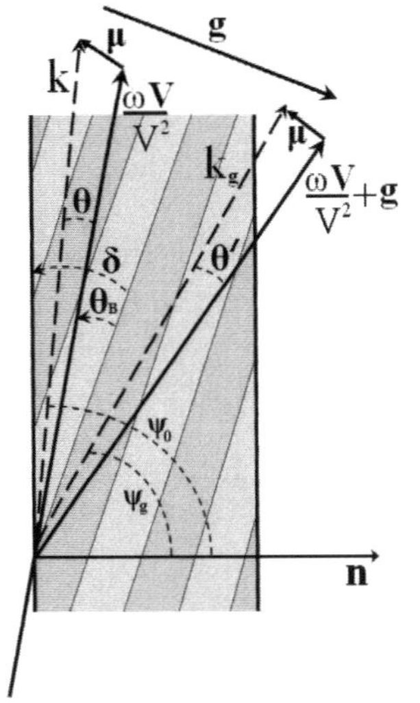

Figure 1. Geometry of the radiation process and the system of the using parameters notations, θ and θ' are the radiation angles, θ_B is Bragg angle, $\mathbf{k}$ and $\mathbf{k}_g$ are wave vectors of incident and diffracted photons.

We consider the equation for the Fourier transform of the electromagnetic field

$$\mathbf{E}(\mathbf{k},\omega) = \int dt\, d^3\mathbf{r}\, \mathbf{E}(\mathbf{r},t)\exp(i\omega t - i\mathbf{kr}). \qquad (1)$$

We use the two-wave approximation of dynamic diffraction theory, in which the incident and diffracted wave are considered on equal grounds. Since the electromagnetic field associated with a relativistic particle can accurately be considered as transverse both incident $\mathbf{E}_0(\mathbf{k},\omega)$ and diffracted $\mathbf{E}_\mathbf{g}(\mathbf{k},\omega)$ electromagnetic waves are determined by pairs of transverse polarization amplitudes:

$$\begin{aligned}\mathbf{E}_0(\mathbf{k},\omega) &= E_0^{(1)}(\mathbf{k},\omega)\mathbf{e}_0^{(1)} + E_0^{(2)}(\mathbf{k},\omega)\mathbf{e}_0^{(2)},\\ \mathbf{E}_\mathbf{g}(\mathbf{k},\omega) &= E_\mathbf{g}^{(1)}(\mathbf{k},\omega)\mathbf{e}_1^{(1)} + E_\mathbf{g}^{(2)}(\mathbf{k},\omega)\mathbf{e}_1^{(2)},\end{aligned} \qquad (2)$$

where the vector $\mathbf{e}_0^{(1)}$ and $\mathbf{e}_0^{(2)}$ are perpendicular to the vector $\mathbf{k}$ and the vectors $\mathbf{e}_1^{(1)}$ and $\mathbf{e}_1^{(2)}$ are perpendicular to the vector $\mathbf{k}_\mathbf{g} = \mathbf{k} + \mathbf{g}$. Vectors $\mathbf{e}_0^{(2)}$ and $\mathbf{e}_1^{(2)}$ lie in the plane of the vectors $\mathbf{k}$ and $\mathbf{k}_\mathbf{g}$ (π-polarization), and vectors $\mathbf{e}_0^{(1)}$ and $\mathbf{e}_1^{(1)}$ are perpendicular to this plane (σ-polarization). The vector $\mathbf{g}$ is defined similarly to the reciprocal lattice vector in the crystal - it is perpendicular to the layers of protection, and its length is equal to $g = \frac{2\pi}{T}n$, $n = 0, \pm1, \pm2, \ldots$

The equations for the Fourier transform of the electromagnetic field in a two-wave approximation of dynamical diffraction theory have the form [19]:

$$\begin{cases}(\omega^2(1+\chi_0)-k^2)E_0^{(s)}+\omega^2\chi_{-\mathbf{g}}C^{(s)}E_{\mathbf{g}}^{(s)}=8\pi^2 ie\omega\theta VP^{(s)}\delta(\omega-\mathbf{kV}),\\ \omega^2\chi_{\mathbf{g}}C^{(s)}E_0^{(s)}+(\omega^2(1+\chi_0)-k_{\mathbf{g}}^2)E_{\mathbf{g}}^{(s)}=0,\end{cases} \tag{3}$$

where $\chi_{\mathbf{g}}$, $\chi_{-\mathbf{g}}$ are coefficients of the Fourier expansion of the periodic structure dielectric susceptibility over the reciprocal vectors $\mathbf{g}$:

$$\chi(\omega,\mathbf{r})=\sum_{\mathbf{g}}\chi_{\mathbf{g}}(\omega)\exp(i\mathbf{gr})=\sum_{\mathbf{g}}\left(\chi_{\mathbf{g}}'(\omega)+i\chi_{\mathbf{g}}''(\omega)\right)\exp(i\mathbf{gr}). \tag{4}$$

The values $C^{(s)}$ and $P^{(s)}$ in the system (3) are defined as follows

$$C^{(s)}=\mathbf{e}_0^{(s)}\mathbf{e}_1^{(s)},\ C^{(1)}=1,\ C^{(2)}=\cos 2\theta_B,\qquad P^{(s)}=\mathbf{e}_0^{(s)}(\boldsymbol{\mu}/\mu),\ P^{(1)}=\sin\varphi,\ P^{(2)}=\cos\varphi, \tag{5}$$

where $\boldsymbol{\mu}=\mathbf{k}-\omega\mathbf{V}/V^2$ is the component of the virtual photon momentum perpendicular to the particle velocity $\mathbf{V}$, $\mu=\omega\theta/V$, $\theta<<1$ is the angle between vectors $\mathbf{k}$ and $\mathbf{V}$, θ_B is Bragg angle, φ is the radiation azimuth angle measured from the plane formed by the velocity vector $\mathbf{V}$ and $\mathbf{g}$.

The vector $\mathbf{g}$ length can be also expressed through the Bragg angle θ_B and the Bragg frequency ω_B: $g=2\omega_B\sin\theta_B/V$. The angle between the vector $\frac{\omega\mathbf{V}}{V^2}$ and the wave vector $\mathbf{k}$ of the incident wave is marked θ and the angle between the vector $\frac{\omega\mathbf{V}}{V^2}+\mathbf{g}$ and the diffracted wave vector $\mathbf{k}_{\mathbf{g}}$ is indicated as θ'. The system (3) under $s=1$ describes the fields of σ - polarization, and under $s=2$ the fields of π -polarization.

The values χ_0 and $\chi_{\mathbf{g}}$ are defined as follows:

$$\chi_0(\omega)=\frac{a}{T}\chi_a+\frac{b}{T}\chi_b,\ \chi_{\mathbf{g}}(\omega)=\frac{\exp(-iga)-1}{igT}(\chi_b-\chi_a),\ \chi_0'=\frac{a}{T}\chi_a'+\frac{b}{T}\chi_b',\qquad \chi_0''=\frac{a}{T}\chi_a''+\frac{b}{T}\chi_b'',$$

$$\operatorname{Re}\sqrt{\chi_{\mathbf{g}}\chi_{-\mathbf{g}}}=\frac{2\sin\left(\frac{ga}{2}\right)}{gT}(\chi_b'-\chi_a'),\quad \operatorname{Im}\sqrt{\chi_{\mathbf{g}}\chi_{-\mathbf{g}}}=\frac{2\sin\left(\frac{ga}{2}\right)}{gT}(\chi_b''-\chi_a'') \tag{6}$$

By solving the dispersion equation following from the system (3)

$$(\omega^2(1+\chi_0)-k^2)(\omega^2(1+\chi_0)-k_{\mathbf{g}}^2)-\omega^4\chi_{-\mathbf{g}}\chi_{\mathbf{g}}C^{(s)2}=0 \tag{7}$$

with the use of standard methods of dynamical theory[18], we find the expression for k and $k_{\mathbf{g}}$:

$$k=\omega\sqrt{1+\chi_0}+\lambda_0,\qquad k_{\mathbf{g}}=\omega\sqrt{1+\chi_0}+\lambda_{\mathbf{g}}. \tag{8}$$

$$\lambda_{\mathbf{g}}^{(1,2)}=\frac{\omega}{4}\left(\beta\pm\sqrt{\beta^2+4\chi_{\mathbf{g}}\chi_{-\mathbf{g}}C^{(s)^2}\frac{\gamma_{\mathbf{g}}}{\gamma_0}}\right),\ \lambda_0^{(1,2)}=\omega\frac{\gamma_0}{4\gamma_{\mathbf{g}}}\left(-\beta\pm\sqrt{\beta^2+4\chi_{\mathbf{g}}\chi_{-\mathbf{g}}C^{(s)^2}\frac{\gamma_{\mathbf{g}}}{\gamma_0}}\right), \tag{9}$$

where $\beta=\alpha-\chi_0\left(1-\frac{\gamma_{\mathbf{g}}}{\gamma_0}\right)$, $\alpha=\frac{1}{\omega^2}(k_{\mathbf{g}}^2-k^2)$, $\gamma_0=\cos\psi_0$, $\gamma_{\mathbf{g}}=\cos\psi_{\mathbf{g}}$, ψ_0 – the angle between the wave vector of the incident wave $\mathbf{k}$ and vector normal to the surface of the plate $\mathbf{n}$, $\psi_{\mathbf{g}}$ is the angle between the wave vector $\mathbf{k}_{\mathbf{g}}$ and the vector $\mathbf{n}$ (see figure 1). Dynamic additions λ_0 and $\lambda_{\mathbf{g}}$ for X-ray wave vectors are related by formula

$$\lambda_{\mathbf{g}}=\frac{\omega\beta}{2}+\lambda_0\frac{\gamma_{\mathbf{g}}}{\gamma_0} \tag{10}$$

Since the dynamic additions are small: $|\lambda_0|<<\omega$, $|\lambda_{\mathbf{g}}|<<\omega$, one can show that $\theta\approx\theta'$ (see figure 1), therefore further we will use the notation θ for both of these angles.

We represent the solution of the system of equations (3) for the diffracted field in a periodic structure in such a form:

$$E_{\mathbf{g}}^{(s)medium}=-\frac{8\pi^2 ieV\theta P^{(s)}}{\omega}\frac{\omega^2\chi_{\mathbf{g}}C^{(s)}}{4\frac{\gamma_0^2}{\gamma_{\mathbf{g}}^2}\left(\lambda_{\mathbf{g}}-\lambda_{\mathbf{g}}^{(1)}\right)\left(\lambda_{\mathbf{g}}-\lambda_{\mathbf{g}}^{(2)}\right)}\delta\left(\lambda_{\mathbf{g}}-\lambda_{\mathbf{g}}^{*}\right)+E_{\mathbf{g}}^{(s)^{(1)}}\delta(\lambda_{\mathbf{g}}-\lambda_{\mathbf{g}}^{(1)})+E_{\mathbf{g}}^{(s)^{(2)}}\delta(\lambda_{\mathbf{g}}-\lambda_{\mathbf{g}}^{(2)}),\qquad(11a)$$

where $\lambda_0^*=\omega\left(\frac{\gamma^{-2}+\theta^2-\chi_0}{2}\right)$, $\lambda_{\mathbf{g}}^*=\frac{\omega\beta}{2}+\frac{\gamma_{\mathbf{g}}}{\gamma_0}\lambda_0^*$, $\gamma=1/\sqrt{1-V^2}$ – Lorentz factor of the particle, $E_{\mathbf{g}}^{(s)^{(1)}}$ and $E_{\mathbf{g}}^{(s)^{(2)}}$ are free diffracted fields in the multilayer target.

For the field in vacuum in front of the radiator the solution of (3) has the form:

$$E_0^{(s)vac\,I}=\frac{8\pi^2 ieV\theta P^{(s)}}{\omega}\frac{1}{-\chi_0-\frac{2}{\omega}\lambda_0}\delta\left(\lambda_0-\lambda_0^*\right)=\frac{8\pi^2 ieV\theta P^{(s)}}{\omega}\frac{1}{\frac{\gamma_0}{\gamma_{\mathbf{g}}}\left(-\chi_0-\frac{2}{\omega}\frac{\gamma_0}{\gamma_{\mathbf{g}}}\lambda_{\mathbf{g}}+\beta\frac{\gamma_0}{\gamma_{\mathbf{g}}}\right)}\delta\left(\lambda_{\mathbf{g}}-\lambda_{\mathbf{g}}^*\right)\qquad(11b)$$

where we use the relation $\delta\left(\lambda_0-\lambda_0^*\right)=\frac{\gamma_{\mathbf{g}}}{\gamma_0}\delta\left(\lambda_{\mathbf{g}}-\lambda_{\mathbf{g}}^*\right)$.

The diffracted field behind the radiator in vacuum is as follows:

$$E_{\mathbf{g}}^{(s)vac}=E_{\mathbf{g}}^{(s)Rad}\delta\left(\lambda_{\mathbf{g}}+\frac{\omega\chi_0}{2}\right),\qquad(12)$$

where $E_{\mathbf{g}}^{(s)Rad}$ is the field of coherent radiation in the direction close to the Bragg direction. From the second equation of the system (3) we can derive the expression relating the incident and diffracted fields in the medium:

$$E_0^{(s)medium}=\frac{2\omega\lambda_{\mathbf{g}}}{\omega^2\chi_{\mathbf{g}}C^{(s)}}E_{\mathbf{g}}^{(s)medium}.\qquad(13)$$

To determine the amplitude of the field $E_{\mathbf{g}}^{(s)Rad}$, we use the boundary conditions at the entrance and exit surfaces of the multilayer plate:

$$\int E_0^{(s)vacI}d\lambda_0=\int E_0^{(s)medium}d\lambda_0,\quad \int E_{\mathbf{g}}^{(s)medium}d\lambda_0=0,$$
$$\int E_{\mathbf{g}}^{(s)medium}\exp\left(i\frac{\lambda_{\mathbf{g}}}{\gamma_{\mathbf{g}}}L\right)d\lambda_{\mathbf{g}}=\int E_{\mathbf{g}}^{(s)vac}\exp\left(i\frac{\lambda_{\mathbf{g}}}{\gamma_{\mathbf{g}}}L\right)d\lambda_{\mathbf{g}}.\qquad(14)$$

We will present the radiation field in the form of two terms:

$$E_{\mathbf{g}}^{(s)Rad}=E_{PXR}^{(s)}+E_{DTR}^{(s)},\qquad(15a)$$

$$E_{PXR}^{(s)}=-\frac{8\pi^2 ieV\theta P^{(s)}}{\omega}\frac{\omega^2\chi_{\mathbf{g}}C^{(s)}}{8\frac{\gamma_0}{\gamma_{\mathbf{g}}}\sqrt{\beta^2+4\chi_{\mathbf{g}}\chi_{-\mathbf{g}}C^{(s)^2}\frac{\gamma_{\mathbf{g}}}{\gamma_0}}}\cdot\frac{1}{\lambda_0^*}\times\left[\left(\beta+\sqrt{\beta^2+4\chi_{\mathbf{g}}\chi_{-\mathbf{g}}C^{(s)^2}\frac{\gamma_{\mathbf{g}}}{\gamma_0}}\right)\left(\frac{1-\exp\left(-i\frac{\lambda_{\mathbf{g}}^*-\lambda_{\mathbf{g}}^{(2)}}{\gamma_{\mathbf{g}}}L\right)}{\lambda_{\mathbf{g}}^*-\lambda_{\mathbf{g}}^{(2)}}\right)\right.$$
$$\left.-\left(\beta-\sqrt{\beta^2+4\chi_{\mathbf{g}}\chi_{-\mathbf{g}}C^{(s)^2}\frac{\gamma_{\mathbf{g}}}{\gamma_0}}\right)\left(\frac{1-\exp\left(-i\frac{\lambda_{\mathbf{g}}^*-\lambda_{\mathbf{g}}^{(1)}}{\gamma_{\mathbf{g}}}L\right)}{\lambda_{\mathbf{g}}^*-\lambda_{\mathbf{g}}^{(1)}}\right)\right]\exp\left[i\left(\frac{\omega\chi_0}{2}+\lambda_g^*\right)\frac{L}{\gamma_g}\right],\qquad(15b)$$

$$E_{DTR}^{(s)}=\frac{8\pi^2 ieV\theta P^{(s)}}{\omega}\frac{\chi_{\mathbf{g}}C^{(s)}}{\frac{\gamma_0}{\gamma_{\mathbf{g}}}\sqrt{\beta^2+4\chi_{\mathbf{g}}\chi_{-\mathbf{g}}C^{(s)^2}\frac{\gamma_{\mathbf{g}}}{\gamma_0}}}\left(\frac{\omega}{-\omega\chi_0-2\lambda_0^*}+\frac{\omega}{2\lambda_0^*}\right)$$

$$\times\left[\exp\left(-i\frac{\lambda_{\mathbf{g}}^*-\lambda_{\mathbf{g}}^{(1)}}{\gamma_{\mathbf{g}}}L\right)-\exp\left(-i\frac{\lambda_{\mathbf{g}}^*-\lambda_{\mathbf{g}}^{(2)}}{\gamma_{\mathbf{g}}}L\right)\right]\exp\left[i\left(\frac{\omega\chi_0}{2}+\lambda_g^*\right)\frac{L}{\gamma_g}\right]. \quad (15c)$$

The expression (15b) and (15c) represent the amplitudes of the radiation fields, similar to the amplitudes of PXR and DTR in a crystal. The DTR is the result of diffraction by a periodically layered artificial structure of the transition radiation, which is generated on the front surface of the target.
For further analysis of the radiation, the dynamic addition (9) can be represented as follows:

$$\lambda_g^{(1,2)}=\frac{\omega|\chi'_{\mathbf{g}}|C^{(s)}}{2}\left(\xi^{(s)}-\frac{i\rho^{(s)}(1-\varepsilon)}{2}\pm\sqrt{\xi^{(s)2}+\varepsilon-2i\rho^{(s)}\left(\frac{(1-\varepsilon)}{2}\xi^{(s)}+\kappa^{(s)}\varepsilon\right)-\rho^{(s)^2}\left(\frac{(1-\varepsilon)^2}{4}+\kappa^{(s)^2}\varepsilon\right)}\right), \quad (16)$$

where

$$\xi^{(s)}(\omega)=\eta^{(s)}(\omega)+\frac{1-\varepsilon}{2\nu^{(s)}},\ \eta^{(s)}(\omega)=\frac{\alpha}{2\left|\mathrm{Re}\sqrt{\chi_{\mathbf{g}}\chi_{-\mathbf{g}}}\right|C^{(s)}}\equiv\frac{\sin^2\theta_B}{V^2C^{(s)}}\frac{gT}{|\chi'_b-\chi'_a|\left|\sin\left(\frac{ga}{2}\right)\right|}\left(1-\frac{\omega(1-\theta\cos\varphi\cot\theta_B)}{\omega_B}\right),$$

$$\nu^{(s)}=\frac{C^{(s)}\,\mathrm{Re}\sqrt{\chi_{\mathbf{g}}\chi_{-\mathbf{g}}}}{\chi'_0}\equiv\frac{2C^{(s)}\left|\sin\left(\frac{ga}{2}\right)\right|}{g}\left|\frac{\chi'_b-\chi'_a}{a\chi'_a+b\chi'_b}\right|,\ \rho^{(s)}=\frac{\chi''_0}{\left|\mathrm{Re}\sqrt{\chi_{\mathbf{g}}\chi_{-\mathbf{g}}}\right|C^{(s)}}\equiv\frac{a\chi''_a+b\chi''_b}{|\chi'_b-\chi'_a|C^{(s)}}\frac{g}{2\left|\sin\left(\frac{ga}{2}\right)\right|},$$

$$\kappa^{(s)}=\frac{\chi''_{\mathbf{g}}C^{(s)}}{\chi''_0}\equiv\frac{2C^{(s)}\left|\sin\left(\frac{ga}{2}\right)\right|}{g}\left|\frac{\chi''_b-\chi''_a}{a\chi''_a+b\chi''_b}\right|,\ \varepsilon=\frac{\gamma_{\mathbf{g}}}{\gamma_0}. \quad (17)$$

An important parameter in (17) is the parameter ε that determines the degree of the field reflection asymmetry relative to the target surface, which can be represented as

$$\varepsilon=\frac{\sin(\delta+\theta_B)}{\sin(\delta-\theta_B)}, \quad (18)$$

where θ_B is the angle between the electron velocity and reflective layers, δ – the angle between the target surface and reflective layers. Note that the angle of electron incidence on the target surface increases when the parameter ε decreases, and vice versa (see figure 2). But in the approach used in this work the calculations will be correct only when the condition $\lambda_g^{(1,2)}(\varepsilon)<<\omega\sqrt{1+\chi_0}$ (see (8)) is provided, i.e. the parameter ε must be limited in accordance with this condition and with formula (16).

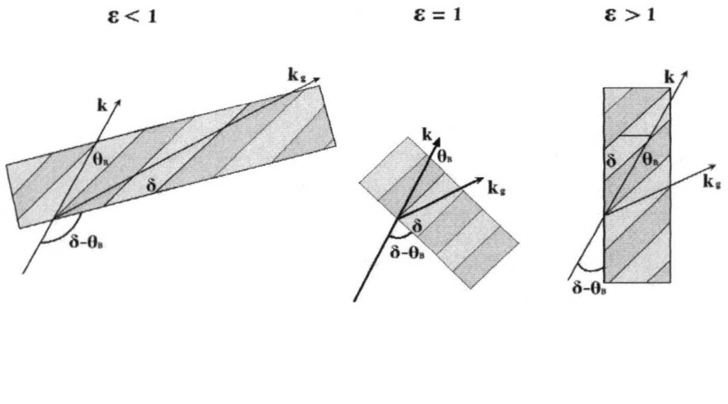

Figure 2. Schema of the asymmetric ($\varepsilon>1$ and $\varepsilon<1$ cases) reflection of the radiation from the multilayered plate. The case ($\varepsilon=1$) corresponds to the symmetric reflection.

3. Spectral-angular radiation density

Substituting (16) for $\lambda_g^{(1,2)}$ into (15b) and (15c), then substituting (15b) for $E_{PXR}^{(s)}$ and (15c) for $E_{DTR}^{(s)}$ in the well-known [19] expression for the spectral-angular density of X-rays

$$\omega\frac{d^2N}{d\omega d\Omega}=\omega^2(2\pi)^{-6}\sum_{s=1}^{2}\left|E_{\mathbf{g}}^{(s)Rad}\right|^2, \tag{19}$$

we will obtain the expression for summands, describing the contributions to the spectral-angular density of the radiation of the mechanisms PXR, DTR and of the summand, which is the result of the interference of these radiation mechanisms.

$$\omega\frac{d^2N_{\mathrm{PXR}}^{(s)}}{d\omega d\Omega}=\frac{e^2}{4\pi^2}P^{(s)^2}\frac{\theta^2}{(\theta^2+\gamma^{-2}-\chi_0')^2}R_{\mathrm{PXR}}^{(s)}, \tag{20a}$$

$$R_{\mathrm{PXR}}^{(s)}=\left(1-\frac{\xi}{\sqrt{\xi^2+\varepsilon}}\right)^2\frac{1+\exp\left(-2b^{(s)}\rho^{(s)}\Delta^{(1)}\right)-2\exp\left(-b^{(s)}\rho^{(s)}\Delta^{(1)}\right)\cos\left(b^{(s)}\left(\sigma^{(s)}+\frac{\xi-\sqrt{\xi^2+\varepsilon}}{\varepsilon}\right)\right)}{\left(\sigma^{(s)}+\frac{\xi-\sqrt{\xi^2+\varepsilon}}{\varepsilon}\right)^2+\rho^{(s)^2}\Delta^{(1)^2}}. \tag{20б}$$

$$\omega\frac{d^2N_{\mathrm{DTR}}^{(s)}}{d\omega d\Omega}=\frac{e^2}{4\pi^2}P^{(s)^2}\theta^2\left(\frac{1}{\theta^2+\gamma^{-2}}-\frac{1}{\theta^2+\gamma^{-2}-\chi_0'}\right)^2R_{DTR}^{(s)}, \tag{21a}$$

$$R_{DTR}^{(s)}=\frac{4\varepsilon^2}{\xi^2+\varepsilon}\exp\left(-b^{(s)}\rho^{(s)}\frac{1+\varepsilon}{\varepsilon}\right)\left[\sin^2\left(b^{(s)}\frac{\left(\sqrt{\xi^2+\varepsilon}\right)}{\varepsilon}\right)+sh^2\left(b^{(s)}\rho^{(s)}\frac{(1-\varepsilon)\xi^{(s)}+2\varepsilon\kappa^{(s)}}{2\varepsilon\sqrt{\xi^2+\varepsilon}}\right)\right], \tag{21б}$$

where

$$\Delta^{(1)}=\frac{\varepsilon+1}{2\varepsilon}-\frac{1-\varepsilon}{2\varepsilon}\frac{\xi^{(s)}}{\sqrt{\xi^{(s)^2}+\varepsilon}}-\frac{\kappa^{(s)}}{\sqrt{\xi^{(s)^2}+\varepsilon}},\ \sigma^{(s)}=\frac{1}{|\chi_{\mathbf{g}}'|C^{(s)}}\left(\theta^2+\gamma^{-2}-\chi_0'\right)\equiv\frac{1}{\nu^{(s)}}\left(\frac{\theta^2}{|\chi_0'|}+\frac{1}{\gamma^2|\chi_0'|}+1\right),$$

$$b^{(s)}=\frac{\omega\left|\mathrm{Re}\sqrt{\chi_{\mathbf{g}}\chi_{-\mathbf{g}}}\right|C^{(s)}}{2}\frac{L}{\gamma_0}. \tag{22}$$

The expressions (20)–(21) constitute the main result of this work. They are obtained in two-wave approximation of dynamic diffraction theory, taking into account the absorption of radiation in the layered plate substances and the orientation of the diffracting layers relative to the surface of the plate. These expressions allow us to investigate the spectral and angular characteristics of radiation depending on the energy of relativistic electrons and on the parameters of the artificial periodic structure of the target

4. Bormann effect in DTR

Since two X-ray waves determine the DTR yield, for the analysis of their contributions to the radiation spectral density it is convenient to represent the expression (21b) in such a form:

$$R_{DTR}^{(s)}=\frac{\varepsilon^2}{\xi(\omega)^2+\varepsilon}\left[e^{-b^{(s)}\rho^{(s)}\left[\frac{1+\varepsilon}{\varepsilon}-\frac{(1-\varepsilon)\xi^{(s)}+2\varepsilon\kappa^{(s)}}{\varepsilon\sqrt{\xi^2+\varepsilon}}\right]}+e^{-b^{(s)}\rho^{(s)}\left[\frac{1+\varepsilon}{\varepsilon}+\frac{(1-\varepsilon)\xi^{(s)}+2\varepsilon\kappa^{(s)}}{\varepsilon\sqrt{\xi^2+\varepsilon}}\right]}-2e^{-b^{(s)}\rho^{(s)}\frac{1+\varepsilon}{\varepsilon}}\cdot cos\left(\frac{2b^{(s)}\sqrt{\xi^2+\varepsilon}}{\varepsilon}\right)\right] \tag{23}$$

When consider the expression (23), one can see that the terms in brackets successively describe the waves belonging to the first and second fields, and their interference. Next we write the expression (23) in a more demonstrable form of

$$R_{DTR}^{(s)} = \frac{\varepsilon^2}{\xi(\omega)^2+\varepsilon}\left[e^{-L_f\mu_1^{(s)}} + e^{-L_f\mu_2^{(s)}} - 2\cdot e^{-L_f\mu_0\left(\frac{1+\varepsilon}{2}\right)}\cdot cos\left(\frac{L_f}{L_{ext}^{(s)}}\sqrt{\xi^2+\varepsilon}\right)\right], \quad (24)$$

where

$$\mu_1^{(s)} = \mu_0\left[\frac{1+\varepsilon}{2} - \frac{(1-\varepsilon)\xi^{(s)}+2\varepsilon\kappa^{(s)}}{2\sqrt{\xi^2+\varepsilon}}\right], \qquad \mu_2^{(s)} = \mu_0\left[\frac{1+\varepsilon}{2} + \frac{(1-\varepsilon)\xi^{(s)}+2\varepsilon\kappa^{(s)}}{2\sqrt{\xi^2+\varepsilon}}\right], \quad (25)$$

L_f is the path of a photon in a crystal, $\mu_0 = \omega\chi_0''$ – the linear coefficient of X-waves absorption in the averaged amorphous medium, $L_{ext}^{(s)} = \dfrac{1}{\omega\left|\mathrm{Re}\sqrt{\chi_{\mathbf{g}}\chi_{-\mathbf{g}}}\right|C}$ – the length of the X-waves extinction in a periodic medium.

The formula (24) clearly demonstrates the dynamic Borrmann effect arising during the passage of x-rays DTR through a periodic medium. Namely, in the X-ray scattering in a periodical medium the abnormal weak absorption is observed for the first wave field $\mu_1^{(s)} << \mu_0$ (i.e. anomalous transmission of the first field X-rays) and abnormal strong absorption for the second one $\mu_2^{(s)} > \mu_0$. By this reason, for the sufficiently large photon path in the substance of the plate the DTR only by one of the fields in a periodic structure will be formed, namely, by the field with effective absorption coefficient $\mu_1^{(s)}$.

Physics of the Borrmann effect [20] consists in the formation of the standing waves from the incident and scattered waves, whose antinodes are localized in the regions of space with a lower electron density for one of the waves (first term in (23) and (24)) and in the regions of space with a higher electron density for second wave (second term in (23) and (24)). Parameter $\kappa^{(s)}$ appearing in (25) determines the degree of manifestation of the Borrmann effect in the anomalous X-ray waves passing through a periodic structure. As in the case of free X-ray waves in crystals, a prerequisite for manifestation of the effect of DTR in layered medium is the condition $\kappa^{(s)} \approx 1$, that corresponds to the minimal value of the linear absorption coefficient $\mu_1^{(s)}$.

Next, we will carry out a numerical analysis for each of the waves and of their interference term separately. For this purpose the expression (23) we write in the following form

$$R_{DTR}^{(s)} = R_1^{(s)} + R_2^{(s)} + R_{\mathrm{int}}^{(s)}, \quad (26a)$$

$$R_1^{(s)} = \frac{\varepsilon^2}{\xi(\omega)^2+\varepsilon} e^{-b^{(s)}\rho^{(s)}\left[\frac{1+\varepsilon}{\varepsilon} - \frac{(1-\varepsilon)\xi^{(s)}+2\varepsilon\kappa^{(s)}}{\varepsilon\sqrt{\xi^2+\varepsilon}}\right]}, \quad (26b)$$

$$R_2^{(s)} = \frac{\varepsilon^2}{\xi(\omega)^2+\varepsilon} e^{-b^{(s)}\rho^{(s)}\left[\frac{1+\varepsilon}{\varepsilon} + \frac{(1-\varepsilon)\xi^{(s)}+2\varepsilon\kappa^{(s)}}{\varepsilon\sqrt{\xi^2+\varepsilon}}\right]}, \quad (26c)$$

$$R_{\mathrm{int}}^{(s)} = -\frac{2\varepsilon^2}{\xi(\omega)^2+\varepsilon} e^{-b^{(s)}\rho^{(s)}\frac{1+\varepsilon}{\varepsilon}} \cdot cos\left(\frac{2b^{(s)}\sqrt{\xi^2+\varepsilon}}{\varepsilon}\right). \quad (26c)$$

We will carry out the calculations for σ – polarized waves, i.e. for $s = 1$. In order to get demonstrable results, we will consider the case when the layers are of equal thickness $a = b = T/2$. We will consider the reflections, that correspond to $g = \dfrac{2\pi}{T}$. In this case, the parameters in the expressions (26) will take the following values:

$$\xi(\omega)=\frac{2\pi\sin^2(\theta_B)}{|\chi_b'-\chi_a'|}\cdot\left(1-\frac{\omega}{\omega_B}\right)+\frac{1-\varepsilon}{2\nu^{(1)}},\ \kappa^{(1)}=\frac{2}{\pi}\cdot\left|\frac{\chi_b''-\chi_a''}{\chi_b''+\chi_a''}\right|,\ \ \rho^{(1)}=\frac{\pi}{2}\cdot\left|\frac{\chi_b''+\chi_a''}{\chi_b'-\chi_a'}\right|,\ \nu^{(1)}=\frac{2}{\pi}\cdot\left|\frac{\chi_b'-\chi_a'}{\chi_b'+\chi_a'}\right|,$$

$$b^{(1)}=\frac{\omega_B|\chi_b'-\chi_a'|}{2\pi\sin(\delta-\theta_B)}L\,. \tag{27}$$

For a thin target ($b^{(1)}=5$), the curves drawn by (26), are shown in figure 3 describing the spectral density of the DTR (for $\omega_B=8\ \kappa eV$) in the artificial periodic structure consisting of amorphous layers of beryllium (Be) and tungsten (W). We see in this case, that the DTR is formed by the fields of two waves in a periodic structure, whose contributions in the spectral distribution are of comparable magnitude which will cause a strong interference of these waves. The interference term brings oscillations in the spectral density.

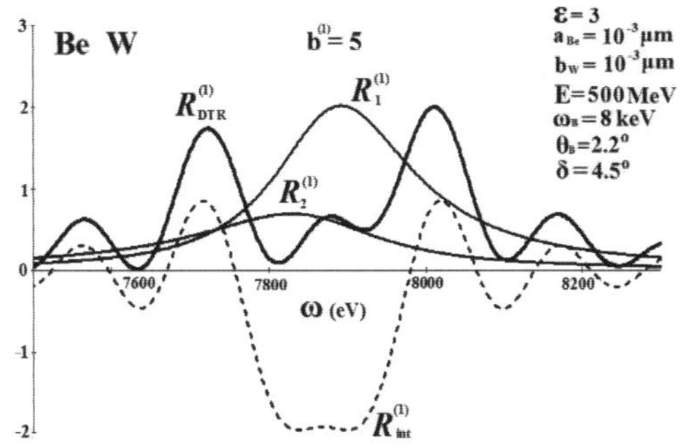

Figure 3. The contributions of the two fields, $R_1^{(1)}$ and $R_2^{(1)}$, and of their interference term $R_{\text{int}}^{(1)}$ into the total spectral density of DTR $R_{DTR}^{(1)}=R_1^{(1)}+R_2^{(1)}+R_{\text{int}}^{(1)}$

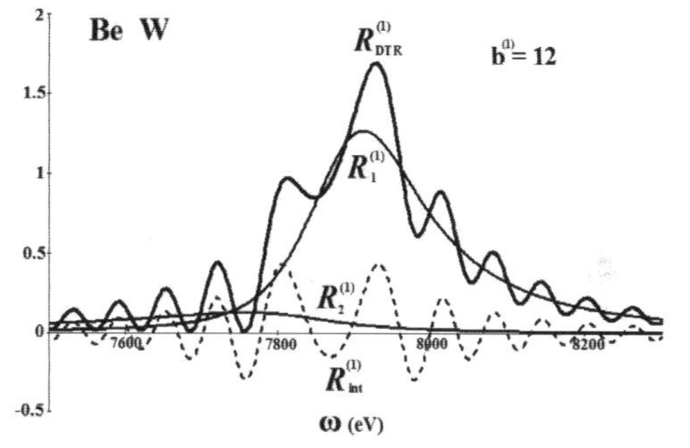

Figure 4. The same as figure 3 for bigger target thickness.

With increase of the target thickness one of the waves decays rapidly (figure 4 and figure 5), while the other one traverses the target without a significant decrease in amplitude. Under these conditions the contribution of the interference term markedly decreases and the spurious peaks in the spectrum are attenuated and then completely disappear (figure 6).

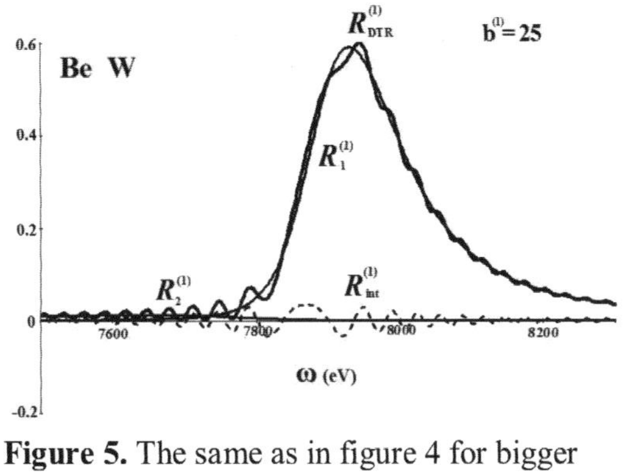

Figure 5. The same as in figure 4 for bigger target thickness.

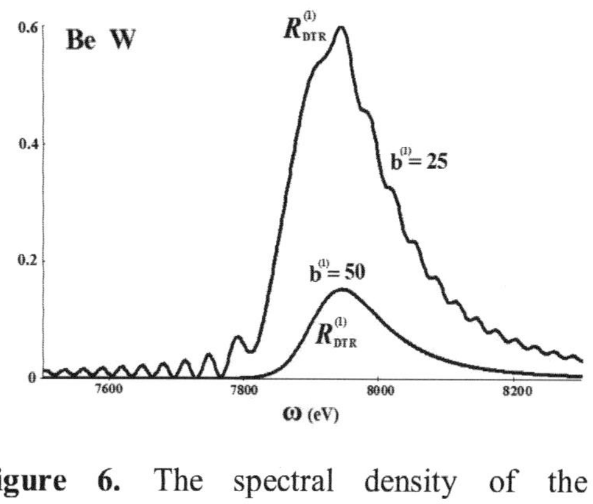

Figure 6. The spectral density of the relativistic electron DTR for different values of the target thickness.

It should be noted that the spectral curves in figure 5 and Figure 6 are constructed for a large target thickness, when the photon path length is longer than the average photo-absorption in an amorphous medium $l_{abs}=\frac{1}{\mu_0}$, which corresponds to the conditions of the Borrmann effect manifestation in an artificial periodic structure.

5. Conclusion

A theory for the coherent radiation of the relativistic electron crossing an artificial periodic structure is constructed for the case of Laue scattering geometry. The expressions for spectral-angular characteristics of the radiation in Bragg direction are derived and investigated.

The contributions to the DTR yield of two X-ray waves, which are responsible for DTR formation, are studied. It is shown that with increase of the target thickness, one of the waves is absorbed anomalously strongly and the other wave abnormally weakly, i.e. the Borrmann effect is manifested in DTR in an artificial periodic structure in the Laue geometry.

Based on these expressions it is shown that the angular density of diffracted transition radiation in layered target is more than one order higher than the density for a single crystal radiator under similar conditions.

References

[1] Garibian G M, Yang C 1983 *X- ray Transition Radiation*, Erevan, USSR, (in Russian)
[2] Caticha A 1989 *Phys. Rev. A.* **40** 4322
[3] Nasonov N N 1998 Phys. Lett. A. **246** 148
[4] Artru X Rullhusen P 1998 *Nucl. Instr. Meth. B* **145** 1
[5] Ter-Mikaelian M 1972 High-Energy Electromagnetic Process in Condensed Media (New York : Wiley)
[6] Garibian G, Yang C 1971 *J. Exp. Theor. Phys.* **61** 930
[7] Baryshevsky V, Feranchuk I 1971 *J. Exp. Theor. Phys.* **61** 944
[8] Nasonov N, Noskov A 2003 Nucl. Instr. Meth. In Phys. Res. B **201** 67
[9] Kubankin A.S., Nasonov N.N., Sergienko V.I., Vnukov I.E. 2003 *Nucl. Instr. Meth. B* **201** 97
[10] Adischev Y N, Arishev S N, Vnukov A V, et al. 2003 *Nucl. Instr. Meth. B* **201** 114
[11] Nasonov N N, Kaplin V V, Uglov S.R., et al. *2005 Nucl. Instr. Meth. B* **227** 41.
[12] Blazhevich S, Noskov A 2006 *Nucl. Instr. Meth. B.* **252** 69
[13] Blazhevich S, Noskov A 2008 *Nucl. Instr. Meth. In Phys. Res. B* **266** 3777
[14] Blazhevich S, Noskov A 2009 *J. Exp. Theor. Phys.* **136** 1043
[15] Blazhevich S, Noskov A *2008 Nucl. Instr. Meth. In Phys. Res. B* **266** 3770
[16] M.A. Piestrup, D.G. Boyers, C.I. Pincus et al. 1992 *Phys.Rev. A.* **45** 1183.
[17] Nasonov N N, KaplinV V, Uglov S R, Piestrup M A, Garyт C K 2003 *Phys. Rev. E.* **68** 3604
[18] Pinsker Z 1984 *Dynamic Scattering of X-rays in Crystals* (Berlin: Springer)
[19] Bazylev V, Zhevago N 1987 *Emission From Fast Particles Moving in a Medium and External Fields* Moscow Nauka, 1987
[20] Borrmann G 1941 *Zh. Phys.* **42** 157

Setup for study of radiation in VUV region generated by 5.7 MeV electrons in a multilayer mirror

S R Uglov, V N Zabaev, V V Kaplin, S I Kuznetsov

Tomsk Polytechnic University, P.O. Box 25, Lenin Ave. 2-a, 634050 Tomsk, Russia

E-mail: uglov@tpu.ru

Abstract. The experimental setup created for investigating the radiation in the vacuum ultraviolet (VUV) region generated by 5.7 MeV electrons in a multilayer X-ray mirror at different angles of the electron beam incidence on the X-ray mirror surface is described. The method proposed for parametric X-rays separation from the generated radiation is analyzed. The method is based on the K-edge effect of photon absorption in fluorine at transmission of generated radiation through a thin Teflon filter. The results of the first test measurements are discussed.

1. Introduction

It is known that in the spectrum of radiation generated by an electron beam in a periodic radiator, in contrast to a uniform flat one, the appearance of additional quasi-monochromatic components should be expected due to the radiation diffraction on a multilayer structure of the radiator. A number of studies [1–4] have experimentally shown that the artificial periodic structures (APS) and the crystals are suitable for generation of quasi-monochromatic X-ray radiation at passing charged particles through them. The mechanisms of radiation providing the main contributions to the generated radiation are transition x-ray radiation (TR) and bremsstrahlung (Br) with subsequent diffraction of their photons in the periodic structure of the radiator, as well as the mechanism of generation known as parametric x-ray radiation (PXR). PXR is coherent radiation for a given direction and the wavelength emitted by periodic structure excited by a relativistic charged particle. In the experiments, these types of radiation manifest itself as a number of coherent spectral peaks with the position, width and intensity depending on the mechanism of radiation. Further, for convenience, these types of radiation will be referred to be a general term – the coherent radiation from a periodic structure (CRPS).

In contrast to the crystals, the technology of artificial periodic structures, such as multilayer X-ray mirrors, allow to create a structure with the specified parameters, such as the atomic composition of the layers and the law of its distribution. One from the attractive features of such structures for the generation of radiation is that at least they have a greater efficiency of the generation of radiation in comparison with crystals [3], and therefore have a less destructive effect on the electron beam in the case of a thin radiator, that is important for possible further use of this electron beam for any additional applications or for the repeated penetration of the internal radiator for the radiation generation in circular accelerators.

X-ray emission from APS was previously observed in the photon energy range of 4–20 keV in the experiments at 500 MeV electron beam of the Tomsk synchrotron [1, 3] and at betatron internal beam of 15–35 MeV electrons [4]. Since APS period is greater than that of the crystals, they can be used for radiation generation in soft X-rays and vacuum ultraviolet (VUV) ranges [5]. Currently, there is a gap in experimental studies of parametric radiation in the VUV range. However, this might serve as promising base of tunable VUV source for some applications.

In this paper the experimental setup created for investigating the radiation in the VUV region generated by microtron electrons in periodic structures is described. The results of the first test experiment carried out on the created setup are presented. In addition, the detailed discussion is provided for the method for extracting the parametric component of the total VUV radiation generated by electrons interacting with the APS. The first attempt of experimental observation of the parametric component in VUV radiation generated by electrons with energy of 5.7 MeV in APS is also described. The orientation dependences of the angular distributions of radiation generated both on the surfaces of a homogeneous thin plate and multilayer X-ray mirror were measured and compared.

2. Experimental setup

The experimental setup was established on the basis of the Tomsk microtron M-5. The scheme of the experimental chamber with a detection unit is shown in figure 1. The frequency of the accelerator is 25

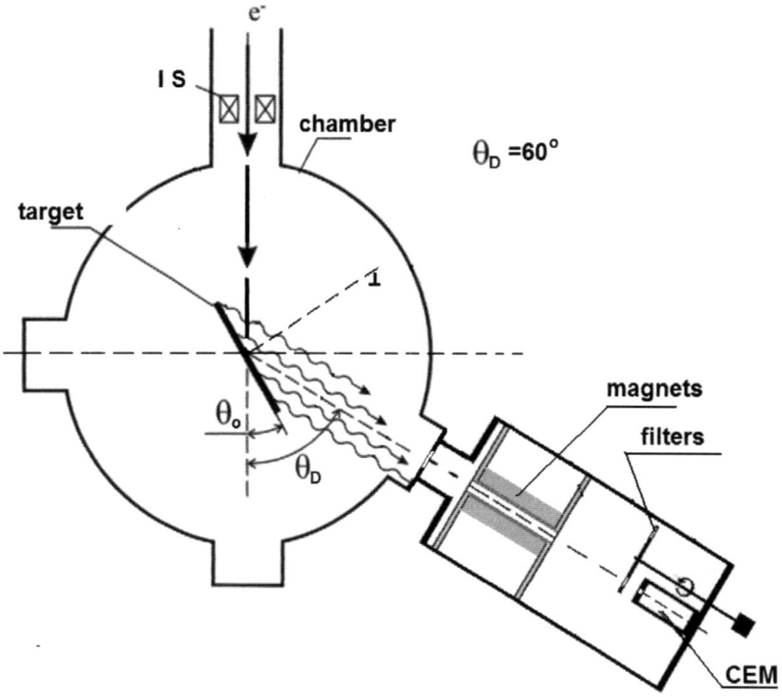

Figure 1. Scheme of experimental chamber with a detection unit.

Hz, the duration of the extracted electron beam is 0.4 μs. Electron beam diameter is defined by a collimator with the diameter of 3 mm and length of 60 mm located at the distance of 600 mm from the mirror. Monitoring the current of electrons accelerated up to 5.7 MeV is carried out by an induction sensor (IS) placed between the collimator and the radiator. VUV yield is measured by a channel electron multiplier (CEM) VEU-6. The detector is located in the median plane of the accelerator at the distance of 64 cm from the radiator and at an angle of 60° to the direction of the electron beam. Permanent magnets and diaphragms are set along the direction of VUV registration for cleaning the flow of the generated photons from electrons scattered by the target. CEM works in the counting mode. The measuring VUV system maintains linearity of photon registration at the loading of about 2 pulses during the electron dumping on the radiator. The registration system loading is checked during

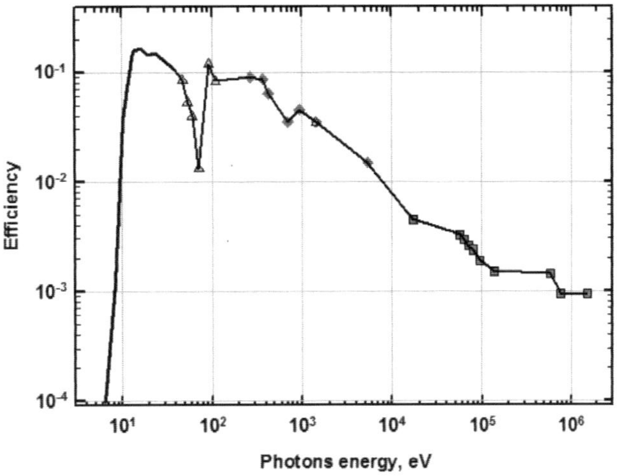

Figure 2. Efficiency of CEM detector.

the experiment for every 100 repetitions of the accelerator. Figure 2 shows a typical [6] dependence of the efficiency of photon detection on the photon energy. Spectral range of detected radiation is additionally regulated with absorption filters installed before the CEM's window. The absorbing filters are replaced by means of the turret mounted in front of the detector.

The multilayer structure of the X-ray mirror radiator consisted of 400 pairs of W and B_4C layers with period $d = 1.8$ nm is placed on Si substrate with a thickness of 35 μm. The horizontal and vertical dimensions of the radiator used in the first experiment were 6 and 9 mm, respectively. The radiator was placed in the vacuum chamber on the end of the vertical holder. Vertical movement of the holder and its rotation around vertical axis allowed setting the radiator on the trajectory of the electron beam and changing the orientation angle θ_O of electron incidence on its surface. Collection of information, its processing and control of the radiator attitude was carried out with the use of the specialized units of the computer-running CAMAC through a special program.

3. On study of CRPS and TR

According to our previous experiments, a back transition radiation (BTR) generated by electrons on the input Si-plate surface is effectively recorded with CEM. BTR yield is mainly located in the spectral range of 10 – 50 eV. BTR is also generated on the input surface of APS. The measured orientation dependences (OD) of BTR generated in APS and Si plate are shown in figure 3, curves 1 and 2, respectively. The figure shows that ODs measured for homogeneous Si plate and for multilayer structure are largely identical. Minima in the centers of the ODs are at the angle $\theta_O = \theta_B = \theta_D / 2$ that corresponds to symmetric Bragg orientation of the radiator relative to the electron beam and a collimated detector. Maxima in both ODs are located symmetrically with respect to the dips in its centers. Angular distance $\Delta\theta_O$ between the ODs peaks is comparable to the value of $\gamma^{-1} = 89.6$ mrad, where γ is the Lorentz factor of accelerated electrons. In studies involving the detection of CRPS generated in APS, back transition radiation was seen as a background which should be taken into account at determining the contribution of a CRPS quasi-monochromatic component. To detect CRPS in the VUV range it was chosen the method based on the property of CRPS described by the dependence (1) defining the relation between photon energy E_γ, angle θ_D of photon emission and angle θ_O of the electron incidence on the radiator layers

$$E_\gamma = \hbar\omega = \frac{2\pi\hbar c}{d}\frac{\sin\theta_O}{\beta^{-1} - \sqrt{\varepsilon(\omega)}\,\cos\theta_D}, \tag{1}$$

where $\beta = v/c$ – relative velocity of the electrons; d – period of APS;

$\varepsilon(\omega) = 1 - \left[a(1-\varepsilon(\omega)_W) + b(1-\varepsilon(\omega)_C)\right]/d$ – dielectric constants averaged over the period d;

$\varepsilon(\omega)_W$ and $\varepsilon(\omega)_C$ – dielectric constants of tungsten and carbon.

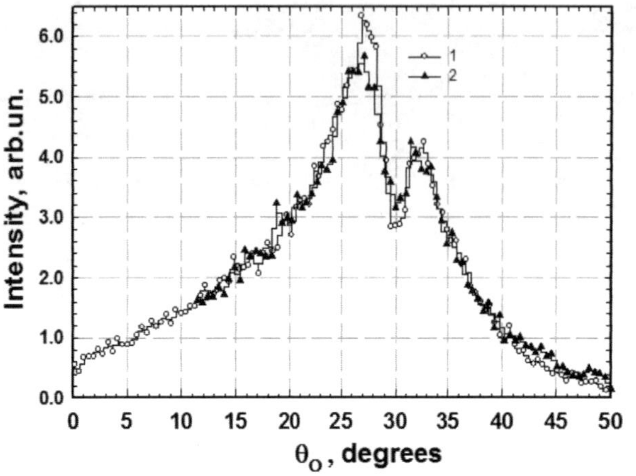

Figure 3. Orientation dependence of BTR generated in APS, curve 1, and in Si plate, curve 2. θ_o is an angle between electron beam and radiator surface, $\theta_B = 30^o$.

The principle of the method for finding the parametric X-rays is illustrated in figure 4. For discussion we used CRPS orientation dependence, curve 1, with its shape being typical for the

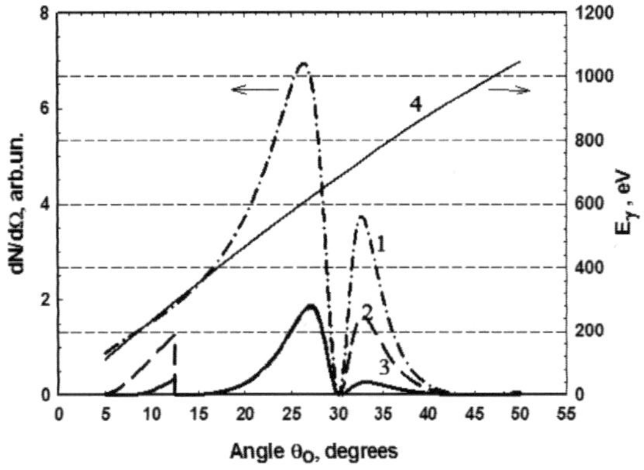

Figure 4. Illustration of changes in the shape of the orientation dependence (curve 1) after transmission of radiation through the 1 μm thick polyethylene *$(C_2H_4)_n$* and Teflon *$(C_2F_4)_n$* absorption filters, curve 2 and 3, respectively. Curve 4 – the photon energy in dependence with the angle θ_O of electron incidence on the target surface.

crystals. Curve 4 represents the dependence of angular position of CRPS spectral peak on angle θ_O for the structure with a period of 1.8 nm, angle of photon emission $\theta_D = 60^o$ and electron energy of 5.7 MeV. The use of a thin-foil filter from the material with the absorption edge in the range of investigated energies will lead to a significant distortion of the specific shape of the CRPS angular distribution. A similar method with the use of a thin nickel foil was used in [7] to estimate the "natural" width of the spectral line of parametric X-ray radiation produced by a Si crystal in the region of photon energy of about 8 keV. In our case, the radiation energy in the center of the CRPS angular distribution, according to the formula (1), is about 600 eV. Therefore, it is advisable to use a filter containing fluorine, for example, Teflon *$(C_2F_4)_n$* film or LiF crystal. The *K* absorption-edge energy of fluorine $E_K = 688.37$ eV. In addition to the absorption of radiation near the *K* edge, the 1 μm thick filter provides almost complete suppression of transition radiation in the range of 10–50 eV.

Orientation dependence of CRPS at using the thin $(C_2F_4)_n$ filter will be compared with orientation dependence at using another filter which has no absorption edge in the investigated range of photon energies, for example, polyethylene $(C_2H_4)_n$ filter. The coefficients of radiation transmission through $(C_2H_4)_n$ and $(C_2F_4)_n$ filters with a thickness of 1 μm, calculated using the database [8], are shown in figure 5.

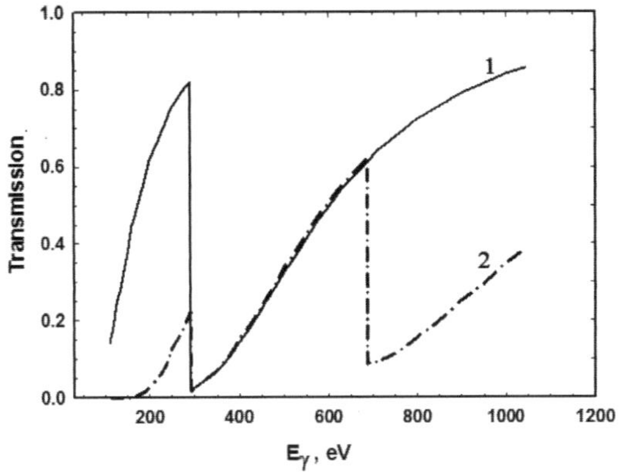

Figure 5. Coefficients of radiation transmission through 1 μm polyethylene and 1 μm Teflon filters, curves 1 and 2, respectively, in dependence with the photon energy.

Figure 4 also shows the changes in the shape of CRPS orientation dependence at the radiation penetration of the polyethylene and Teflon filters, curves 2 and 3, respectively. As seen, the radiation intensity in the right peak is reduced by about 1.5 and 7.2 times at transmission of radiation through polyethylene and Teflon filters, respectively. Left maximum of intensity in OD is reduced by about 2 times in both cases. When using the Teflon filter, a stronger suppression of the right maximum is sustained, approximately 3.6 times higher as compared to the left one, which can be easily seen in the experiment. Thus, the analysis of the calculations allows us to conclude that by observing the predicted differences in OD, it will be possible to detect the presence of quasi-monochromatic components in the investigated radiation, the nature of which can be explained by a CRPS mechanism. Note that the calculated ODs submitted by curves 2 and 3 have the additional maxima on its left sides due to the influence of the absorption edge of carbon – a component of both filters. Their position $\theta_0 = 12.5^\circ$ corresponds to the energy of generated radiation of 291 eV, according to formula (1).

4. Test measurements

The first experiments for testing the setup and on the monochromatic component separation in the radiation generated in the layered structure were carried out according to the above considered method.

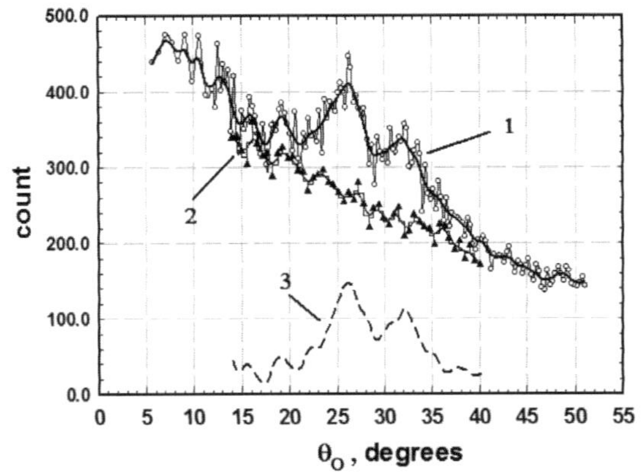

Figure 6. Orientation dependence of radiation yield from APS detected with CEM after transmission of radiation through the absorption filters. Curves 1 and 2 – for the filters consisted of single and double Teflon films with thickness of 1.7 μm. Curve 3 – smoothed difference between the yield values shown by curves 1 and 2.

Measurements of the orientation dependences of the radiation were done with the use of the absorbing filters made of Teflon films. Figure 6 shows the orientation dependence obtained with using a 1.7 μm

Teflon filter, curve 1, which has a double-maximum form and a strong enough background substrate, with its intensity increasing with the decrease of angle θ_O. Curve 2 in figure 6 shows the orientation dependence obtained using a double-film Teflon filter with a full thickness of 3.4 μm, which has not component of a double-maximum form. Curve 3 in figure 6 shows difference between the yield values shown by curves 1 and 2. The ratio of intensities of the OD maxima, curve 3, is about 2/3, instead of the expected value of 1/7. It means that the predicted peculiarities of the radiation angular distribution due to selective absorption of the Teflon filter are not observed. Double-peak form of the orientation dependence is preserved, and the expected suppression of the right peak is absent.

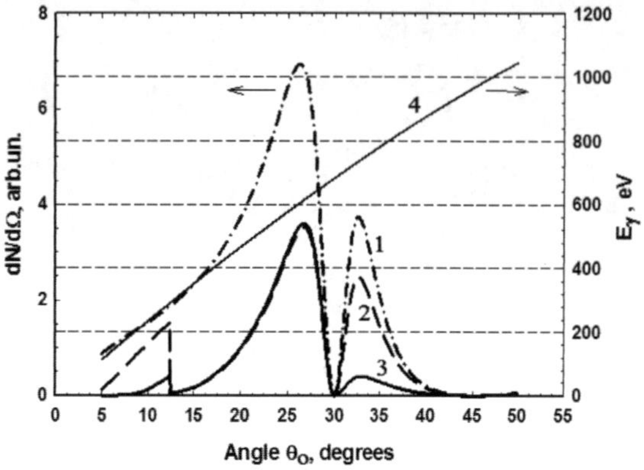

Figure 7. Same as in figure 4. Curve 2 – for polyethylene filter with thickness of 2 μm, curve 3 – for a combined (1 μm Teflon + 1 μm polyethylene films) filter.

One of the possible reasons for the result is that one-film filter used in the experiment has pores through which a soft transition radiation falls into the detector. This conclusion is supported by the fact that after installing the filter consisting of two identical 1.7 micron Teflon films, the peaks in OD were not observed (curve 2 in figure 6). Hence, the second Teflon filter completely absorbed soft transition radiation. Therefore, a composite filter consisting of thin polyethylene and Teflon films should be used because the dual-filter ensures the mutual overlap of the pores in the filters that can reliably suppress soft background of transition radiation. Figure 7 presents the orientation dependences of the radiation from a multilayer structure, calculated for the case of using the composite filters. Curve 2 – for the composite 2 μm polyethylene filter consisted of two films with thicknesses of 1 μm, curve 3 – for a combined (1 μm Teflon + 1 μm polyethylene) filter. As can be seen, the use of an additional filter does not violate the principle of the method. The asymmetry of the peaks of the orientation dependence is retained (the ratio of about 1/7). Although, the radiation yield will be lower by a factor of about 2. Another possible reason may be that the shape of the orientation dependence can be different from that used to discuss the method of separating the monochromatic component in the radiation generated in the layered structure, curve 1, figure 4. Our effort to calculate OD for x-ray mirror with using the existing theories has given ambiguous results.

5. Conclusion

The first test experiments have shown that the created experimental setup is workable. However, the analysis of the experimental results after testing the method for the observation of CRPS provided several suggestions for its improvement. First, it is necessary to reduce background induced by bremsstrahlung formed on the chamber walls by electrons scattered by the target at large angles. To this end, the radiator-to-detector distance should be increased. Narrower diaphragms for tighter collimation of radiation might also be useful. Secondly, it is also necessary to improve quality of the filters in order to exclude influence of its imperfections on experimental results. For further experiments it will be made a set of more perfect filters for different regions of the radiation cone. Finally, for the ultimate justification of the method of CRPS observation it is necessary to have the exact theoretical form of the CRPS orientation dependence. Unfortunately, the existing theories give

Journal of Physics: Conference Series **357** (2012) 012012 doi:10.1088/1742-6596/357/1/012012

conflicting results so far. But, in any case, CRPS component will be seen due to a "jump" in the distribution of radiation passed through a *K* edge filter because of the relation between photon energy and direction of its emission given by equation (1).

Acknowledgments
This work was partly supported by the Russian Foundation for Basic Research, project 10-02-00942, "The nonuniform photonic crystal as a new source of coherent submicrofocus X-rays from relativistic electrons".

References
[1] Kaplin V V, Uglov S R, Zabaev V N, Piestrup M A, Gary C K and Nasonov N N 2000 *Appl. Phys. Lett.* 76 3647
[2] Kaplin V V, Kuznetsov S I, Timchenko N A, Uglov S R and Zabaev V N 2001 *Nucl. Instrum. Meth. Phys. Res.* B **173** 238
[3] Nasonov N, Kaplin V V, Uglov S R, Piestrup M A and Gary C 2003 *Phys. Rev.* E **68** 036504
[4] Kaplin V V, Uglov S R, Sohoreva V V, Bulaev O F, Voronin A A, Piestrup M, Gary C and Fuller M 2009 *Nucl. Instrum. Meth. Phys. Res.* B 276 777
[5] Nasonov N N, Kubankin A S, Zhukova P N, Goldstein M, Williams D L, Piestrup M A and Park H 2007 *Nucl. Instrum. Meth. Phys. Res.* B **254** 259
[6] Ainbund M R and Polenov B V 1981 *Secondary electron multipliers and their use* (Moscow: Energoatomizdat) p 140
[7] Brenzinger K H, Limburg B, Backe H, Dambach S, Euteneuer H, Hagenbuck F, Herberg C,Kaiser K H, Kettig O, Kube G, Lauth W, Schöpe H and Walche Th 1997 *Phys. Rev. Lett.* **79** 2462
[8] http://ftp.esrf.eu/pub/scisoft/xop2.3/DabaxFiles/

Optical transition radiation in presence of acoustic waves for an oblique incidence

A R Mkrtchyan, V V Parazian and A A Saharian

Institute of Applied Problems in Physics, 25 Nersisyan Street, 0014 Yerevan, Armenia

E-mail: vparazian@gmail.com, saharian@ysu.am

Abstract. Forward transition radiation is considered in an ultrasonic superlattice excited in a finite thickness plate under oblique incidence of relativistic electrons. We investigate the influence of acoustic waves on both the intensity and polarization of the radiation. In the quasi-classical approximation, formula is derived for the spectral-angular distribution of the radiation intensity. It is shown that the acoustic waves generate new resonance peaks in the spectral and angular distributions. The heights and the location of the peaks can be controlled by choosing the parameters of the acoustic wave. The numerical examples are given for a plate of fused quartz.

1. Introduction

Transition radiation is produced when a uniformly moving charged particle crosses an interface between two media with different dielectric properties. Such radiation has a number of remarkable properties and at present it has found many important applications (see, for instance, Refs. [1]-[6]). In particular, the transition radiation is widely used for particle identification, for the measurement of transverse size, divergence and energy of electron and proton beams. An enhancement for the transition radiation intensity may be achieved by using the interference between the radiation emitted by many interfaces in a multilayer structure. From the point of view of controlling the parameters of various radiation processes in a medium, it is of interest to investigate the influence of external fields, such as acoustic waves, temperature gradient etc., on the corresponding characteristics.

In Refs. [7, 8] we have considered the X-ray and optical transition radiation from ultrarelativistic electrons in an ultrasonic superlattice excited in melted quartz plate. The radiation from a charged particle for a semi-infinite laminated medium has been recently discussed in [9]. In these references the transition radiation is considered under normal incidence of a charged particle upon the interface of the plate. In the present paper we generalize the corresponding results for oblique incidence. The angle between the particle velocity and the normal to the interface is an additional parameter which can be used for the control of angular-frequency distribution and the polarization of the radiation.

We have organized the paper as follows. In the next section the intensity of the radiation in forward direction is investigated for both parallel and perpendicular polarizations by using the quasi-classical approximation. In section 3, numerical examples are presented for the radiation intensity in the case of a plate made of fused quartz. The main results are summarized in section 4.

2. Spectral-angular distribution of the radiation

Consider the transition radiation from a charged particle in a plate of thickness l with dielectric permittivity ε_0 which is immersed into a homogeneous medium with permittivity ε_1. We assume that the z axis of the Cartesian coordinate system (x, y, z) is directed along the normal to the plate and the boundaries of the plate are located at $z = -l$ and $z = 0$. The trajectory of the particle with velocity $\mathbf{v}$ is in the (x, z) plane and forms with the z axis a given angle α. For the particle coordinates one has

$$x(t) = v(t - t_0)\sin\alpha, \;\; y(t) = 0, \;\; z(t) = -l + v(t - t_0)\cos\alpha. \tag{1}$$

Here we do not take multiple scattering effects into account (for the discussion of these effects see [1]). So, we can write $\mathbf{v} = (v\sin\alpha, 0, v\cos\alpha)$. In the presence of longitudinal ultrasonic vibrations, excited in the plate along the normal to its surface, the dielectric permittivity can be written in the form

$$\varepsilon(z) = \begin{cases} \varepsilon_0 + \Delta\varepsilon\cos(k_s z + \omega_s t + \phi), & -l \leqslant z \leqslant 0, \\ \varepsilon_1, & z < -l, \; z > 0, \end{cases} \tag{2}$$

where ω_s, k_s are the cyclic frequency and the wave number of the ultrasound, ϕ is the initial phase. In the discussion below we will assume that during the transit time of the particle the dielectric permittivity in the superlattice is not notably changed. For relativistic particles and for the plate thickness $l \lesssim 1$ cm this condition is satisfied for $\omega_s \ll 10^{12}$ Hz.

For the radiation in the spectral range $\omega \gg k_s c$, the presence of the small parameter $k_s c/\omega$ allows us to employ the quasi-classical approximation for the investigation of the radiation field in the forward direction. The latter may be evaluated in a way similar to that discussed in [1] for the case of normal incidence (see also [10]). Considering the radiation in the region $z > 0$, we denote by θ and φ the polar and azimuthal angles for the radiated photon with respect to the axis z. For the corresponding wave vector one has

$$\mathbf{k} = \frac{\omega}{c}\sqrt{\varepsilon_1}(\sin\theta\cos\varphi, \sin\theta\sin\varphi, \cos\theta). \tag{3}$$

Unlike to the case of the normal incidence, discussed in [7, 8], in the case of oblique incidence we have radiations with two different polarizations. For the first one the electric field is parallel to the observation plane (plane formed by the vectors $\mathbf{k}$ and $\mathbf{v}$) and for the second one the electric field is perpendicular to the observation plane. These two polarizations are referred as parallel and perpendicular polarizations, respectively.

First we consider the spectral-angular density of the total radiation intensity averaged over the phase of particle flight into the plate. In the angular region $\sin\theta < \sqrt{\varepsilon_0/\varepsilon_1}$, it is given by the expression

$$\begin{aligned} I(\omega, \theta, \varphi) &= \frac{dI(\omega, \theta, \varphi)}{d\omega d\theta d\varphi} = \frac{e^2\sin^3\theta\cos^2\alpha}{\pi^2 c\sqrt{\varepsilon_1}} \sum_{m=-\infty}^{+\infty} J_m^2(a(\theta)) \\ &\times \left[\frac{\mathbf{P}(\theta, \varphi, \alpha)}{V(\theta, \varphi, \alpha)} - \frac{\mathbf{Q}(\theta, \varphi, \alpha)}{U_m(\theta, \varphi, \alpha)} \frac{\cos^{1/2}\theta}{(\varepsilon_0/\varepsilon_1 - \sin^2\theta)^{1/4}}\right]^2 \\ &\times \sin^2\left[\frac{\omega l\sqrt{\varepsilon_1}}{2c\cos\alpha} U_m(\theta, \varphi, \alpha)\right], \end{aligned} \tag{4}$$

where e is the charge of the particle, $J_m(x)$ is the Bessel function, and

$$a(\theta) = \frac{\omega\Delta\varepsilon/(2ck_s)}{\sqrt{\varepsilon_0 - \varepsilon_1\sin^2\theta}}. \tag{5}$$

Journal of Physics: Conference Series **357** (2012) 012013 doi:10.1088/1742-6596/357/1/012013

In (4) we have defined the functions

$$\begin{aligned} U_m(\theta, \varphi, \alpha) &= 1/\beta_1 - \sin\theta\cos\varphi\sin\alpha - \cos\alpha\sqrt{\varepsilon_0/\varepsilon_1 - \sin^2\theta} - \frac{mk_s c}{\omega\sqrt{\varepsilon_1}}\cos\alpha, \\ V(\theta, \varphi, \alpha) &= 1/\beta_1 - \sin\theta\cos\varphi\sin\alpha - \cos\theta\cos\alpha, \end{aligned} \tag{6}$$

with $\beta_1 = v\sqrt{\varepsilon_1}/c$ and

$$\begin{aligned} \mathbf{P}(\theta, \varphi, \alpha) &= (\sin\varphi, \cot\theta\tan\alpha - \cos\varphi, -\sin\varphi\tan\alpha), \\ \mathbf{Q}(\theta, \varphi, \alpha) &= (\sqrt{\varepsilon_1/\varepsilon_0}\sin\varphi, \cot\theta\tan\alpha - \sqrt{\varepsilon_1/\varepsilon_0}\cos\varphi, -\sin\varphi\tan\alpha), \end{aligned} \tag{7}$$

In general, the dielectric permittivities ε_0 and ε_1 are functions of ω. In what follows we assume that $\beta_1 < 1$. For the case of normal incidence, $\alpha = 0$, the general formula (4) coincides with the result previously derived in [8].

For the spectral-angular density of the radiation with the perpendicular polarization we have the following expression:

$$\begin{aligned} I_\perp(\omega, \theta, \varphi) &= \frac{e^2\sin^2(2\alpha)}{4\pi^2 c}\frac{\left(1 - \sqrt{\varepsilon_1/\varepsilon_0}\right)^2}{\sqrt{\varepsilon_0 - \varepsilon_1\sin^2\theta}}\sin^3\theta\cos\theta\sin^2\varphi \\ &\times \sum_{m=-\infty}^{+\infty}\frac{J_m^2(a(\theta))}{U_m^2(\theta, \varphi)}\sin^2\left[\frac{\omega l\sqrt{\varepsilon_1}}{2c\cos\alpha}U_m(\theta, \varphi)\right]. \end{aligned} \tag{8}$$

The radiation intensity for the parallel polarization is found from the relation

$$I_\parallel(\omega, \theta, \varphi) = I(\omega, \theta, \varphi) - I_\perp(\omega, \theta, \varphi), \tag{9}$$

with the total intensity given by (4). In the case of normal incidence, the intensity for the perpendicular polarization vanishes, $I_\perp(\omega, \theta, \varphi) = 0$, and the radiation is polarized in the observation plane.

Various special cases of formulas (4) and (8) have been considered in the literature. In the case of the normal incidence we recover the results of Refs. [7] and [8] for the X-ray and optical transition radiations respectively. In the absence of the acoustic wave we have $\Delta\varepsilon = 0$ and in formulas (4) and (8) the $m = 0$ term contributes only. In this case we obtain the quasi-classical approximation for the radiation intensity in a finite thickness plate for oblique incidence. The corresponding exact expression for the radiation intensity in this problem is well known from the literature [11] (see also Refs. [1]-[6]). The features of the optical transition radiation in a finite thickness plate for an oblique incidence have been discussed in Refs. [12] on the base of Pafomov's formulas. For a transparent material in the over-threshold case and under the condition $l\omega/v \gg 1$, the dominant contribution comes from the term in the exact formula with the resonant factor $\sin^2(y)/y^2$, with y given by the argument of the sin function in (4) with $m = 0$. Now, for simplicity considering the case of the radiation into the vacuum, it can be seen that for a relativistic particle with $1 - v/c \ll 1$, the radiation intensity near the Cherenkov peaks is well approximated by the formulas obtained from (4) and (8) in the limit $\Delta\varepsilon = 0$.

As it is seen from formulas (4) and (8), the radiation intensities for both polarizations have peaks the angular location of which is determined from the equation

$$U_m(\theta, \varphi, \alpha) = 0, \tag{10}$$

with the function $U_m(\theta, \varphi, \alpha)$ defined in (6). In the approximation under consideration the location of the peaks does not depend on the amplitude of the acoustic oscillations. In the presence of acoustic waves we have a set of peaks specified by m. The angular location of the peak with $m = 0$ coincides with that for the peak in the absence of acoustic waves. The angular distance between the peaks induced by the acoustic waves is of the order $(\omega_s/\omega)(c/v_s)$. In order to present the condition (10) in a physically more transparent form, we introduce the wave vector $\mathbf{k}_0$ for the photon inside the plate (in the absence of the acoustic wave):

$$\mathbf{k}_0 = \frac{\omega}{c}\sqrt{\varepsilon_0}\,(\sin\theta_0\cos\varphi, \sin\theta_0\sin\varphi, \cos\theta_0)\,, \tag{11}$$

where $\sin\theta_0 = \sqrt{\varepsilon_1/\varepsilon_0}\sin\theta$. In terms of this vector, the condition for the peaks is written in the form

$$\mathbf{k}_{0m}\mathbf{v} = \omega,\ \mathbf{k}_{0m} \equiv \mathbf{k}_0 + m\mathbf{k}_s, \tag{12}$$

with $\mathbf{k}_s = (0, 0, k_s)$. The latter is the condition for the Cherenkov radiation in the medium of the plate. Hence, we conclude that the peaks in the region $z > 0$, determined by the condition (10), correspond to the Cherenkov radiation emitted inside the plate and refracted from the boundary. We can have a situation where the Cherenkov radiation emitted inside the plate is completely reflected from the boundary and in the exterior region we have no peaks. In this case the equation (10) has no solutions. There are also cases then the Cherenkov radiation is confined inside the plate in the absence of the acoustic excitations and the peaks defined by (10) appear as a result of the influence of the acoustic waves. The location of the peaks can be controlled by choosing the incidence angle α.

We denote by $\theta = \theta^{(m)}(\alpha, \varphi)$ the location of the peaks with respect to the polar angle θ, for given values of α and φ. For the radiation intensity at the peaks we get the expression

$$\begin{aligned} I(\omega, \theta^{(m)}, \varphi) &= \frac{e^2\varepsilon_1 l^2\omega^2\sin^3\theta^{(m)}\cos\theta^{(m)}}{4\pi^2c^3(\varepsilon_0 - \varepsilon_1\sin^2\theta^{(m)})^{1/2}} \\ &\times \mathbf{Q}^2(\theta^{(m)}, \varphi, \alpha)J_m^2\left(\frac{\omega\Delta\varepsilon/(2ck_s)}{\sqrt{\varepsilon_0 - \varepsilon_1\sin^2\theta^{(m)}}}\right). \end{aligned} \tag{13}$$

The relative contribution of the component with the perpendicular polarization is given by

$$\frac{I_\perp(\omega, \theta^{(m)}, \varphi)}{I(\omega, \theta^{(m)}, \varphi)} = \left(1 - \sqrt{\varepsilon_1/\varepsilon_0}\right)^2 \frac{\sin^2\varphi\sin^2\alpha}{\mathbf{Q}^2(\theta^{(m)}, \varphi, \alpha)}. \tag{14}$$

In the absence of the ultrasonic vibrations the location of the peak is given by $\theta^{(0)}$. The ultrasound reduces the height of this peak by the factor $J_0^2\left(a(\theta^{(0)})\right)$. In particular, for a given radiation frequency, the frequency or the amplitude of the ultrasound can be tuned to eliminate this peak.

3. Radiation in a plate of fused quartz

In this section we consider the optical transition radiation in a plate of fused quartz. For the velocity of longitudinal ultrasonic vibrations one has $v_s \approx 5.6 \times 10^5$ cm/s. For the dielectric permittivity of fused quartz we use the Sellmeier dispersion formula

$$\varepsilon_0 = 1 + \sum_{i=1}^{3}\frac{a_i\lambda^2}{\lambda^2 - l_i^2} \tag{15}$$

with the parameters $a_1 = 0.6961663$, $a_2 = 0.4079426$, $a_3 = 0.8974794$, $l_1 = 0.0684043$, $l_2 = 0.1162414$, $l_3 = 9.896161$. In (15), λ is the wavelength of the radiation measured in micrometers. Formula (15) well describes the dispersion properties of fused quartz in the range $0.2\mu\text{m} \leqslant \lambda \leqslant 6.7\mu\text{m}$. In this spectral range fused quartz is very weakly absorbing.

In figures below we plot the spectral-angular density of the total radiation intensity in the forward direction, $I(\omega, \theta, \varphi)/\hbar$, and the spectral-angular density for the component with perpendicular polarization, $I_{\perp}(\omega, \theta, \varphi)/\hbar$, for electrons with the energy 2 MeV and for the plate thickness $l = 1$ cm. For the oscillation amplitude we have taken the value $\Delta n/n_0 = 0.05$, where n_0 is the number of electrons per unit volume for fused quartz.

In accordance with general features described above, the presence of the acoustic wave leads to the appearance of new peaks in both angular and spectral distributions of the radiation intensity. The height of the peaks can be tuned by choosing the parameters of the acoustic wave. In particular, the peak in the radiation intensity which is present in the absence of the acoustic wave is reduced by the factor $J_0^2(a)$. This peak can be completely removed by taking the parameters of the acoustic wave in such a way to have $a = j_{0,s}$, $s = 1, 2, 3, \ldots$, where $z = j_{0,s}$ are the zeroes of the function $J_0(z)$.

In figure 1, for separate values of the angle α (numbers near the curves), we display the radiation intensity, defined by (4), as a function of the cyclic frequency for the fixed value the polar angle $\theta = 1$ (left panel) and as a function of θ for the fiexed value of $\omega = 3.23 \times 10^{14}$ Hz (left panel). The graphs in figure 1 are plotted for the frequency of acoustic wave $\nu_s = 1.5$ MHz and for the azimuthal angle $\varphi = 0.1$. Similar graphs for the radiation with the perpendicular polarization are displayed in figure 2 for the same values of the parameters. In figure 3 we show the spectral distribution for the total intensity for different values of the acoustic wave frequency ν_s (numbers near the curves). The graphs are plotted for $\theta = 1$, $\varphi = 0.1$, $\alpha = 0.1$.

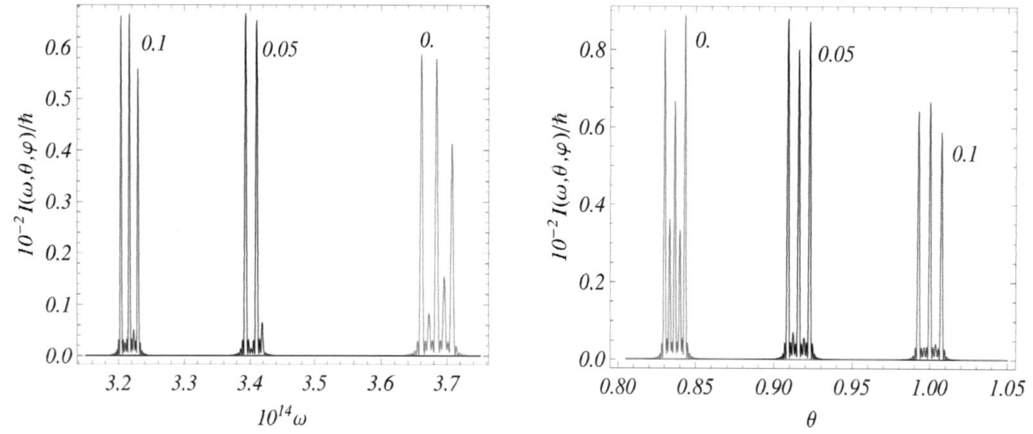

Figure 1. The spectral and angular distributions of the radiation intensity for different values of the incidence angle α (numbers near the curves). The values of the parameters are as follows: $\varphi = 0.1$, $\nu_s = 1.5$ MHz. For the left panel $\theta = 1$ and for the right one $\omega = 3.23 \times 10^{14}$ Hz.

4. Conclusion

In the present paper we have studied the influence of acoustic waves, excited in a plate, on the spectral and angular characteristics of the transition radiation for the oblique incidence of a charged particle. For the optical transition radiation, the presence of the small parameter λ/λ_0, with λ and λ_s being the wavelengths for the radiation and acoustic wave, respectively, allows us to apply the quasi-classical approximation for the evaluation of the radiation intensity in the forward direction. The spectral-angular distribution is described by formula (4) for the total

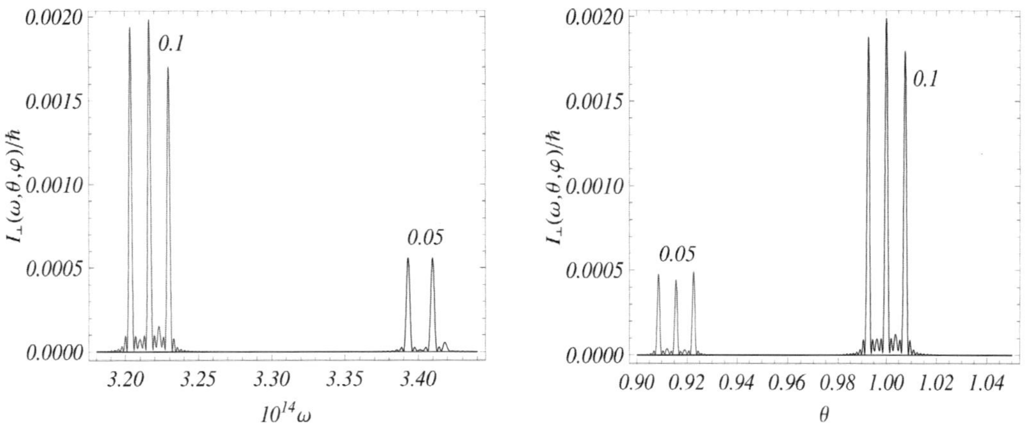

Figure 2. The same as in figure 1 for the perpendicular polarization.

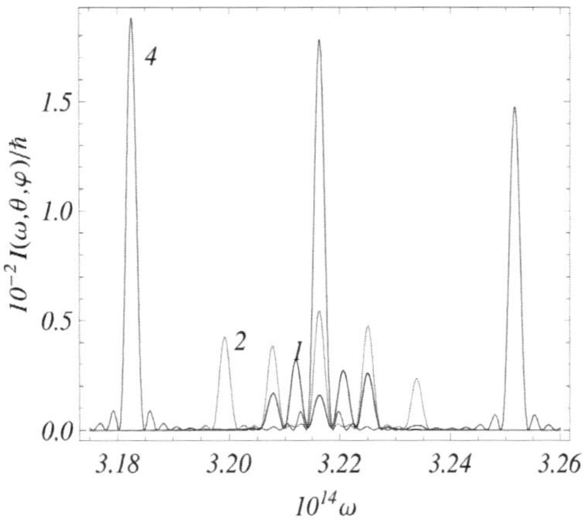

Figure 3. Spectral distribution of the total intensity for separate values of the acoustic wave frequency ν_s (numbers near the curves).

radiated energy and by (8) for the component with perpendicular polarization. The radiation intensities for both polarizations have strong peaks with the angular location determined from the condition (10). These peaks correspond to the Cherenkov radiation emitted inside the plate and refracted from the boundary. In the presence of acoustic waves we have a set of peaks with the angular separation of the order $(\omega_s/\omega)(c/v_s)$. The numerical examples for the optical transition radiation in a plate of fused quartz show that the acoustic waves allow to control the both angular and spectral parameters of the radiation. In particular, new resonance peaks appear in the spectral-angular distribution of the radiation intensity.

Acknowledgment

The authors are grateful to Professors Babken Khachatryan and Levon Grigoryan for valuable discussions and suggestions. A.R.M and V.V.P. acknowledge the Organizers of the RREPS11 Conference for a support.

References

[1] Ter-Mikaelian M L 1972 *High Energy Electromagnetic Processes in Condensed Media* (New York: Wiley Interscience)

[2] Gharibian G M and Yan S 1983 *Rentgenovskoye Perekhodnoye Izluchenie* (Yerevan: Izdatelstvo AN Arm. SSR)

[3] Ginzburg V L and Tsytovich V N 1990 *Transition Radiation and Transition Scattering* (Bristol: Adam Hilger)

[4] Baier V N, Katkov V M and Strakhovenko V M 1998 *Electromagnetic Processes at High Energies in Oriented Single Crystals* (Singapore: World Scientific)

[5] Rullhusen P, Artru X and Dhez P 1998 *Novel Radiation Sources Using Relativistic Electrons* (Singapore: World Scientific)

[6] Potylitsin A P 2011 *Electromagnetic Radiation of Electrons in Periodic Structures* (Berlin: Springer)

[7] Grigoryan L Sh, Mkrtchyan A H and Saharian A A 1998 *Nucl. Instr. Meth.* B **145** 197

[8] Mkrtchyan A R, Parazian V V and Saharian A A 2010 *Mod. Phys. Lett.* B **24** 2693

[9] Grigoryan L Sh, Mkrtchyan A R, Khachatryan H F, Arzumanyan S R and Wagner W 2010 *J. Phys.: Conf. Series* **236** 012012

[10] Khachatryan B V, 2004 *Doctor of Sciences Thesis* (Yerevan)

[11] Pafomov V E 1969 *Proc. FIAN SSSR, Nuclear Physics and Particle Interaction with Matter* **44** 90 (Moscow: Nauka)

[12] Ružička J and Mehes J 1986 *Nucl. Instr. Meth.* A **250** 491; Hrmo A and Ruzicka J 2000 *Nucl. Instrum. Meth.* A **451** 506

Journal of Physics: Conference Series **357** (2012) 012014 doi:10.1088/1742-6596/357/1/012014

High frequency Transition Radiation from uniaxial crystal

E Glazkova, M Ryazanov, M Strikhanov

National Research Nuclear University "MEPhI", Moscow, Kashirskoe sh. 31, 115409, Russia

E-mail: elenaglazkova2289@gmail.com

Abstract. Transition radiation from an ultrarelativistic particle flying out of the crystal surface at an arbitrary angle is considered. The expression for the spectral-angular density of radiation in the high frequencies has been obtained. The dependence of the resulting radiation distribution depending on the directions of the crystal main axis relative to the surface and the angle of the emission particle velocity is investigated.

1. Introduction

Transition radiation arises when a charged particle is passing through an interface between two media. Polarization currents that are produced in the material by moving charged particle field are the transition radiation sources. Usually transition radiation is considered in the frames of macroscopic electrodynamics when boundary conditions at the interface between two media is used [1]. If we use boundary conditions when surface shape is complex, it leads to the calculating difficulties. However, in the high frequencies area

$$\left(e^2/hc\right)^2 mc^2 << h\omega << \left(e^2/hc\right)mc^2, \quad (1)$$

when the radiation field frequency is much more higher than atomic frequencies and wavelength more than the atom size, so the permittivity $\varepsilon(\mathbf{r},\omega)$ is close to unity. This gives us an opportunity to use the method of successive approximation by degree of the difference $1-\varepsilon(\mathbf{r},\omega)$ for transition radiation in the high frequencies area. We can neglect the difference $\varepsilon(\mathbf{r},\omega)$ from unity in the expression for the field in the calculation of the polarization current in linear approximation, because the Fourier transform of the polarization current is proportional to the difference $1-\varepsilon(\mathbf{r},\omega)$, i.e. we can replace the field $\mathbf{E}(\mathbf{r},\omega)$, that is produced by a charged particle in medium by the field $\mathbf{E}^0(\mathbf{r},\omega)$, that is created by the same charged particle motion in a vacuum. In this approximation, medium boundaries are taken into account as boundaries when polarization current existence and at the same time boundaries conditions for the fields are automatically satisfied. If linear approximation by $1-\varepsilon(\mathbf{r},\omega)$ is used, it makes consideration of the transition radiation from complex surface much simpler. In general, transition radiation from interface between anisotropic medium and vacuum depends on the crystal surface orientation. It is of interest to evaluate the high frequencies transition radiation intensity when a fast charged particle shoot through from a uniaxial crystal.

2. Polarization currents in uniaxial crystal

It is known [7] that considering monocrystal to a crystal optics approximation one should take into account orientational ordering of crystals molecules and neglect the positional molecules ordering. So, in the crystal optics approximation the crystal is considered as an anisotropic homogeneous medium. The unit vector **e** is the vector that shows the direction of a uniaxial crystal principal axis. Then we can write its permittivity in the crystal optics approximation as:

$$\varepsilon_{is} = \varepsilon_0(\omega)(\delta_{is} - e_i e_s) + \varepsilon_e(\omega) e_i e_s. \tag{2}$$

In frequencies region equation (1), permittivity distinction from unity is small, i. e.

$$\delta_{is} - \varepsilon_{is}(\omega) = (\delta_{is} - e_i e_s)\{1 - \varepsilon_0(\omega)\} + e_i e_s \{\varepsilon_0(\omega) - \varepsilon_e(\omega)\} << 1. \tag{3}$$

Inequality equation (3) allows one to find an approximate solution by using the method of successive approximation in powers of the small difference $\delta_{is} - \varepsilon_{is}(\omega)$. If the medium is at $z < Z(x, y)$, the Fourier transforms of polarization current density $\mathbf{J}(\mathbf{r},\omega)$ and electric field in medium $\mathbf{E}(\mathbf{r},\omega)$ are connected with permittivity by the ratio:

$$J_i(\mathbf{r},\omega) = \frac{i\omega}{4\pi} \cdot (\delta_{is} - \varepsilon_{is}(\omega)) \cdot E_s(\mathbf{r},\omega)\theta(Z(x,y) - z), \tag{4}$$

where $\theta(u) = (u + |u|)/2|u|$.

In the high-frequency region Fourier transform of the polarization current density is proportional to the small difference equation (3). Therefore, in the first approximation it is possible to replace in equation (4) exact value of electric field in medium $\mathbf{E}(\mathbf{r},\omega)$ the electric field $\mathbf{E}_0(\mathbf{r},\omega)$ of a charged particle in a vacuum:

$$J_i(\mathbf{r},\omega) = \frac{i\omega}{4\pi} \cdot (\delta_{is} - \varepsilon_{is}(\omega)) \cdot E_{0s}(\mathbf{r},\omega)\theta(Z(x,y) - z). \tag{5}$$

In an isotropic medium the Fourier transform of polarization currents density is:

$$J_{0i}(\mathbf{r},\omega) = \frac{i\omega}{4\pi} \cdot (1 - \varepsilon_0(\omega)) \cdot E_{0i}(\mathbf{r},\omega)\theta(Z(x,y) - z). \tag{6}$$

Comparing equations (5) and (6) it can be seen that for the same medium geometry and the same particle motion Fourier transform for polarization current density in a uniaxial crystal $\mathbf{J}(\mathbf{q},\omega)$ in the linear on $\delta_{is} - \varepsilon_{is}(\omega)$ is connected to Fourier transform approximation of polarization current density in an isotropic medium with permittivity $\varepsilon_0(\omega)$, $\mathbf{J}_0(\mathbf{r},\omega)$ by ratio:

$$J_i(\mathbf{r},\omega) = \frac{\delta_{is} - \varepsilon_{is}(\omega)}{[1 - \varepsilon_0]} J_{0s}(\mathbf{r},\omega). \tag{7}$$

In the linear approximation by $\delta_{is} - \varepsilon_{is}(\omega)$ polarization current density equation (5) can be assumed to be known, so, the problem reduces to finding the polarization current equation (6). The angular and frequency radiation distribution in vacuum is created by a current $\mathbf{J}(\mathbf{k},\omega)$ can be represented as:

$$\frac{d^2W}{d\omega d\Omega} = \frac{(2\pi)^6}{c} \left|[\mathbf{k},\mathbf{J}(\mathbf{k},\omega)]\right|^2 = \frac{(2\pi)^6}{c} \left\{k^2 |\mathbf{J}(\mathbf{k},\omega)|^2 - |(\mathbf{k},\mathbf{J}(\mathbf{k},\omega))|^2\right\}, \tag{8}$$

where

$$\mathbf{J}(\mathbf{q},\omega) = \int d^3r \mathbf{J}(\mathbf{r},\omega) \exp(-i\mathbf{q}\mathbf{r}). \tag{9}$$

It followed from equations (3) and (5) that the Fourier transform of the polarization current density can be represented as the sum of the part $J_i^{tr}(\mathbf{q},\omega)$ that is perpendicular to a principal crystal optical axis and the part $J_i^l(\mathbf{q},\omega)$ that is parallel to this axis :

$$J_i(\mathbf{q},\omega) = J_i^{tr}(\mathbf{q},\omega) + J_i^{l}(\mathbf{q},\omega); J_i^{tr}(\mathbf{q},\omega) = \left(\delta_{is} - e_i e_s\right) J_{0s}(\mathbf{q},\omega),$$
$$J_i^{l}(\mathbf{q},\omega) = e_i e_s \varsigma(\omega) J_{0s}(\mathbf{q},\omega); \varsigma(\omega) = \frac{\{\varepsilon_0(\omega) - \varepsilon_e(\omega)\}}{1-\varepsilon_0(\omega)}. \quad (10)$$

The contributions of each of these parts depends on the monocrystal surface orientation relative to the crystal main axis. Using equation (7), it can be written:

$$\left|\left[\mathbf{k},\mathbf{J}(\mathbf{k},\omega)\right]\right|^2 = \left|\left[\mathbf{k},\mathbf{J}^{tr}(\mathbf{k},\omega)\right]\right|^2 + \left|\left[\mathbf{k},\mathbf{J}^{l}(\mathbf{k},\omega)\right]\right|^2 + 2\,\mathrm{Re}\left(\left[\mathbf{k},\mathbf{J}^{l}(\mathbf{k},\omega)\right],\left[\mathbf{k},\mathbf{J}^{tr*}(\mathbf{k},\omega)\right]\right). \quad (11)$$

Considering the radiation along the crystals main axis direction $\mathbf{k} \parallel \mathbf{e}$, so that $\left[\mathbf{k},\mathbf{J}^{l}(\mathbf{k},\omega)\right] = 0$, it is easy to see that $\left|\left[\mathbf{k},\mathbf{J}(\mathbf{k},\omega)\right]\right|^2 = \left|\left[\mathbf{k},\mathbf{J}^{tr}(\mathbf{k},\omega)\right]\right|^2 = k^2\left|\mathbf{J}(\mathbf{k},\omega)\right|^2$. This implies that there is only radiation to the parallel to the main axis part of polarization current with any crystal surface orientation. Using equation (11), one can express the distribution of the transition radiation in a uniaxial crystal though the polarization current density in an isotropic medium $\mathbf{J}_0(\mathbf{r},\omega)$ .This will determine us the transition radiation intensity ratio in a crystal and in isotropic medium:

$$\frac{d^2W}{d\omega d\Omega} = \frac{(2\pi)^6}{c}\Big(k^2\left|\mathbf{J}^0(\mathbf{k},\omega)\right|^2 - \left|(\mathbf{k},\mathbf{J}^0(\mathbf{k},\omega))\right|^2 + k^2\xi^2(\omega)\left|(\mathbf{e},\mathbf{J}^0(\mathbf{k},\omega))\right|^2$$
$$-\xi^2(\omega)\left|(\mathbf{k},\mathbf{e}),(\mathbf{e},\mathbf{J}^0(\mathbf{k},\omega))\right|^2 + 2\xi(\omega)\,\mathrm{Re}\left\{\left(\left[\mathbf{k},\mathbf{J}^0(\mathbf{k},\omega)\right]^*,[\mathbf{k},\mathbf{e}]\right)\cdot\left(\mathbf{e},\mathbf{J}^0(\mathbf{k},\omega)\right)\right\}\Big). \quad (12)$$

Polarization current in an isotropic medium $\mathbf{J}_0(\mathbf{k},\omega)$ is not dependent on the uniaxial crystal axis direction. That's why, some of the orientational dependences features of transition radiation that is created by charged particle, flying out from an unaxial crystal, can be seeing without using the explicit form of $\mathbf{J}_0(\mathbf{k},\omega)$, which will be demonstrated below.

3. Angles and frequencies distribution of transition radiation

If charge moves with the constant velocity $\mathbf{v} \parallel z$ at an angle ξ to the crystal surface, where ξ is the angle between velocity direction and surface normal (see figure 1), the polarization current density in isotropic medium has the form:

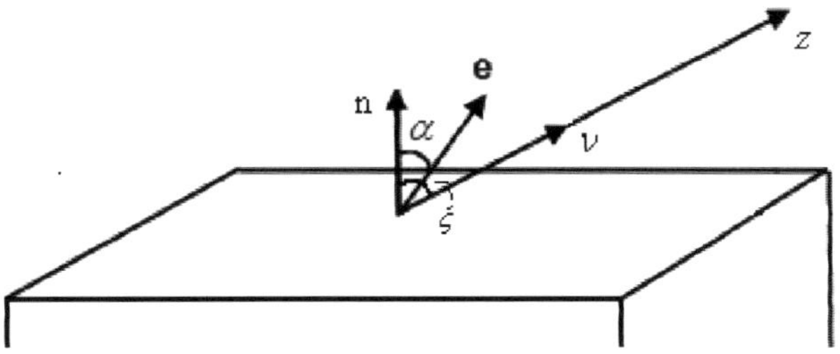

Figure 1. Scheme of a charged particle emission.

$$J_{0i}(\mathbf{k},\omega)=\frac{i\omega}{16\pi^3 v}\cdot(1-\varepsilon_0(\omega))\int_{-\infty}^{+\infty}dx\int_{-\infty}^{xtg\xi}dz\cdot e^{i(\frac{\omega}{v}-k_z)z}e^{-ik_x x}\int_{-\infty}^{+\infty}dq_x e^{ixq_x}E_i^0\left(q_x,k_y,\frac{\omega}{v}\right). \tag{13}$$

The field is produced by such a particle can be represented as:

$$\mathbf{E}_0(\mathbf{r},t)=\int d^3q\int d\omega\mathbf{E}_0(\mathbf{q},\omega)\exp(i\mathbf{qr}-i\mathbf{qv}t),$$
$$\mathbf{E}_0(\mathbf{q},\omega)=-\frac{ie}{2\pi^2}\frac{\mathbf{v}\omega/c^2-\mathbf{q}}{q^2-\omega^2/c^2}\delta(\omega-q_z v). \tag{14}$$

Furthermore, a case when a charged particle moves through a crystal with an ultrarelativistic speed was considered. In ultrarelativistic case the radiation concentrates at small angles ϑ near the particle velocity, so that one can write $k_x\simeq k\vartheta\cos\varphi$; $k_y\simeq k\vartheta\sin\varphi$; $k_z\simeq k\left(1-\frac{\vartheta^2}{2}\right)$. Therefore $\left(\frac{\omega}{v}-k_z\right)\simeq\frac{k}{2}L$; $\left(\frac{\omega}{v}-k_z\right)tg\xi+k_x\simeq k\left(\frac{L}{2}tg\xi+n_x\sqrt{\varepsilon_0}\right)$, where $L=\gamma^{-2}+\vartheta^2+\frac{\omega_p^2}{\omega^2}$, $\mathbf{n}=(n_x,n_y,n_z)=\left(\vartheta\cos\varphi,\vartheta\sin\varphi,1-\frac{\vartheta^2}{2}\right)$ and when $\omega>>\omega_p$ one can assume that $\sqrt{\varepsilon}\simeq 1-\omega_p^2/2\omega^2$.

According to the relations given in [2],

$$\int_0^{+\infty}dxe^{-px}\sin(qx+\lambda)=\frac{1}{p^2+q^2}(q\cos\lambda+p\sin\lambda),$$
$$\int_0^{+\infty}dxe^{-px}\cos(qx+\lambda)=\frac{1}{p^2+q^2}(p\cos\lambda-q\sin\lambda) \tag{15}$$

and neglecting the amendments of the order of γ^{-2}, the expression for the polarization current density components can be obtained:

$$J_{0x}(\mathbf{k},\omega)=\frac{\omega e}{i8\pi^4 vk^2}\cdot\frac{1}{L}\cdot\frac{1}{F^2+\left(\frac{L}{2}tg\xi+n_x\sqrt{\varepsilon_0}\right)^2}\left(\frac{L}{2}tg\xi+n_x\sqrt{\varepsilon_0}\right)(1-\varepsilon_0),$$
$$J_{0y}(\mathbf{k},\omega)=\frac{-n_y\sqrt{\varepsilon_0}(1-\varepsilon_0)\omega e}{i8\pi^4 vk^2}\cdot\frac{1}{L}\cdot\frac{1}{F^2+\left(\frac{L}{2}tg\xi+n_x\sqrt{\varepsilon_0}\right)^2}, \tag{16}$$
$$J_{0z}(\mathbf{k},\omega)=-\frac{\gamma^{-2}\omega e}{i8\pi^4 vk^2}\cdot\frac{1}{L}\cdot\frac{1}{F^2+\left(\frac{L}{2}tg\xi+n_x\sqrt{\varepsilon_0}\right)^2}(1-\varepsilon_0).$$

where $F=\sqrt{n_y^2\varepsilon_0+\gamma^{-2}}$.

Using equations (12) and (16), it is easy to obtain the following expression for angles and frequencies radiation distribution in vacuum:

$$\frac{d^2W}{d\omega d\Omega}=\frac{e^2}{\pi^2v^2}\frac{1}{L^2}\cdot\frac{1}{\left(F^2+\left(\frac{L}{2}tg\xi+n_x\sqrt{\varepsilon_0}\right)^2\right)^2}(1-\varepsilon_0)^2$$
$$\times\Big\langle 2\varsigma(\omega)\Big[\frac{L^2}{4}tg^2\xi\left(\left(n_y^2-\vartheta^2\right)\sin\alpha+\sin\alpha-n_x\cos\alpha\right)$$
$$+Ltg\xi\left(n_x\sqrt{\varepsilon_0}\sin\alpha-n_x^2\sqrt{\varepsilon_0}\cos\alpha+\frac{1}{2}n_y^2\sqrt{\varepsilon_0}\cos\alpha\right)+n_x^2\varepsilon_0\sin\alpha\Big]\cdot\sin\alpha \tag{17}$$
$$+\frac{L^2}{4}tg^2\xi\cdot\left(1-n_x^2\right)+Ltg\xi\cdot n_x\sqrt{\varepsilon_0}+\left(n_x^2+n_y^2\right)\varepsilon_0\Big\rangle.$$

We would like to note that this distribution is obtained in the first approximation to the difference $\Delta\varepsilon(\omega)=\varepsilon_e(\omega)-\varepsilon_o(\omega)$.

4. Discussion of the results

Let us see, how angles and frequencies radiation distribution in a vacuum, given by equation (17) depends on the angle α between the direction of the charged particles velocity and the crystal main axis. Ad hoc, the following ratio was considered:

$$\frac{d^2W}{d\omega d\Omega}(\alpha)=\frac{\frac{d^2W}{d\omega d\Omega}-\frac{d^2W}{d\omega d\Omega}_{izotr}}{\frac{d^2W}{d\omega d\Omega}_{izotr}}$$
$$=2\varsigma(\omega)\Big[\frac{L^2}{4}tg^2\xi\left(\sin\alpha\left(n_y^2-\vartheta^2\right)+\sin\alpha-n_x\cos\alpha\right) \tag{18}$$
$$+Ltg\xi\left(n_x\sqrt{\varepsilon_0}\sin\alpha-n_x^2\sqrt{\varepsilon_0}\cos\alpha+\frac{1}{2}n_y^2\sqrt{\varepsilon_0}\cos\alpha\right)+n_x^2\varepsilon_0\sin\alpha\Big]\cdot\sin\alpha.$$

In the special case when radiation quanta are emitted at an angle close to the surface normal one can find that:

$$\cos\xi>>\gamma^{-2}+\frac{\omega_p^2}{\omega^2}.$$

Introducing polar and azimuthal emission angles by the relations $\mathbf{n}=(\sin\vartheta\cos\varphi,\sin\vartheta\sin\varphi,\cos\vartheta)$ and taking into account that the radiation is emitted at small angles to the charged particle motion direction $\vartheta<<1$, one gets from equation (17):

$$\frac{d^2W}{d\omega d\Omega}=\frac{e^2}{\pi^2v^2}\frac{1}{\left(\gamma^{-2}+\vartheta^2+\frac{\omega_p^2}{\omega^2}\right)^2}\cdot\frac{1}{\left(\vartheta^2+\gamma^{-2}\right)^2}(1-\varepsilon_0)^2\vartheta^2\left\langle 2\varsigma(\omega)\cos^2\varphi\cdot\sin^2\alpha+1\right\rangle. \tag{19}$$

In a cubic crystal case, in which $\Delta\varepsilon(\omega)=\varepsilon_e(\omega)-\varepsilon_o(\omega)$, the well-known result for particles that emitting perpendicular to crystal surface can be obtained:

$$\frac{d^2W}{d\Omega d\omega}\simeq\frac{ce^2}{\pi^2v^2}\cdot\frac{(1-\varepsilon_0)^2}{\left(\gamma^{-2}+\vartheta^2+\omega_p^2/\omega^2\right)^2}\cdot\frac{\vartheta^2}{\left(\vartheta^2+\gamma^{-2}\right)^2}. \tag{20}$$

The result (8) was obtained theoretically in [3] and experimentally confirmed in [4].

Thus, the dependence of the resulting distribution radiation on the crystal main axis direction and relative to the surface was considered. Also, the dependence of the resulting distribution radiation on the angle of particle emission relative to the normal to the surface was investigated.

Acknowledgements
This work was supported by Russian Ministry of Education and Science under contract № P681 and grant № 2.1.1.10054.

References

[1] Ginzburg V L and Frank I M Theory of transition radiation 1946 *JETP* **16** 15
[2] Grandshteyn I S and Ryzhik I M *Tables of Integrals, Series and Products* (Fizmatgiz, Moscow, 1962)
[3] Garibyan G The theory of transition radiation and ionization losses particle energy 1959 *JETP* **37** 527
[4] Yuan L, Wang C, Uto H, Prunster S Energy Dependence of X-Ray Transition Radiation from Ultrarelativistic Charged Particles 1970 *Phys. Rev. Lett.* **31** 603
[5] Potylitsyn A P, Ryazanov M I, Strikhanov M N and Tishchenko A A *Diffraction radiation from relativistic particles* (Springer, 2010)
[6] Bolotovskii B M and Voskresensky G V Diffraction radiation 1968 *UFN* **94** 3 377
[7] Vasiliev D M *Crystallography* (St. P., St. Petersburg, 2003)

Transition radiation from graphene

V M Sukharev, M N Strikhanov and A A Tishchenko
National Research Nuclear University "MEPHI", 115409, Moscow, Russia

E-mail: sukharef@gmail.com

Abstract. Transition radiation from graphene monolayer is studied theoretically. The cases of normal, oblique and grazing incidence are considered in ultrarelativistic case. The spectral-angular radiation distribution is studied in detail, both for the forward and for backward radiation. The radiation along the monolayer is predicted to exist. The radiation at grazing incidence is demonstrated to be considerably more intensive than at normal and moderately oblique incidence.

Introduction

Discovering of graphene gave an impulse to wide research of this material. The experimental studies [1, 2] demonstrate that graphene conductivity is higher than conductivity of the other two-dimensional structures. On the other hand, graphene is a monolayer in its natural form, whereas usually transition radiation is considered for rather thick targets. This difference can lead to the changing the angular radiation distribution comparatively with standard TR models. Besides, having the TR from graphene studied, one can investigate more complicated structures made of graphene. So, from theoretical point of view graphene can be of interest as an object of exploration. We consider the case when conductivity of the graphene monolayer is set a priori [3], which allows us to make numerical calculations of TR characteristics from graphene.

1. Field of TR from monolayer

It is necessary to find TR characteristics from infinite graphene monolayer, which is placed in vacuum ($z=0$) (figure 1). The source of TR are polarization currents, which are produced by the field of a charged particle e moving uniformly through the monolayer.

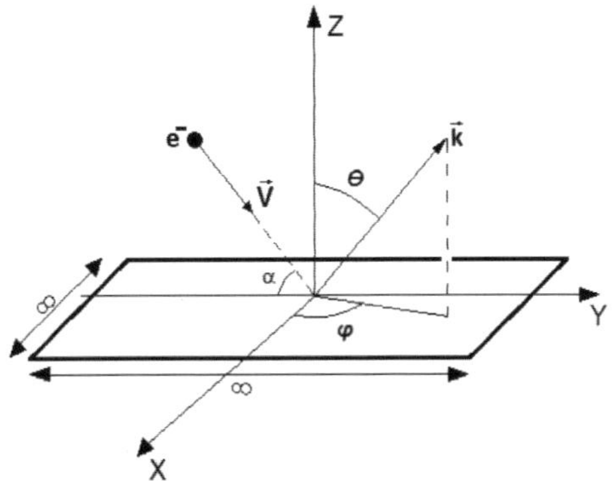

Figure 1. Principle scheme.

Published under licence by IOP Publishing Ltd

Journal of Physics: Conference Series **357** (2012) 012015 doi:10.1088/1742-6596/357/1/012015

Maxwell equations for Fourier images can be written in form:

$$rot\mathbf{H}(\mathbf{r},\omega)=\frac{4\pi}{c}\mathbf{j}(\mathbf{r},\omega)-\frac{i\omega}{c}\mathbf{E}(\mathbf{r},\omega),$$
$$rot\mathbf{E}(\mathbf{r},\omega)=\frac{i\omega}{c}\mathbf{H}(\mathbf{r},\omega). \tag{1}$$

From (1) one can get expression for magnetic field of our system:

$$\left(\Delta+k^2\right)\mathbf{H}(\mathbf{r},\omega)=-\frac{4\pi}{c}rot\mathbf{j}(\mathbf{r},\omega), \tag{2}$$

where $k^2=\omega^2/c^2$. Solution of this equation in the wave zone $|\mathbf{r}|$? $\left\{|\mathbf{r}'|,\lambda\right\}$ can be written as:

$$\mathbf{H}(\mathbf{r},\omega)=\frac{1}{cr}\exp(ikr)\int d^3r'\left(\mathbf{k}\times\mathbf{j}(\mathbf{r}',\omega)\right)\exp(-i\mathbf{k}\mathbf{r}'). \tag{3}$$

where $\mathbf{k}=\mathbf{n}k=\frac{\mathbf{r}}{r}k$ – a wave vector of the radiation field.

Spectral and angular radiation distribution is given by the expression [4]:

$$\frac{d^2W(\mathbf{n},\omega)}{d\Omega d\omega}=\frac{1}{c}\left|\int d^3\mathbf{r}\exp(-i\mathbf{k}\mathbf{r})\left(\mathbf{k}\times\mathbf{j}(\mathbf{r},\omega)\right)\right|^2. \tag{4}$$

where $\mathbf{j}(\mathbf{k},\omega)$ is the polarization currents density. To describe the polarization currents density we will use an expression:

$$\mathbf{j}(\mathbf{r},\omega)=\sigma(\omega)\delta(z)\mathbf{E}_0(\mathbf{r},\omega), \tag{5}$$

where $\sigma(\omega)$– is the graphene monolayer conductivity, $\mathbf{E}_0(\mathbf{r},\omega)$ – the own moving charge field. The form of the graphene conductivity can be found, for example, in [3].

The radiation distribution from the graphene monolayer can be obtained from (5) and (4):

$$\frac{d^2W(\mathbf{n},\omega)}{d\Omega d\omega}=\frac{(2\pi)^4}{c}|\sigma(\omega)|^2\left|\mathbf{k}\times\mathbf{E}_0(\mathbf{k}_\perp,z\quad 0,\omega)\right|^2= \tag{6}$$

The Fourier image of the own charge field of the in (6) can be found as:

$$\mathbf{E}_0(\mathbf{k}_\perp,z=0,\omega)=-\frac{ie}{2\pi^2}\frac{A}{|v_z|}\left(\omega\mathbf{v}-c^2\mathbf{Q}\right),$$
$$\mathbf{Q}=\mathbf{k}_\perp+a\mathbf{e}_z,$$
$$A=\left[c^2\left(\left(\mathbf{k}_\perp^2+a^2\right)\quad\left(\frac{\omega}{c}\right)^2\right)\right]^{-1}, \tag{7}$$
$$a=(\omega-\mathbf{v}_\perp\mathbf{k}_\perp)v_z^{-1}.$$

Therefore one can get a common expression for the distribution of the TR from graphene monolayer over the angles and frequencies:

$$\frac{d^2W(\mathbf{n},\omega)}{d\Omega d\omega}=\frac{(2e)^2}{c}\left|\sigma(\omega)\right|^2\left(\frac{A}{v_z}\right)^2\left|\mathbf{k}\times\left(\omega\mathbf{v}-c^2\mathbf{Q}\right)\right|^2. \tag{8}$$

2. Normal incidence case

First we consider normal incidence of charged particle on the monolayer. In this case velocity vector has only one component:

$$\mathbf{v}=\left\{v_x,v_y,v_z\right\}=\left\{0,0,\ v\right\} \tag{9}$$

The radiation distribution takes the form:

$$\frac{d^2W(\mathbf{n},\omega)}{d\Omega d\omega}=\frac{(2e)^2}{c}\left|\sigma(\omega)\right|^2 A^2k_\perp^2\left(\frac{c^2}{v}\left(\frac{\omega}{v}+k_z\right)\ \ \omega\right)^2, \tag{10}$$

where we use the definitions from (7).

In spherical coordinates (see figure 1) equation (10) gets the form:

$$\frac{d^2W(\mathbf{n},\omega)}{d\Omega d\omega}=\frac{(2e)^2}{c}\left|\sigma(\omega)\right|^2 A^2\left(\frac{\omega^2}{c}\right)^2\left(\beta^{-1}\left(\beta^{-1}+\cos\theta\right)\ \ 1\right)^2\sin^2\theta, \tag{11}$$

where

$$A=\left[\omega^2\left(\left(\sin^2\theta+\beta^{-2}\right)\ \ 1\right)\right]^{-1}. \tag{12}$$

For the ultrarelativistic case one can easily get:

$$\frac{d^2W(\mathbf{n},\omega)}{d\Omega d\omega}=\frac{(2e)^2}{c}\frac{\left|\sigma(\omega)\right|^2}{c^2}\frac{\sin^2\theta\cos^2\theta}{\left(\sin^2\theta+\gamma^{-2}\right)^2}. \tag{13}$$

In optical frequency range the graphene conductivity takes the simple form [3]:

$$\sigma(\omega)=\frac{e^2}{4\mathrm{h}}. \tag{14}$$

and the radiation distribution expression is:

$$\frac{d^2W(\mathbf{n},\omega)}{d\Omega d\omega}=\frac{e^2}{c}\frac{1}{4}\left(\frac{e^2}{\mathrm{h}c}\right)^2\frac{\sin^2\theta\cos^2\theta}{\left(\sin^2\theta+\gamma^{-2}\right)^2}. \tag{15}$$

where $\frac{e^2}{\mathrm{h}c}=\frac{1}{137}$ is the fine structure constant.

The radiation distribution as a function of the angle θ (figure 1) is demonstrated by figure 2.

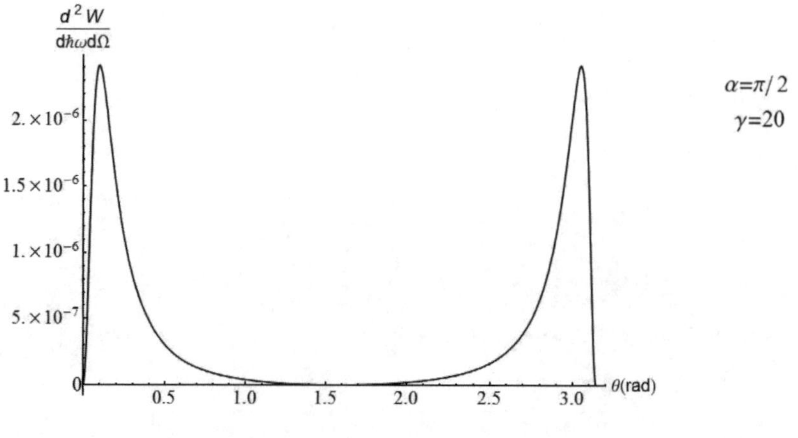

Figure 2. TR distribution in the normal incidence case, where $\sigma(\omega)$ is taken from (14).

It is worth noting that (15) resembles the one from the work [5], where TR from the monolayer was also investigated. The cosine in (15) is not usual for TR from half-infinite space; however, the same cosine was obtained in [4] in a different way, by means of microscopic theory. This cosine defines a shape of TR along the monolayer. In spite of weakness of this radiation, it might be of use for diagnostic purposes – of course, only if it is possible to register this radiation experimentally.

The other thing that differs (15) from the standard TR expressions [6] is a square of fine structure constant. The presence of such a small parameter is expectable, it defines the distinctions between the radiation from monolayer and from macroscopically thick targets.

As one can see, in the normal incidence case the spectral-angular density of radiation does not depend on the azimuth angle φ. The peaks are equally high and distribution is symmetrical. It is notable also, that radiation is concentrated within the two narrow ranges near $\theta = 0$ and $\theta = \pi$. These ranges characterize "backward" and "forward" radiation emitted in the directions that can be obtained according to expression [6]:

$$\Delta = \pi/2 \pm \alpha, \tag{16}$$

which in case of normal incidence $\alpha = \pi/2$ are equal to π and 0.

The radiation along the monolayer, i.e. at $\theta = \pi/2$ and near $\theta = \pi/2$, is extremely weak and can be negligible in comparison with the radiation maxima.

The dependence of radiation maximums from Lorentz factor is usual: Lorentz factor increasing leads to radiation increasing and to narrowing of radiation distribution peaks: the higher Lorentz factor, the narrower ranges for distribution peaks.

3. Oblique incidence case

Let us consider an oblique incidence in a relativistic case. In this case velocity vector has two components:

$$\mathbf{v} = \{v_x, v_y, v_z\} = \{0, v\cos\alpha, -v\sin\alpha\}. \tag{17}$$

The distribution of TR over the angles and frequencies can be written in form:

$$\frac{d^2W(\mathbf{n},\omega)}{d\Omega d\omega}=\frac{e^2}{c}\frac{4|\sigma(\omega)|^2}{c^2}\left(\omega^2 A\right)^2\times\left\{n_x^2\left[\left(\sin\alpha-\left(\frac{1}{\sin\alpha}\left(1-n_y\cos\alpha\right)+n_z\right)\right)^2+\cos^2\alpha\right]+\right.$$
$$\left.+n_y^2\left[\left(\sin\alpha-\left(\frac{1}{\sin\alpha}\left(1-n_y\cos\alpha\right)+n_z\right)\right)+\left(n_z/n_y\right)\cos\alpha\right]^2\right\}, \tag{18}$$

where

$$A=\left[\omega^2\left(\left(\sin^2\theta+\frac{\left(1-\sin^2\alpha\right)}{\sin^2\alpha}+\gamma^{-2}\right)-n_y\frac{ctg\alpha}{\sin\alpha}\left(2-n_y\cos\alpha\right)\right)\right]^{-1}. \tag{19}$$

In the case $\alpha=\pi/2$ (18) transforms into (13).
The distribution shape over the angles is demonstrated in figure 3.

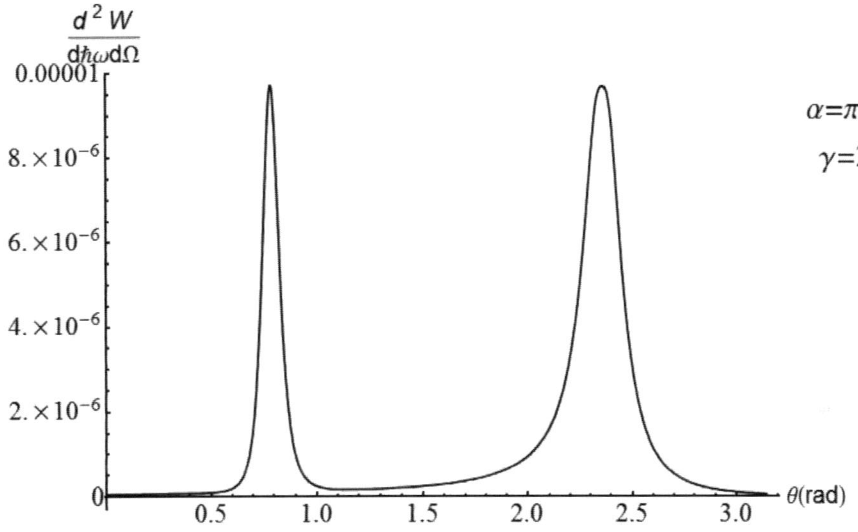

Figure 3. TR distribution in the oblique incidence case $\varphi=\pi/2-\delta$, $\delta=0.07$; $\sigma(\omega)$ is taken from (14).

As well as in the normal incidence case it is seen that the radiation is concentrated in two narrow peaks, the angular directions of which are described by (16). These peaks characterize "forward" and "backward" radiation. Asymmetry of the shape reflects asymmetry of initial problem (geometry) in the oblique incidence case. The radiation peaks are equally high as in for normal incidence, but in the oblique case peaks are higher than in the normal incidence case.

Radiation directed along the monolayer near $\theta=\pi/2$ is not always extremely low and negligible in comparison with distribution maxima as opposed to normal incidence case. For example, at $\alpha=\pi/4$ one can neglect the radiation along the monolayer ($\varphi=1.14$), but for $\alpha=\pi/5$ the radiation along the monolayer becomes comparable with the radiation emitted in forward and backward direction. The Lorentz factor does not play important role for the shape of the radiation distribution.

It is also interesting to notice, that if the incidence angle is small enough ($\alpha=\pi/10$), the radiation concentrates near $\theta=\pi/2$, that means that radiation is also directed along the monolayer.

Journal of Physics: Conference Series **357** (2012) 012015 doi:10.1088/1742-6596/357/1/012015

4. Grazing incidence case

Let us transform angular part of (18) for the case when the incidence angle α is extremely small, but not equal to zero. The radiation distribution for this case is shown in figures 4. One can see that for incidence angle $\alpha = \pi/100$, the distribution asymmetry can be seen still (figure 4 (a)), and radiation along (in) the monolayer is considerable ($\theta = \pi/2$). Figures 4 (b) and 4 (c) demonstrate that the asymmetry dies away when the incidence angle is much less than the previous quantity. Also, we can see that in grazing incidence case considerable part of the radiation is directed along the monolayer.

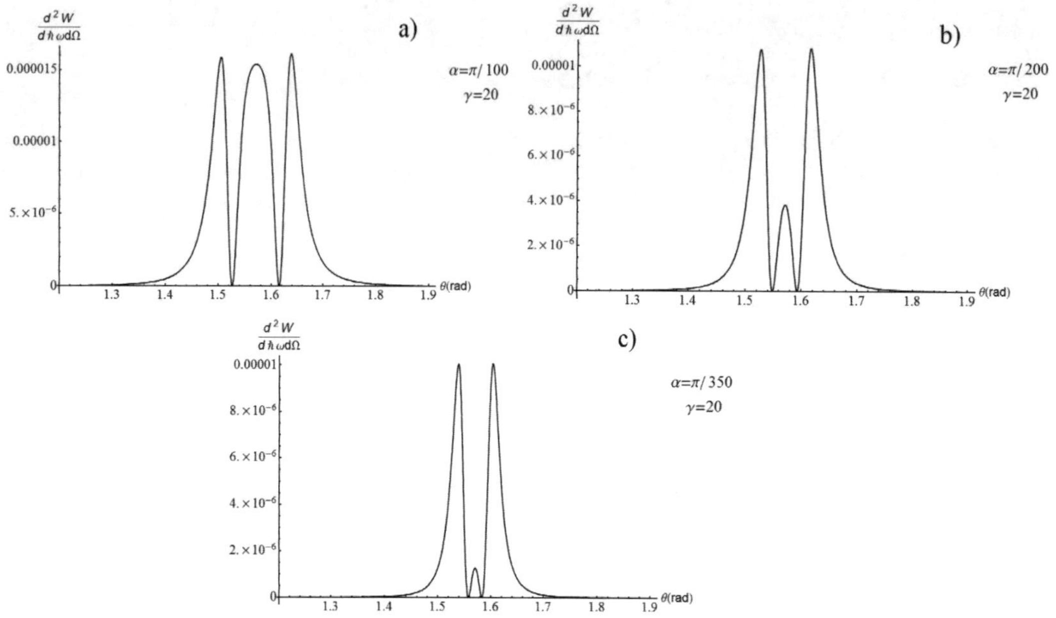

Figure 4. TR distribution in case of grazing incidence ($\varphi = \pi/2$); $\sigma(\omega)$ is taken from (14).

Also in grazing case Lorentz factor influences on the distribution shape more intensively. Figures 5 demonstrate that Lorentz factor increasing leads to increasing of the radiation along the monolayer, and the maximum of TR is even higher than in oblique incidence case. The distribution of TR over the angles and frequencies at $\alpha << 1$ can be written as

$$\frac{d^2W(\mathbf{n},\omega)}{d\Omega d\omega} = \frac{e^2}{c}\left(\frac{e^2}{2hc}\right)^2 \frac{1}{\left[\left((\sin^2\theta+\gamma^{-2})\sin^2\alpha+1-\sin^2\alpha\right)-\sin\theta\sin\varphi(2-\sin\theta\sin\varphi)\right]^2}\times$$
$$\times\left\{\sin^2\theta\left(\sin^3\alpha-\sin^2\alpha\left(\frac{1}{\sin\alpha}(1-n_y)+n_z\right)\right)^2+\right. \qquad (20)$$
$$\left.+(n_x^2+n_z^2)\sin^4\alpha+2\sin^4\alpha\left(\sin\alpha-\left(\frac{1}{\sin\alpha}(1-n_y)+n_z\right)\right)n_zn_y\right\}.$$

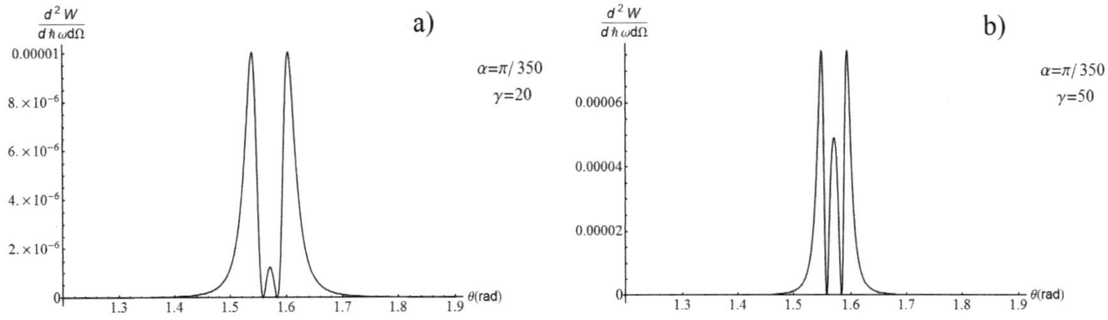

Figure 5. Radiation distribution for grazing incidence case ($\varphi = \pi/2$); $\sigma(\omega)$ is taken from (14).

6. Conclusion

TR characteristics from graphene monolayer in ultrarelativistic case is considered. The cases of normal and oblique incidence, including the grazing one, are studied.

In normal incidence case it is shown that:

1) the spectral-angular radiation distribution is independent of the azimuth angle;
2) highness of both peaks are equal, and the radiation distribution has clear-cut symmetry;
3) radiation is concentrated in two narrow ranges: near $\theta = 0$ and $\theta = \pi$. These ranges characterize "backward" and "forward" radiation;
4) with the Lorentz factor increasing the radiation maximum increases.

In oblique incidence case it is shown:

1). radiation distribution over the angles is asymmetrical;

2). radiation directed along the monolayer near $\theta = \pi/2$ depends on the incidence angle and does not depend on the Lorentz factor;

3). if the incidence angle is small enough ($\alpha = \pi/10$), radiation is concentrated near $\theta = \pi/2$, i.e. directed along the monolayer.

In grazing incidence case the variety of possible ways of radiation propagation is demonstrated. It is also shown, that the Lorentz factor influences on the distribution radiation shape more intensively than for both previous incidence cases, and more complicated behavior for radiation propagation along the monolayer takes place. The radiation at grazing incidence proves to be more intensive than at moderately oblique and normal incidence.

Acknowledgments

This work is supported by the Ministry of Education and Science of the Russian Federation (contract No. 14.740.11.0472 and, partially, P1093 and P681).

References

[1] Kuzmenko A B, van Heumen E, Carbone F and van der Marel D 2008 *Phys. Rev. Lett.* **100** 117401
[2] Gusynin V P, Sharapov S G and Carbotte J P 2006 *Phys. Rev. Lett.* **96** 256802
[3] Falkovsky L A 2008 Phys. Usp. **51** 9 887–897 [**178** 923–934]
[4] Potylitsyn A P, Ryazanov M I, Strikhanov M N, Tishchenko A A 2010 *Diffraction Radiation from Relativistic Particles* (Springer-Verlag)
[5] Ryazanov M I, Tishchenko A A, 2002 *Laser Physics* **12** 1442
[6] Garibyan G M, Yan Shi 1983 *X-ray Transition Radiation* (Yerevan: AN ArmSSR)

Coherent X-radiation along the velocity of a relativistic electron in a bounded periodic multilayer medium

S Blazhevich[1], I Kolosova[2], A Noskov[2]
[1] National Research University Belgorod State University, Belgorod, Russia
[2] Belgorod University of Cooperation, Economics and Low, Belgorod, Russia

E-mail: blazh@bsu.edu.ru

Abstract. In the present work a theory of the coherent X-radiation generated by the relativistic electron crossing an artificial periodic layered structure in Laue scattering geometry is built. The expressions are derived for spectral-angular distribution of the radiation in direction of the particle velocity. The comparative analysis of the spectral-angular density of the radiation is carried out for an artificial periodic structure and a single crystal in similar conditions. It is shown, that the spectral peak of the parametric X-radiation in the artificial periodic layered structure in the forward direction (FPXR) will be considerably wider than the one in a single crystal target which can make its experimental detection and investigation appreciably easier.

1. Introduction

Traditionally the radiation of the relativistic particle in periodical multilayer structure is considered only in the Bragg scattering geometry for the case, when the reflecting layers are situated across the target entrance surface, i.e. for the case of symmetric reflection. This radiation is usually considered as the resonance transition radiation [1-2]. In the work [3] this radiation was represented as a sum of the diffracted transition radiation (DTR) and parametric X-radiation (PXR). Recently in our work [4] a theory of coherent X-radiation in Bragg direction generated by a relativistic electron crossing the artificial periodic layered structure has been constructed in Laue scattering geometry for general asymmetry of the electron coulomb field reflection, when the reflecting layered structure is situated at any angle to target surface.

The dynamic theory of relativistic particle in the crystal predicts the radiation not only in Bragg scattering direction, but also in the direction along of incident particle velocity [5-7] (FPXR). This radiation is a result of the manifestation of dynamic diffraction in PXR. The detailed theoretical decryption of the dynamical effect of FPXR together with accompanying background of transition radiation in the case of symmetric reflection was given in works [8-10]. Later, in our woks [11-12] the coherent X-radiation along the velocity of the relativistic electron crossing a crystal was considered both in the Laue geometry and in the Bragg geometry in the general case of asymmetric reflection, i.e. in the case when the atomic planes in the crystal can be situated at any angular δ to the target entrance surface.

In the present wok the dynamic theory of coherent radiation generated by relativistic electron crossing artificial periodic structure bounded by two parallel planes is constructed in Laue scattering geometry for radiation directed along the particle velocity (FPXR). Based on the two-wave approach of dynamic theory of diffraction the expressions of spectral-angular density of the radiation are derived. The calculated values of the spectral-angular densities of the FPXR in an artificial periodic structure and in a single crystal target are compared in similar radiation conditions.

2. Radiation amplitude

Let us consider the radiation of a relativistic electron crossing with the velocity $\mathbf{V}$ a multilayer structure consisted of the periodically situated amorphous layers with thicknesses a and b ($T = a + b$ – period of the structure) of dielectric susceptibility χ_a and χ_b, respectively.

To derive the expressions describing the spectral and angular characteristics FPXR and transition radiation (TR), we used the notation and techniques similar to those shown in [13].

The solution of the combined equations (3) (see [13]) gives us the relativistic particle field in the periodic structure in such a form:

$$E_0^{(s)medium} = \frac{8\pi^2 ieV\theta P^{(s)}}{\omega} \frac{-\omega^2\beta - 2\omega\frac{\gamma_g}{\gamma_0}\lambda_0}{4\frac{\gamma_g}{\gamma_0}\left(\lambda_0 - \lambda^{(1)}\right)\left(\lambda_0 - \lambda^{(2)}\right)} \delta\left(\lambda_0 - \lambda_0^*\right)$$
$$+ E_0^{(s)^{(1)}} \delta\left(\lambda_0 - \lambda^{(1)}\right) + E_0^{(s)^{(2)}} \delta\left(\lambda_0 - \lambda^{(2)}\right), \tag{1}$$

where $\lambda_0^* = \omega\left(\dfrac{\gamma^{-2} + \theta^2 - \chi_0}{2}\right)$, $\gamma = 1/\sqrt{1 - V^2}$ is Lorentz factor of the electron, $E_0^{(s)^{(1)}}$, $E_0^{(s)^{(2)}}$ are free incident fields in the considered medium.

The solution of the combined equations (see [13]) gives us the relativistic particle field in the vacuum in front of the radiator as

$$E_0^{(s)vac\,I} = \frac{8\pi^2 ieV\theta P^{(s)}}{\omega} \frac{1}{-\chi_0 - \frac{2}{\omega}\lambda_0} \delta\left(\lambda_0 - \lambda_0^*\right). \tag{2}$$

The expression for the field in the vacuum behind the target we write as

$$E_0^{(s)vac\,II} = \frac{8\pi^2 ieV\theta P^{(s)}}{\omega} \frac{1}{-\chi_0 - \frac{2\lambda_0}{\omega}} \delta(\lambda_0 - \lambda_0^*) + E_0^{(s)Rad} \delta\left(\lambda_0 + \frac{\omega\chi_0}{2}\right), \tag{3}$$

where $E_0^{(s)Rad}$ is the amplitude of coherent field along the electron velocity.

The expression connecting the incident and diffracted fields in the layered periodic medium follows from the second equation of the system (3):

$$E_0^{(s)medium} = \frac{2\omega\lambda_{\mathbf{g}}}{\omega^2 \chi_{\mathbf{g}} C^{(s)}} E_{\mathbf{g}}^{(s)medium}. \tag{4}$$

Using the standard boundary conditions at entrance and exit surfaces of the target

$$\int E_0^{(s)vacI} d\lambda_0 = \int E_0^{(s)medium} d\lambda_0\,, \quad \int E_{\mathbf{g}}^{(s)medium} d\lambda_0 = 0\,,$$

$$\int E_0^{(s)medium} e^{i\frac{\lambda_0}{\gamma_0}L} d\lambda_0 = \int E_0^{(s)vac\,II} e^{i\frac{\lambda_0}{\gamma_0}L} d\lambda_0\,, \tag{5}$$

we obtain the expression for the radiation field amplitude:

$$E_{Rad}^{(s)} = -\frac{4\pi^2 ieV\theta P^{(s)}}{\omega} e^{i\frac{\lambda_0^* + \frac{\omega\chi_0}{2}}{\gamma_0}L} \left[\left(1 - \frac{\beta}{\sqrt{\beta^2 + 4\chi_{\mathbf{g}}\chi_{-\mathbf{g}} C^{(s)^2} \frac{\gamma_{\mathbf{g}}}{\gamma_0}}}\right)\left(\frac{\omega}{-\omega\chi_0 - 2\lambda_0^*} + \frac{\omega}{2(\lambda_0^* - \lambda_0^{(2)})}\right)\left(1 - e^{i\frac{\lambda_0^{(2)} - \lambda_0^*}{\gamma_0}L}\right)\right.$$

$$+\left(1+\frac{\beta}{\sqrt{\beta^2+4\chi_{\mathbf{g}}\chi_{-\mathbf{g}}C^{(s)^2}\frac{\gamma_{\mathbf{g}}}{\gamma_0}}}\right)\left(\frac{\omega}{-\omega\chi_0-2\lambda_0^*}+\frac{\omega}{2(\lambda_0^*-\lambda_0^{(1)})}\right)\left(1-e^{i\frac{\lambda_0^{(1)}-\lambda_0^*}{\gamma_0}L}\right)\Bigg] \tag{6}$$

The formula (6) allows the description of spectral and angular characteristics of the radiation. Prior to the beginning of the spectral-angular characteristics analysis it is necessary to note that three mechanisms make a contribution to the total yield of the relativistic electron coherent radiation in multilayer periodic structure in the forward direction: bremsstrahlung, transition radiation and parametric radiation (FPXR).

The radiation amplitude $E_{Rad}^{(s)}$ contains the contributions of the radiation mechanisms analogue to FPXR and TR in the single crystal target. So far as the transition radiation is the main underground noise in experimental studying of FPXR, let us represent $E_{Rad}^{(s)}$ as a sum of FPXR and TR amplitudes. Such a presentation allows the estimation of the FPXR and TR contributions to the total radiation yield.

So we write the expression for the radiation field (6) as

$$E_0^{(s)Rad}=E_0^{(s)FPXR}+E_0^{(s)TR}, \tag{7}$$

$$E_0^{(s)FPXR}=\frac{4\pi^2 ieV\theta P^{(s)}}{\omega}e^{i\frac{\lambda_0^*+\frac{\omega\chi_0}{2}}{\gamma_0}L}\frac{\omega^2\chi_{\mathbf{g}}\chi_{-\mathbf{g}}C^{(s)^2}}{2\sqrt{\beta^2+4\chi_{\mathbf{g}}\chi_{-\mathbf{g}}C^{(s)^2}\frac{\gamma_{\mathbf{g}}}{\gamma_0}}}\frac{1}{\lambda_0^*}\left[\frac{1-e^{i\frac{\lambda_0^{(2)}-\lambda_0^*}{\gamma_0}L}}{\lambda_0^*-\lambda_0^{(2)}}-\frac{1-e^{i\frac{\lambda_0^{(1)}-\lambda_0^*}{\gamma_0}L}}{\lambda_0^*-\lambda_0^{(1)}}\right] \tag{8}$$

$$E_0^{(s)TR}=\frac{4\pi^2 ieV\theta P^{(s)}}{\omega}e^{i\frac{\lambda_0^*+\frac{\omega\chi_0}{2}}{\gamma_0}L}\left(\frac{\omega}{\omega\chi_0+2\lambda_0^*}-\frac{\omega}{2\lambda_0^*}\right)$$

$$\times\left[\left(1-\frac{\beta}{\sqrt{\beta^2+4\chi_{\mathbf{g}}\chi_{-\mathbf{g}}C^{(s)^2}\frac{\gamma_{\mathbf{g}}}{\gamma_0}}}\right)\left(1-e^{i\frac{\lambda_0^{(2)}-\lambda_0^*}{\gamma_0}L}\right)+\left(1+\frac{\beta}{\sqrt{\beta^2+4\chi_{\mathbf{g}}\chi_{-\mathbf{g}}C^{(s)^2}\frac{\gamma_{\mathbf{g}}}{\gamma_0}}}\right)\left(1-e^{i\frac{\lambda_0^{(1)}-\lambda_0^*}{\gamma_0}L}\right)\right]. \tag{9}$$

The dynamical corrections corresponding to the two branches of dispersion relation depend not only on photon frequency and target properties, but also on ε. Thus, the dispersion of real photons in the crystal depends on the asymmetry:

$$k^{(1,2)}=\omega\sqrt{1+\chi_0}+\lambda_0^{(1,2)}(\omega,\varepsilon). \tag{10}$$

The dispersion of virtual photons associated with the Coulomb field is described by the formula

$$k^*=\omega\sqrt{1+\chi_0}+\lambda_0^*=\omega+\frac{\omega}{2}\left(\theta^2+\gamma^{-2}\right). \tag{11}$$

A FPXR reflex can be observed only if at least one denominator in the bracketed expression in (7b) vanishes:

$$\mathrm{Re}\left(\lambda_0^*-\lambda_0^{(1)}\right)=0,\qquad \mathrm{Re}\left(\lambda_0^*-\lambda_0^{(2)}\right)=0. \tag{12}$$

Substituting (8) into (7b) and (7c), we represent the FPXR and TR field amplitudes as

$$E_0^{(s)FPXR}=\frac{4\pi^2 ieV\theta P^{(s)}}{\omega}\frac{e^{i\frac{\omega(\gamma^{-2}+\theta^2)}{2\gamma_0}L}}{\gamma^{-2}+\theta^2-\chi_0}\frac{1}{\sqrt{\xi^{(s)2}+\varepsilon}}\times$$
$$\times\left(\frac{1-\exp(-ib^{(s)}\Omega^{(2)}(\omega)-b^{(s)}\rho^{(s)}\Delta^{(2)})}{\Omega^{(2)}(\omega)-i\rho^{(s)}\Delta^{(2)}}-\frac{1-\exp\left(-ib^{(s)}\Omega^{(1)}(\omega)-b^{(s)}\rho^{(s)}\Delta^{(1)}\right)}{\Omega^{(1)}(\omega)-i\rho^{(s)}\Delta^{(1)}}\right), \tag{13a}$$

$$E_0^{(s)TR}=\frac{4\pi^2 ieV\theta P^{(s)}}{\omega}e^{i\frac{\omega(\gamma^{-2}+\theta^2)}{2\gamma_0}L}\left(\frac{1}{\gamma^{-2}+\theta^2}-\frac{1}{\gamma^{-2}+\theta^2-\chi_0}\right)\times$$
$$\left(\left(1-\xi^{(s)}/\sqrt{\xi^{(s)^2}+\varepsilon}\right)\left(1-\exp\left(-ib^{(s)}\Omega^{(2)}(\omega)-b^{(s)}\rho^{(s)}\Delta^{(2)}\right)\right)+\right.$$
$$\left.+\left(1+\xi^{(s)}/\sqrt{\xi^{(s)^2}+\varepsilon}\right)\left(1-\exp(-ib^{(s)}\Omega^{(1)}(\omega)-b^{(s)}\rho^{(s)}\Delta^{(1)})\right)\right), \tag{13b}$$

where

$$\Delta^{(1)}=\frac{\varepsilon+1}{2\varepsilon}-\frac{1-\varepsilon}{2\varepsilon}\frac{\xi^{(s)}}{\sqrt{\xi^{(s)^2}+\varepsilon}}-\frac{\kappa^{(s)}}{\sqrt{\xi^{(s)^2}+\varepsilon}},\quad \Delta^{(2)}=\frac{\varepsilon+1}{2\varepsilon}+\frac{1-\varepsilon}{2\varepsilon}\frac{\xi^{(s)}}{\sqrt{\xi^{(s)^2}+\varepsilon}}+\frac{\kappa^{(s)}}{\sqrt{\xi^{(s)^2}+\varepsilon}},$$

$$\sigma^{(s)}=\frac{1}{\nu^{(s)}}\left(\frac{\theta^2}{|\chi_0'|}+\frac{1}{\gamma^2|\chi_0'|}+1\right),\quad b^{(s)}=\frac{\omega|\chi_g'|C^{(s)}}{2}\frac{L}{\gamma_0},\quad \Omega^{(1,2)}(\omega)=\sigma^{(s)}+\left(\xi(\omega)\mp\sqrt{\xi(\omega)^2+\varepsilon}\right)/\varepsilon. \tag{14}$$

The parameter $b^{(s)}$ can be represented as

$$b^{(s)}=\frac{1}{2\sin(\delta-\theta_B)}\frac{L}{L_{ext}^{(s)}}, \tag{15}$$

i.e., $b^{(s)}$ is half the distance traveled by an electron in the plate, measured in units of extinction length

$$L_{ext}^{(s)}=\frac{1}{2C^{(s)}\omega}\frac{gT}{\left|\sin\left(\frac{ga}{2}\right)\right||\chi_b'-\chi_a'|}. \tag{16}$$

Accordingly, the equation

$$\sigma^{(s)}+\frac{\xi^{(s)}-\sqrt{\xi^{(s)2}+\varepsilon}}{\varepsilon}=0, \tag{17}$$

determines the center frequency ω_* of the spectrum of FPXR photons emitted at a particular observation angle.

3. Spectral-angular density of the radiation

Substituting (7a), (13a), and (13b) into a well known expression for the spectral-angular density of X-ray radiation [14],

$$\omega\frac{d^2N}{d\omega d\Omega}=\omega^2(2\pi)^{-6}\sum_{s=1}^{2}\left|E_0^{(s)Rad}\right|^2, \tag{18}$$

we find the contributions to the spectral-angular density due to FPXR, TR, and their interference:

$$\omega\frac{d^2N_{FPXR}^{(s)}}{d\omega d\Omega}=\frac{e^2}{4\pi^2}P^{(s)^2}\frac{\theta^2}{(\theta^2+\gamma^{-2}-\chi_0')^2}R_{FPXR}^{(s)}, \tag{19a}$$

$$R_{FPXR}^{(s)}=\frac{1}{\xi^{(s)2}+\varepsilon}\frac{1+e^{-2b^{(s)}\rho^{(s)}\Delta^{(1)}}-2e^{-b^{(s)}\rho^{(s)}\Delta^{(1)}}\cos\left(b^{(s)}\Omega^{(1)}(\omega)\right)}{\Omega^{(1)}(\omega)^2+\left(\rho^{(s)}\Delta^{(1)}\right)^2}, \tag{19b}$$

$$\omega\frac{d^2N_{TR}^{(s)}}{d\omega d\Omega}=\frac{e^2}{4\pi^2}P^{(s)^2}\theta^2\left(\frac{1}{\theta^2+\gamma^{-2}}-\frac{1}{\theta^2+\gamma^{-2}-\chi_0'}\right)^2R_{TR}^{(s)}, \tag{20a}$$

$$R_{TR}^{(s)}=\left(1-\xi^{(s)}/\sqrt{\xi^{(s)^2}+\varepsilon}\right)^2\left(1+e^{-2b^{(s)}\rho^{(s)}\Delta^{(2)}}-2e^{-b^{(s)}\rho^{(s)}\Delta^{(2)}}\cos\left(b^{(s)}\Omega^{(2)}(\omega)\right)\right)$$
$$+\left(1+\xi^{(s)}/\sqrt{\xi^{(s)^2}+\varepsilon}\right)^2\left(1+e^{-2b^{(s)}\rho^{(s)}\Delta^{(1)}}-2e^{-b^{(s)}\rho^{(s)}\Delta^{(1)}}\cos\left(b^{(s)}\Omega^{(1)}(\omega)\right)\right)$$
$$+\frac{2\varepsilon}{\xi^{(s)^2}+\varepsilon}\left(1+e^{-b^{(s)}\rho^{(s)}\frac{\varepsilon+1}{\varepsilon}}\cdot\cos\left(\frac{2b^{(s)}\sqrt{\xi^2+\varepsilon}}{\varepsilon}\right)-e^{-b^{(s)}\rho^{(s)}\Delta^{(2)}}\cos\left(b^{(s)}\Omega^{(1)}(\omega)\right)-e^{-b^{(s)}\rho^{(s)}\Delta^{(1)}}\cos\left(b^{(s)}\Omega^{(1)}(\omega)\right)\right), \tag{20b}$$

$$\omega\frac{d^2N_{INT}^{(s)}}{d\omega d\Omega}=\frac{e^2}{4\pi^2}P^{(s)^2}\theta^2\left(\frac{1}{\theta^2+\gamma^{-2}}-\frac{1}{\theta^2+\gamma^{-2}-\chi_0'}\right)\frac{1}{\theta^2+\gamma^{-2}-\chi_0'}R_{INT}^{(s)}, \tag{21a}$$

$$R_{INT}^{(s)}=\frac{-2}{\sqrt{\xi^{(s)2}+\varepsilon}}\mathrm{Re}\left(\frac{1-e^{-ib^{(s)}\Omega^{(1)}(\omega)-b^{(s)}\rho^{(s)}\Delta^{(1)}}}{\Omega^{(1)}(\omega)-i\rho^{(s)}\Delta^{(1)}}\right.$$
$$\left.\times\left(\left(1-\xi^{(s)}/\sqrt{\xi^{(s)^2}+\varepsilon}\right)\left(1-e^{ib^{(s)}\Omega^{(2)}(\omega)-b^{(s)}\rho^{(s)}\Delta^{(2)}}\right)+\left(1+\xi^{(s)}/\sqrt{\xi^{(s)^2}+\varepsilon}\right)\left(1-e^{ib^{(s)}\Omega^{(1)}(\omega)-b^{(s)}\rho^{(s)}\Delta^{(1)}}\right)\right)\right). \tag{21b}$$

Expressions (19)–(21) are the main result of this study. They are obtained in the two-beam approximation of dynamical diffraction theory with taking into account the radiation absorption for an arbitrary orientation of the diffracting planes with respect to the surface of an absorbing crystal. Using the above formulas we will analyze the dynamic effect of reflection asymmetry on spectral-angular characteristics of the radiation.

4. Dynamic scattering parameters of the X-waves

As it was mentioned above, the X-ray scattering in artificial periodic structure is studied traditionally in the Bragg scattering geometry in symmetric reflection case, when the layers of the structure are situated along its entrance surface, while the radiation in crystal target is investigated mainly in the Laue scattering geometry because of well known conditions of the experiment.

The obtained expressions (19–21) allow studying the dependence of radiation characteristics on the values of layer thicknesses a and b for different amorphous substances (with the corresponding dielectric susceptibility χ_a and χ_b) and on the parameter of asymmetry ε, defining the angle between the reflecting layers and the target surface.

The parameter $\nu^{(s)}$ possessing the value $0\leq\nu^{(s)}\leq1$ characterizes the reflecting power of a set of the atomic planes as determined by the interference between waves scattered by atoms in different layers ($\nu^{(s)}\approx1$ and 0 when the interference is constructive and destructive, respectively). Dynamical diffraction phenomena, such as FPXR, can be observed only when the interference is constructive. Under $g=\frac{2\pi}{T}$ it can be presented as

$$\nu^{(s)}=\frac{2C^{(s)}\left|\sin\left(\frac{\pi a}{T}\right)\right|}{g}\left|\frac{\chi_b'-\chi_a'}{a\chi_a'+b\chi_b'}\right|. \tag{22}$$

The parameter $\rho^{(s)} = \frac{L_{ext}^{(s)}}{L_{abs}}$ is the ratio between the extinction length $L_{ext}^{(s)}$ (16) and the absorption length $L_{abs} = \frac{T}{\omega\left|a\chi_a'' + b\chi_b''\right|}$ for X-rays in the layered periodic structure. Note that the energy of the primary wave is completely transferred to the secondary wave propagating in the Bragg direction at the depth equal to the extinction length.

The parameter $\kappa^{(s)}$ quantifies the anomalous transmission effect (Borrmann effect) for X-rays passing through an artificial multilayer periodic structure which is well known in the physics of X-ray scattering in crystal [15]. The Borrmann effect manifestation in a crystal for different asymmetry of scattering was studied in [16]. The necessary condition of the Borrman effect manifestation both in crystal and artificial periodic structures is $\kappa^{(s)} \approx 1$. Under this condition the value of $\Delta^{(1)}$ in (14)

$$\Delta^{(1)} = \frac{\varepsilon+1}{2\varepsilon} - \frac{1-\varepsilon}{2\varepsilon}\frac{\xi^{(s)}}{\sqrt{\xi^{(s)2}+\varepsilon}} - \frac{\kappa^{(s)}}{\sqrt{\xi^{(s)2}+\varepsilon}}, \tag{23}$$

becomes minimal which leads to minimal value of $\rho^{(s)}\Delta^{(1)}$ characterizing absorption of the radiation.

So the Borrmann effect can manifest itself in the coherent radiation of the relativistic electron in a multilayer artificial periodic structure.

5. Radiation amplitude. Comparision of the spectral-angular distributions

To simplify the formulas and to visualize the results we consider the case, when the layers of the periodic structure have the equal thickness, i.e. $T = 2a = 2b$. In this case the formula (19) for the σ-polarized waves takes on the following form:

$$\omega\frac{d^2N_{FPXR}^{(1)}}{d\omega d\Omega} = \frac{e^2}{4\pi^2}\frac{\theta_\perp^2}{\left(\theta_\perp^2+\gamma^{-2}+\frac{\left|\chi_a'+\chi_b'\right|}{2}\right)^2}R_{FPXR}, \tag{24a}$$

$$R_{FPXR} = \frac{1}{\xi(\omega)^2+\varepsilon}\frac{1+e^{-2b^{(1)}\rho^{(1)}\Delta^{(1)}} - 2e^{-b^{(1)}\rho^{(1)}\Delta^{(1)}}\cos\left(b^{(1)}\Omega^{(1)}(\omega)\right)}{\Omega^{(1)}(\omega)^2+\left(\rho^{(s)}\Delta^{(1)}\right)^2}, \tag{24b}$$

where

$$\Omega^{(1)}(\omega) = \sigma(\theta,\gamma)+\left(\xi(\omega)-\sqrt{\xi(\omega)^2+\varepsilon}\right)/\varepsilon, \quad \sigma(\theta,\gamma) = \frac{\pi}{\left|\chi_b'-\chi_a'\right|}\cdot\left(\theta_\perp^2+\gamma^{-2}+\frac{\left|\chi_a'+\chi_b'\right|}{2}\right)$$

$$\Delta^{(1)} = \frac{\varepsilon+1}{2\varepsilon} - \frac{1-\varepsilon}{2\varepsilon}\frac{\xi}{\sqrt{\xi^2+\varepsilon}} - \frac{\kappa^{(1)}}{\sqrt{\xi^2+\varepsilon}}, \quad \xi(\omega) = \frac{2\pi\sin^2(\theta_B)}{\left|\chi_b'-\chi_a'\right|}\cdot\left(1-\frac{\omega}{\omega_B}\right)+\frac{1-\varepsilon}{2\nu^{(1)}},$$

$$\kappa^{(1)} = \frac{2}{\pi}\cdot\left|\frac{\chi_b''-\chi_a''}{\chi_b''+\chi_a''}\right|, \quad \rho^{(1)} = \frac{\pi}{2}\cdot\left|\frac{\chi_b''+\chi_a''}{\chi_b'-\chi_a'}\right|, \quad \nu^{(1)} = \frac{2}{\pi}\cdot\left|\frac{\chi_b'-\chi_a'}{\chi_b'+\chi_a'}\right|, \quad \theta_\perp = \theta\sin\varphi. \tag{25}$$

To compare the spectral-angular density of the radiation in the artificial periodic and crystal structures in similar conditions let us write the expression of FPXR in the crystal [12] for σ-polarized waves:

$$\omega\frac{d^2N_{FPXR}^{(1)Cr}}{d\omega d\Omega} = \frac{e^2}{4\pi^2}\frac{\theta_\perp^2}{\left(\theta_\perp^2+\gamma^{-2}-\chi_0'\right)^2}R_{FPXR}^{Cr}, \tag{26a}$$

$$R_{FPXR}^{Cr} = \frac{1}{\xi(\omega)^2+\varepsilon}\frac{1+e^{-2b^{(1)}\rho^{(1)}\Delta^{(1)}} - 2e^{-b^{(1)}\rho^{(1)}\Delta^{(1)}}\cos\left(b^{(1)}\Omega^{(1)}(\omega)\right)}{\Omega^{(1)}(\omega)^2+\left(\rho^{(s)}\Delta^{(1)}\right)^2}. \tag{26b}$$

Journal of Physics: Conference Series **357** (2012) 012016 doi:10.1088/1742-6596/357/1/012016

Here the notations corresponding to (38) will take on the following view

$$\kappa^{(1)} = \frac{\chi''_{g}}{\chi''_{0}}, \quad \rho^{(1)} = \frac{\chi''_{0}}{|\chi'_{g}|}, \quad \nu^{(1)} = \frac{\chi'_{g}}{\chi'_{0}}, \quad \sigma(\theta,\gamma) = \frac{1}{|\chi'_{g}|}\cdot\left(\theta_{\perp}^{2} + \gamma^{-2} - \chi'_{0}\right),$$

$$\xi(\omega) = \frac{2\sin^{2}(\theta_{B})}{|\chi'_{g}|}\cdot\left(1 - \frac{\omega}{\omega_{B}}\right) + \frac{1-\varepsilon}{2\nu^{(1)}}. \tag{27}$$

Using the formulas (26 б) and (24 б) we charted the curves of the spectral-angular density of FPXR ($\omega_{B} = 8$ кeV) in crystal target of W (see figure 1) and FPXR in multilayer periodic structure composed of amorphous layers of Be and W (see figure 2). The path of the electron in the target $L_{e} = 56$ μm, asymmetry parameter $\varepsilon = 3$ and observation angle $\theta_{\perp} = 5$ мрад are chosen as the same for both the cases.

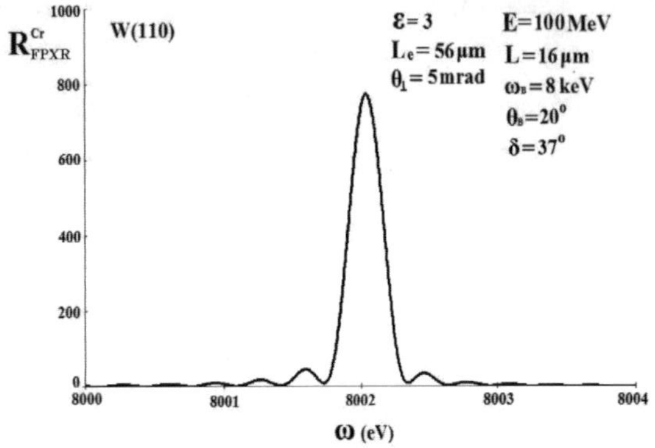

Figure 1. The spectrum of the relativistic electron FPXR in a single crystal (W) medium.

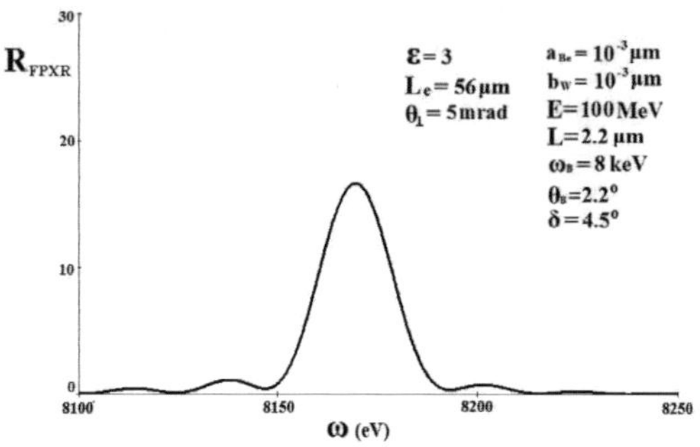

Figure 2. The spectrum of the relativistic electron FPXR in the artificial periodic multilayer (Be-W) medium in the similar figure 3 conditions.

As it followed from figure 1 and figure 2., the spectral amplitude in the crystal is considerably higher than it in a multilayer structure, but the width of the spectral peak of radiation in artificial is much more than in a crystal target because the coherent radiation in this case forms on less number of the medium obstacles. In this connection the experimental investigation of FPXR peak in multilayered periodic structure is interest and easier than in a crystal medium. Let us consider the ratio of the radiation angular densities in these two media. The expressions for angular density of FPXR in artificial periodic structure and crystal medium take on the following view:

$$\frac{dN^{(1)}_{FPXR}}{d\Omega} = \frac{e^{2}}{4\pi^{2}}\frac{\theta_{\perp}^{2}}{\left(\theta_{\perp}^{2} + \gamma^{-2} + \frac{|\chi'_{a} + \chi'_{b}|}{2}\right)^{2}}\int R_{FPXR}\frac{d\omega}{\omega}, \quad \frac{dN^{(1)Cr}_{FPXR}}{d\Omega} = \frac{e^{2}}{4\pi^{2}}\frac{\theta_{\perp}^{2}}{\left(\theta_{\perp}^{2} + \gamma^{-2} - \chi'_{0}\right)^{2}}\int R^{Cr}_{FPXR}\frac{d\omega}{\omega} \tag{28}$$

The curve in the Figure3 and Figure 4 charted by formulas (28) show that the angular density of FPXR in artificial periodic structure as well as the total photons yield are much more than they are in the crystal medium.

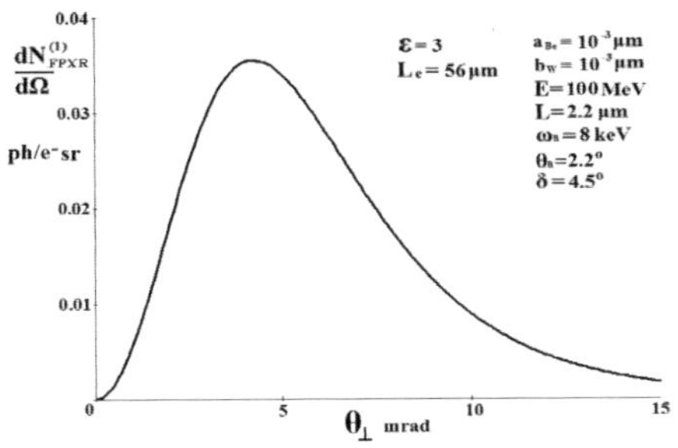

Figure 3. The angular density of relativistic electron FPXR in the artificial periodic multilayer (Be-W) medium

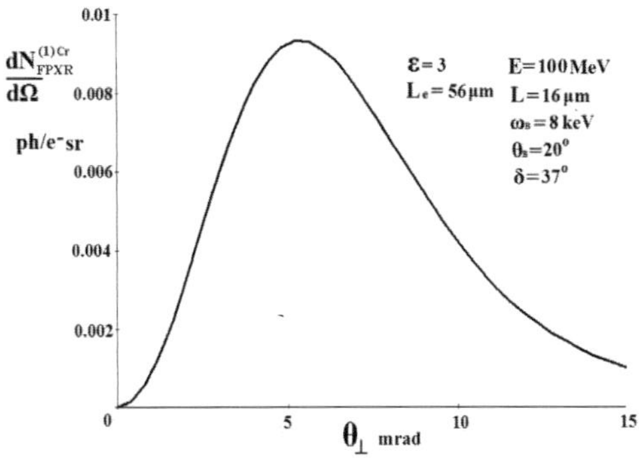

Figure 4. The angular density of the relativistic electron FPXR in the single crystal (W) medium in the similar figure 5 conditions.

6. Conclusion

In the present work the dynamic theory of the coherent X-radiation along the relativistic charged particle in artificial periodic structure are constructed for the Laue scattering geometry and a general case of the asymmetry of the reflection relative to the structure entry surface. In the frame of the two-wave approximation of the diffraction theory the expressions for spectral-angular characteristics of the radiation are obtained based on two different radiation mechanisms: parametric and transition.

It is shown that FPXR spectral peak in artificial periodic structure must be much wider than the analogues peak in the single crystal medium that allows simplifying its experimental investigation. It is shown, that the FPXR angular density in artificial periodic structure expected to be more than an order higher than in a single crystal radiator, under similar conditions.

References

[1] Ter-Mikaelian M 1972 *High-Energy Electromagnetic Process in Condensed Media* (New York: Wiley)
[2] Piestrup M A, Boyers D G, Pincus C I et al. 1992 *Phys.Rev. A*. **45** 1183.
[3] Nasonov N N, Kaplin V V, Uglov S R, Piestrup M A, Garyт C K 2003 *Phys. Rev. E.* **68** 3604.
[4] Blazhevich S V, Noskov A V Book of Abstracts 4th International Conference Charged and Neutral Particles Channeling Phenomena, October 3-8, 2010, Ferrara, Italy.
[5] Garibian G, Yang C 1972 *J. Exp. Theor. Phys.* **63** 1198
[6] Baryshevsky V, Feranchuk I 1976 *Phys. Lett.* A **57** 183
[7] Baryshevsky V G, Feranchuk I D 1983 *J. Physique* (Paris) **44** 913
[8] Kubankin A, Nasonov N, Sergienko V, Vnukov I 2003 *Nucl. Instr. and Meth. In Phys. Res. B* **201** 97
[9] Nasonov N, Noskov A 2003 *Nucl. Instr. Meth. In Phys. Res. B* **201** 67
[10] Kubankin A, Nasonov N, Noskov A, 2002 in Proc. 7 Int. Russian-Japanese Symposium "Interaction of fast charged particles with solids", Nov. 24-30, Kyoto, Japan, p.217-225
[11] Blazhevich S, Noskov A 2010 *J. Phys.: Conf. Ser.* **236** 012013
[12] Blazhevich S, Noskov A 2009 *J.Exp. Theor. Phys.* **136** 1043
[13] Blazhevich S, Kolosova I, Noskov A 2012 J. Phys.: Conf. Ser. (see in this issue)
[14] Bazylev V, Zhevago N 1987 *Emission From Fast Particles Moving in a Medium and External Fields* Moscow Nauka, 1987
[15] Pinsker Z 1984 *Dynamic Scattering of X-rays in Crystals* (Berlin: Springer)
[16] Blazhevich S, Noskov A 2008 *Nucl. Instr. Meth. In Phys. Res. B* **266** 3777

Journal of Physics: Conference Series **357** (2012) 012017 doi:10.1088/1742-6596/357/1/012017

Bremsstrahlung of fast electron on graphene

V A Astapenko[1] and Yu A Krotov[2]
[1] Moscow Institute of Physics and Technology, Moscow, Russia
[2] "Polyus" R&D Institute named after M.F. Stelmakh, Moscow, Russia

E-mail: [1] astval@mail.ru, [2] kryuri@yandex.ru

Abstract. Polarization Bremsstrahlung (PB) [1] and Static (ordinary) Bremsstrahlung (SB) [2] of fast electron scattered on graphene are investigated theoretically for relativistic and non-relativistic projectile velocities. In the framework of the first Born approximation spectral, angular and velocity dependences of PB and SB are calculated and analyzed for various incident electron angles. Coherent and incoherent interactions between electron and two-dimensional graphene are taken into account in the calculations.

1. Introduction

Why graphene? First of all, it is a perfect two-dimensional crystal where a transmitted momentum is discrete in the plane and is arbitrary across it. What can it give in bremsstrahlung processes? How will it be manifested in coherent and incoherent channels? Finally, active work is now underway on obtaining large-size loose-hanging, self-supporting graphene films, which allows looking forward to the performance of an experiment [3].

2. General expression for cross-section of process on ensemble of atoms

The cross-section of a photo process on an ensemble of target atoms looks like (in case of a monatomic target)

$$d\sigma_{target} = \left|\sum_j \exp\left(i\,\mathbf{q}\,\mathbf{r}_j\right)\right|^2 d\sigma_{atom}, \tag{1}$$

where the sum is over all target atoms being in the volume of interaction, $d\sigma_{atom}$ is the differential cross-section of the process on one atom under consideration, $\mathbf{q} = \left(\mathbf{p}_i - \mathbf{p}_f\right)/\hbar - \mathbf{k}$ is the wave vector transferred from an incoming electron to the target, $\mathbf{p}_i, \mathbf{p}_f$ are the initial and finite electron pulses, $\mathbf{k}$ is the wave vector of a photon.

3. Structure factor

The structure factor of a medium in a three-dimensional case (a three-dimensional monocrystal, the angle brackets mean averaging over atom positions):

$$\left\langle \sum_{j,j'} \exp\left(i\mathbf{q}\left(\mathbf{r}_j - \mathbf{r}_{j'}\right)\right) \right\rangle = N\left(1-\exp\left(-u^2 q^2\right)\right) + N n_a (2\pi)^3 \sum_{\mathbf{g}} e^{-u^2 g^2} \left|S(\mathbf{g})\right|^2 \delta^{(3)}(\mathbf{q}-\mathbf{g}), \quad (2)$$

where $N = N_0 N_{cell}$ is the full number of atoms in the volume of interaction, N_0 is the full number of cells in the volume of interaction, N_{cell} is the number of atoms in a cell, $\mathbf{g}$ is the wave vector of a reciprocal lattice, $S(\mathbf{q})$ is the *normalized* structure factor of a unit cell of a crystal on the wave vector $\mathbf{q}$, $S(\mathbf{q}=0)=1$ (M.L. Ter-Mikhaelyan [4] uses a *nonnormalized* structure factor of a cell), $n_a = N_{cell}/\Delta_{cell}$ is the volume concentration of atoms, Δ_{cell} is the volume of a unit cell, $\delta^{(3)}(\mathbf{q}) = \delta(q_x)\delta(q_y)\delta(q_z)$ is the three-dimensional delta function of the wave vector transferred to the target.

In going to a two-dimensional case (we assume that a two-dimensional single crystal lies in the $x\,y$ plane, the z coordinate is fixed: $z = z_0$; graphene), when $\mathbf{q}\,\mathbf{r}_j = q_z z_0 + \mathbf{q}_{\parallel} \mathbf{c}_j$, by analogy with the three-dimensional case for the structure factor of the target we have

$$\left\langle \sum_{j,j'} \exp\left(i\mathbf{q}_{\parallel}\left(\mathbf{c}_j - \mathbf{c}_{j'}\right)\right) \right\rangle = N\left(1-\exp\left(-u^2 q_{\parallel}^2\right)\right) + N n_s (2\pi)^2 \sum_{\mathbf{g}} e^{-u^2 g^2} \left|S(\mathbf{g})\right|^2 \delta^{(2)}(\mathbf{q}_{\parallel}-\mathbf{g}), \quad (3)$$

where $\mathbf{c}_j$ is the radius vector of an atom in the plane of the two-dimensional crystal, $\mathbf{c} = (x, y)$, $\delta^{(2)}(\mathbf{q}_{\parallel}) = \delta(q_x)\delta(q_y)$ is the two-dimensional delta function, n_s is the two-dimensional concentration of atoms, u is the root-mean-square deviation of atoms from the equilibrium position.

For a three-dimensional case we have

$$\left\langle u^2 \right\rangle = \frac{3\hbar^2}{4 M_a T_D}\left[1 + 4\left(\frac{T}{T_D}\right)^2 \int_0^{T_D/T} \frac{y\,dy}{e^y - 1}\right], \quad (4)$$

where T_D is the Debye temperature in energy units, M_a is the mass of atoms of a substance. For a three-dimensional lattice such as diamond $u(C, T = 293^\circ K) = 0.041$ Å ($T_D = 1860^\circ K$). $a = \sqrt{3}\,a_0 = 0.246$ nm is the lattice constant for graphene, $a_0 = 0.142$ nm is the distance between the nearest atoms.

Structure factor of unit cell of graphene is $\left|S(\mathbf{g})\right|^2 = \frac{1}{2}\left[1 + \cos\left(\frac{2\pi}{3}(2n_1 + n_2)\right)\right]$, where the modulus of the graphene reciprocal lattice vector – $g(n_1, n_2) = \frac{4\pi}{\sqrt{3}\,a}\sqrt{n_1^2 + n_2^2 + n_1 n_2}$, $n_{1,2}$ are the integers.

4. Polarization Bremsstrahlung on graphene

Let us consider polarization bremsstrahlung (PB) [1] arising as a result of electron scattering by the two-dimensional plane of graphene. The process geometry is shown in figure 1, where α is the angle of photon emission with respect to the normal to the plane of graphene, ψ is the polar angle of electron incoming with respect to the normal to the plane of graphene), φ is the azimuth angle of electron incoming.

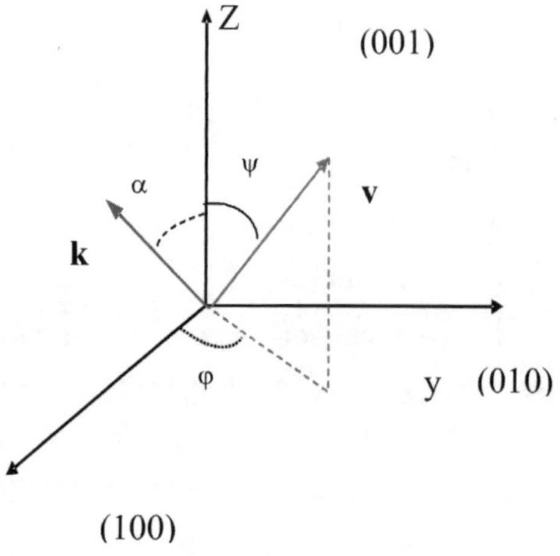

Figure 1. The process geometry.

The cross-section of PB on an atom is

$$\frac{d\sigma_a^{(PB)}}{d\omega d\Omega_{\mathbf{k}}}=\frac{e^2}{\hbar\omega}\frac{c}{\pi \mathrm{v}}\int\delta(\omega-\mathbf{k}\,\mathbf{v}+\mathbf{q}\,\mathbf{v})\frac{\left[\mathbf{s},\omega\mathbf{v}/c^2-\mathbf{q}\right]^2}{\left(\mathbf{q}^2-2\mathbf{k}\mathbf{q}\right)^2}\left|\left(\frac{\omega}{c}\right)^2\alpha(\omega,\mathbf{q})\right|^2 d\mathbf{q}, \tag{5}$$

where $\alpha(\omega,\mathbf{q})$ is the generalized dynamic polarizability of an atom, $\mathbf{s}=\mathbf{k}/|\mathbf{k}|$ is the unit vector in the direction of photon emission. In the multiplicative approximation the generalized dynamic polarizability of an atom is expressed in terms of dipole polarizability and an atomic form factor: $\alpha(\omega,\mathbf{q})=\alpha(\omega)\cdot F(\mathbf{q})$, where $F(\mathbf{q})$ is the atomic form factor normalized by the condition $F(q=0)=1$.

4.1. Incoherent PB on graphene

The cross-section of incoherent PB on a target (in terms of one atom) is

$$\frac{1}{N}\frac{d\sigma_{incoh}^{(PB)}}{d\omega d\Omega_{\mathbf{k}}}=\frac{2e^2}{\pi \mathrm{v}^2 \mathrm{c}^3\hbar\omega}\int_{q_{\min}}^{q_{\max}}\frac{dq}{q}I_{\varphi}(q,\mathrm{v},\omega,\theta)\left(1-e^{-u^2q^2}\right)\left|\omega^2\alpha(\omega,q)\right|^2, \tag{6}$$

where $q_{\min}(\omega,\mathrm{v},\theta)=\frac{\omega}{\mathrm{v}}\left(1-\frac{\mathrm{v}}{c}\cos\theta\right)$, $q_{\max}=2\mu\,\mathrm{v}$, μ is the reduced mass of an electron and an atom of the target, $\theta=angle(\mathbf{k},\mathbf{v})$ is the angle of emission.

$$I_{\varphi}(q,\mathrm{v},\omega,\theta)=\frac{q^3\mathrm{v}}{2\pi}\int d\Omega_{\mathbf{q}}\,\delta(\omega-\mathbf{kv}+\mathbf{qv})\frac{\left[\mathbf{s},\omega\mathbf{v}/c^2-\mathbf{q}\right]^2}{\left(\mathbf{q}^2-2\mathbf{k}\mathbf{q}\right)^2} \tag{7}$$

is the dimensionless kinematic integral, $d\Omega_{\mathbf{q}}$ is the element of the solid angle around the vector $\mathbf{q}$.

4.2. Coherent PB on graphene

The cross-section of coherent PB on a two-dimensional target (in terms of one atom) is

$$\frac{1}{N}\frac{d\sigma_{coh}^{(PB)}}{d\omega d\Omega_{\mathbf{k}}}=\frac{e^2}{\hbar\omega}\frac{c}{\pi \mathrm{v}}\left(\frac{\omega}{c}\right)^4 n_s\left(2\pi\right)^2\times$$
$$\times\sum_{\mathbf{g}} e^{-u^2(\mathbf{g}+\mathbf{q}_\perp)^2}\left|S(\mathbf{g})\right|^2\int\delta(\omega-\mathbf{k}\,\mathbf{v}+\mathbf{g}\,\mathbf{v}+q_z\mathrm{v}_z)\frac{\left[\mathbf{s},\omega\mathbf{v}/c^2-(\mathbf{g}+\mathbf{q}_\perp)\right]^2}{\left((\mathbf{g}+\mathbf{q}_\perp)^2-2\mathbf{k}(\mathbf{g}+\mathbf{q}_\perp)\right)^2}\left|\alpha(\omega,\mathbf{g}+\mathbf{q}_\perp)\right|^2 dq_z. \tag{8}$$

In derivation of this expression it was taken into account that integration over the two-dimensional delta function $\delta^{(2)}(\mathbf{q}_{\|}-\mathbf{g})$ gives $\mathbf{q}_{\|}=\mathbf{g}$, and there remains integration over the component of the wave vector dq_z transferred to the target, this component being normal to the graphene plane, said integration is also "removed" at the expense of the presence of the delta function $\delta(\omega-\mathbf{k}\,\mathbf{v}+\mathbf{g}\,\mathbf{v}+q_z\mathrm{v}_z)$ under the integration sign. As a result, we find a fixed value for the normal (to the graphene plane) component of the wave vector transferred to the target as a function of the problem parameters:

$$q_z=-g\,tg\psi+\omega\frac{1-(\mathrm{v}/c)\cos\theta}{\mathrm{v}\cos\psi}. \tag{9}$$

This value should be substituted in the expression for the cross-section of coherent PB ($|\mathbf{q}_\perp|=q_z$). Taking into account that $(\mathbf{q}_\perp\,\mathbf{g})=0$, we obtain

$$\frac{1}{N}\frac{d\sigma_{coh}^{(PB)}}{d\omega d\Omega_{\mathbf{k}}}=\frac{4\pi n_s}{\cos\psi}\left(\frac{e^2}{\hbar\omega}\right)\frac{\omega^4\left|\alpha(\omega)\right|^2}{c^3\mathrm{v}^2}\sum_{\mathbf{g}} e^{-u^2\left(g^2+q_z^2\right)}\left|S(\mathbf{g})\right|^2\left|F\left(\sqrt{g^2+q_z^2}\right)\right|^2 P(\mathbf{g},\mathbf{k},\mathbf{q}_\perp), \tag{10}$$

where

$$P(\mathbf{g},\mathbf{k},q_z)=\frac{\left(\frac{\omega\mathrm{v}}{c^2}\right)^2+g^2+q_z^2-2\frac{\omega\mathrm{v}}{c^2}(g\sin\psi+q_z\cos\psi)-\left[\frac{\omega\mathrm{v}}{c^2}\cos\theta-g\sin\alpha-q_z\cos\alpha\right]^2}{\left(g^2+q_z^2-2\frac{\omega}{c}(g\sin\alpha+q_z\cos\alpha)\right)^2}, \tag{11}$$

$$\left|S(\mathbf{g})\right|^2=\cos^2\left(\frac{\pi}{3}(2n_1+n_2)\right),\ |\mathbf{g}|=g(n_1,n_2)=\frac{4\pi}{\sqrt{3}\,a}\sqrt{n_1^2+n_2^2+n_1n_2}, \tag{12}$$

$$\cos\theta=\cos\alpha\cos\psi+\cos\varphi\sin\alpha\sin\psi,\ n_s=\frac{4}{\sqrt{3}\,a^2}. \tag{13}$$

Summation over the reciprocal lattice vectors $\mathbf{g}$ implies summation over the integers $n_{1,2}$ determining the modulus of the vector $\mathbf{g}$. Let us note that in contrast to coherent PB in a three-dimensional monocrystal, when the emitted frequency is fixed by the condition $\omega_{\max}^{(3)}=-\frac{\mathbf{g}\,\mathbf{v}}{1-\beta\cos\theta}$, here from we find for a cubic crystal: $\omega_{\max}^{(3)}=-\frac{g\,\mathrm{v}(n_1\sin\psi\cos\varphi+n_2\sin\psi\sin\varphi+n_3\cos\psi)}{1-\beta\cos\theta}$, the frequency of coherent PB in a two-dimensional monocrystal is not a fixed value. Nevertheless, in a two-dimensional case with fulfillment of certain conditions (see below) the PB spectrum has sharp maxima.

For the denominator of equation (11) taking into account the explicit form of q_z, we find

$$\frac{g^4}{\cos^4\psi}\left\{\left(\frac{\omega}{g\,\mathrm{v}}\right)^2\delta\left[\delta-2\beta\cos\alpha\cos\psi\right]+2\beta\frac{\omega}{g\,\mathrm{v}}\cos\psi\left[\cos\alpha\sin\psi-\sin\alpha-\frac{\delta}{\beta}tg\psi\right]+1\right\}^2, \quad (14)$$

where $\delta = 1-\beta\cos\theta$, $\beta = \mathrm{v}/c$. For the zero angle of electron incoming into the graphene plane ($\psi = 0$) the expression for the denominator is simplified to look like

$$\mathrm{Den}\left(\omega,\alpha,\psi=0,\beta,\theta=\alpha\right)=g^4\left\{\left(\frac{\omega}{g\,\mathrm{v}}\right)^2(1-\beta\cos\alpha)(1-3\beta\cos\alpha)-2\beta\frac{\omega}{g\,\mathrm{v}}\sin\alpha+1\right\}^2. \quad (15)$$

The resonance condition in the cross-section of coherent PB in the general case is $\mathrm{Den}(\omega,\alpha,\psi,\beta,\theta)=0$. If this equation is solved with respect to the emission frequency, the following expression for the frequency of a spectral maximum in coherent PB on graphene will be obtained: $\omega_{max} = g\,\mathrm{v}\,F_\omega(\alpha,\psi,\theta,\beta)$. Here the dimensionless function is introduced:

$$F_\omega(\alpha,\psi,\theta,\beta)=\frac{\cos\psi\left(\beta\sin\alpha+\delta\,tg\psi-\beta\sin\psi\cos\alpha\right)+\mathrm{sign}\left(\delta-2\beta\cos\alpha\cos\psi\right)\sqrt{D}}{\delta\left(\delta-2\beta\cos\alpha\cos\psi\right)}, \quad (16)$$

$$D=\cos^2\psi\left(\beta\sin\alpha+tg\psi\,\delta-\beta\sin\psi\cos\alpha\right)^2-\delta\left(\delta-2\beta\cos\alpha\cos\psi\right). \quad (17)$$

In case of the zero angle of electron incoming ($\psi = 0$) into a two-dimensional monocrystal, we have for the function determining the dependence of the resonance frequency of emission on the electron velocity and the angle of photon emission:

$$F_\omega(\psi=0)=\frac{\beta\sin\alpha+\mathrm{sign}(1-3\beta\cos\alpha)\sqrt{(\beta\sin\alpha)^2-(1-\beta\cos\alpha)(1-3\beta\cos\alpha)}}{(1-\beta\cos\alpha)(1-3\beta\cos\alpha)}. \quad (18)$$

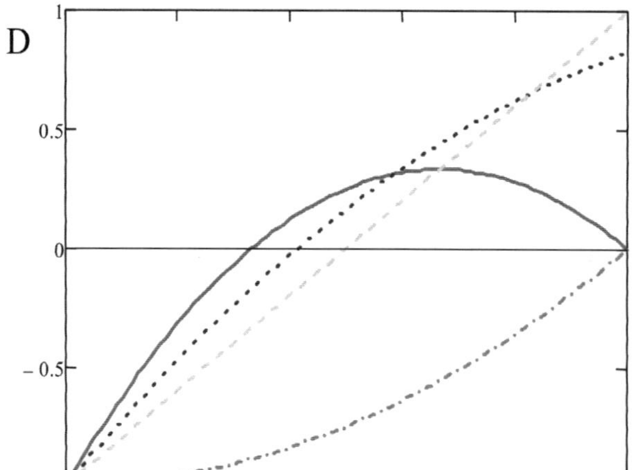

Figure 2. The dependence of the determinant in the expression determining the resonance frequency of PB for the zero angle of electron incoming on the electron velocity for different angles of emission: solid line – α = 0, dotted line – α = π/4, dashed line – α = π/3, dotted-dashed line – α = π/2.

Journal of Physics: Conference Series **357** (2012) 012017 doi:10.1088/1742-6596/357/1/012017

The total cross-section of electron PB on graphene in terms of one atom is equal to the sum of the incoherent and coherent parts. The comparison of the cross-sections (in terms of one atom) of coherent and incoherent PB on graphene and on a carbon atom for the normal incidence of an electron on the graphene plane (ψ=0), and an angle of emission of 30° are given in figures 3 and 4.

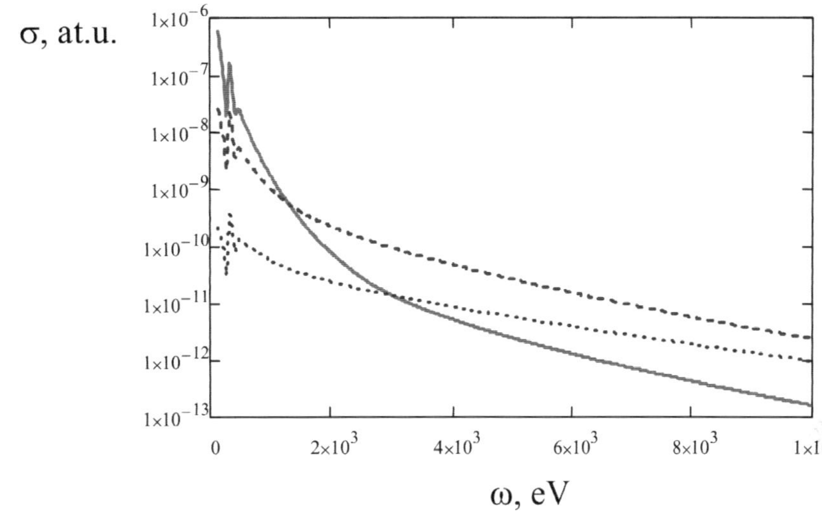

Figure 3. For an electron energy of 30 keV (v = 45 atomic units): solid line – coherent PB, dotted line – incoherent PB, dashed line – atomic PB.

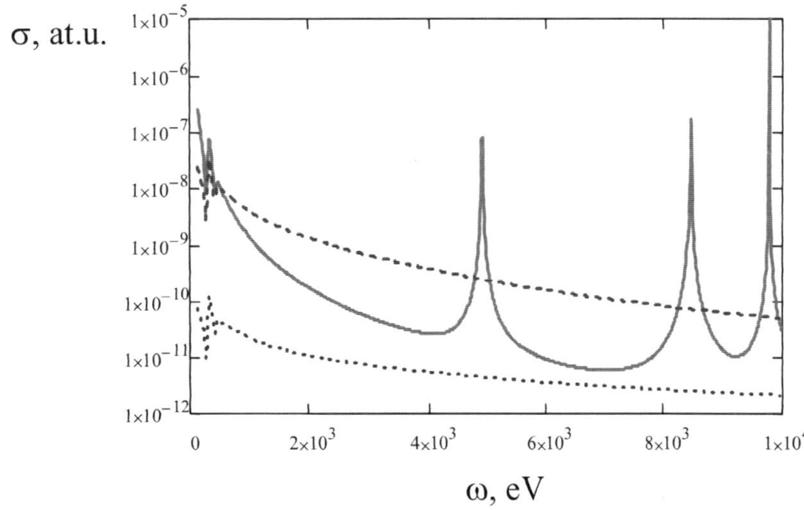

Figure 4. For an electron energy of 4 MeV (v = 136 atomic units): solid line – coherent PB, dotted line – incoherent PB, dashed line – atomic PB.

The dependence of the cross-section of coherent PB on graphene on the angle of electron incoming, for an electron energy of 58 keV (velocity v = 60 atomic units), α = 30° and different photon energies is shown in figure 5.

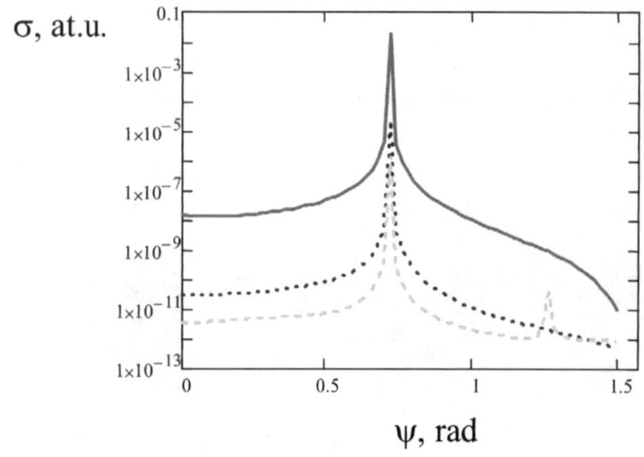

Figure 5. Coherent PB on graphene on the angle of electron incoming, for electron energy of 58 keV (v = 60 at. u.), $\alpha = 30°$: solid line – $\omega = 272$ eV, dotted line – $\omega = 2720$ eV, dashed line – $\omega = 5440$ eV.

5. Static (ordinary) Bremsstrahlung on graphene

For calculation of the coherent part of SB on graphene it is necessary to express the cross-section of the process on an atom [6] in terms of an integral over the transferred wave vector $\mathbf{q}$. Then the expression for the differential cross-section of coherent SB on graphene (in terms of one atom) looks like

$$\frac{1}{N}\frac{d\sigma_{coh}^{(OB)}}{d\omega d\Omega_{\mathbf{k}}} = \frac{2n_s}{\cos\psi}\frac{Z^2 e^6\left(1+\cos^2\theta\right)\left(1-\left(\frac{\mathrm{v}}{c}\right)^2\right)}{\hbar\omega m^2 \mathrm{v}^2 c^3\left(1-\frac{\mathrm{v}}{c}\cos\theta\right)^2}\sum_{\mathbf{g}} e^{-u^2\left(g^2+q_z^2\right)}\left|S(\mathbf{g})\right|^2\frac{\left[1-F\left(\sqrt{g^2+q_z^2}\right)\right]^2}{g^2+q_z^2}. \qquad (19)$$

In view of the last equation summation over the reciprocal lattice vectors $\mathbf{g}$ in this formula reduces to summation over the set of the integers (n_1, n_2) - see formulas (12).

For the incoherent part of the cross-section of SB on graphene (in terms of one atom) we have

$$\frac{1}{N}\frac{d\sigma_{incoh}^{OB}}{d\omega\, d\Omega_{\mathbf{k}}} \approx \frac{1}{\pi}\frac{Z^2 e^6\left(1+\cos^2\theta\right)\left(1-\left(\frac{\mathrm{v}}{c}\right)^2\right)}{m^2\hbar\omega \mathrm{v}^2 c^3\left(1-\frac{\mathrm{v}}{c}\cos\theta\right)^2}\int_{q_{\min}}^{q_{\max}}\frac{\left(1-F(q)\right)^2}{q}\left(1-\exp\left(-u^2 q^2\right)\right)dq. \qquad (20)$$

6. Conclusion

Polarization Bremsstrahlung (PB) and Static (ordinary) Bremsstrahlung (SB) of fast electron scattered on graphene are investigated theoretically. The coherent and incoherent interactions between electron and two-dimensional graphene lattice so as dynamical polarizability and form-factor of carbon atom core are taken into account.

It is shown that spectral features of PB cross section are depended on electron velocity, projectile incident angle and radiation angle. In specified region of the problems parameters sharp maxima in PB spectrum is predicted. For normal electron incidence on graphene plane the analytical description of resonance frequencies on velocity and radiation angles is obtained.

In low frequency range coherent PB is dominated by the process without momentum exchange between projectile and graphene. In this range coherent PB is comparable or greater than the incoherent one. Our analysis show that dominated Bremsstrahlung channels of fast electron on graphene are coherent PB and incoherent SB.

Acknowledgments
One of the authors – Yu Krotov is thankful to professors A Potylitsyn, N Shul'ga and Yu Lozovik for useful discussions.

References

[1] Amusia M Ya, Tsytovich V N, Oiringel I M et al 1992 *Polarization Bremsstrahlung* (New York: Plenum Press)
[2] Heitler W 1954 *The Quantum Theory of Radiation* (Oxford: Clarendon Press)
[3] Geim A K and Novoselov K S 2007 *Nature Mater.* **6** 183
[4] Ter-Mikaelyan M L 1972 *High Energy Electromagnetic Processes in Condensed Media* (New York: Wiley–Interscience)
[5] Lozovik Yu E et al 2008 *Usp. Phys. Nauk* **178** 757
[6] Berestetskii V B, Lifshitz E M and Pitaevskii L P 1982 *Quantum Electrodynamics* (New York, Oxford: Pergamon Press)

Journal of Physics: Conference Series **357** (2012) 012018 doi:10.1088/1742-6596/357/1/012018

A possibility of transverse beam size diagnostics using parametric X-ray radiation

A Gogolev, A Potylitsyn, G Kube*

Tomsk Polytechnic University, Tomsk, Russia
*Deutsches Elektronen-Synchrotron (DESY), Hamburg, Germany

E-mail: alextpuftf@tpu.ru

Abstract. A method of transverse beam size diagnostics with μm space resolution by parametric X-ray radiation is proposed.

1. Introduction

Transverse beam profile diagnostics in modern electron linear accelerators as FELs or injector linacs is mainly based on optical transition radiation (OTR) as standard technique which is observed in backward direction when a charged particle beam crosses the boundary between two media with different dielectric properties. However, the experience from modern linac based light sources shows that OTR diagnostics might fail even for high energetic electron beams because of Coherent OTR emission (COTR). A cause of this emission is the micro-bunching instability, i.e. some unstable micro structures in the electron bunch that compromise the use of OTR monitors as reliable diagnostic scheme.

One possibility to overcome this limitation is to measure at much shorter wavelengths than the visible spectral region, i.e. in the X-ray range. In this contribution it is proposed to use a technique based on the observation of parametric X-radiation (PXR) for beam imaging. Besides the usability in this wavelength region, PXR offers the advantage to be generated at crystal planes oriented under a certain angle to the crystal surface, i.e. OTR and PXR are not reflected in the same direction which allows a spatial suppression of the COTR background.

In the proposed detection scheme a two-dimensional X-ray detector with μm space resolution is placed in the vicinity of a crystalline target that allows to measure a beam profile with transverse sizes down to $\sigma \sim 10$ μm. The detector is situated under a large angle with respect to the beam which offers the possibility to obtain an image of the electron spot with a small distortion if the distance L between target and detector is small ($L/\gamma \leq \sigma$ with γ the Lorentz factor).

2. Simulation

Parametric X-ray radiation (PXR) is emitted by a fast charged particle in a crystal due to diffraction of its virtual-photon field on crystallographic planes. PXR has been theoretically predicted [1–3] and then observed and studied with electron beams at different energies [4, 5].

The energy of PXR photons is determined by the experimental setup in the following way

$$E_n = n\hbar\omega_1 = n\frac{2\pi\hbar c}{d}\frac{\beta\sin\theta_B}{1-\sqrt{\varepsilon}\,\beta\cos\theta_D\,\cos\theta_y},$$

with n the diffraction order, d the inter-planar distance, $\beta = v/c$, ε the dielectric constant of the target material ($\varepsilon \approx 1$ for X-rays), θ_B is the orientation (Bragg) angle of the crystal planes with respect to the particle momentum, and θ_D and θ_y the radiation emission angles. The direction of the PXR maximum is determined by the angle $\theta_D=2\theta_B$ in the diffraction plane, and inclined to the diffraction plane at the angle $\theta_y(\omega)=\sqrt{\gamma^{-2}+(\omega_p/\omega)^2}$ with γ is the Lorentz factor and $\hbar\omega_p$ the plasmon energy which amounts 31 eV for silicon. The diffraction plane is determined by the particle momentum and by vector normal to the crystal planes.

Calculations of the PXR characteristics from 855 MeV electrons in a Si (010) crystal of 10 μm thickness in Laue geometry have been performed according to the kinematic theory [6] by the algorithm described in the [7]. The results of these calculations will be presented in the following for two possible observation geometries and discussed in view of a possible particle beam diagnostics application.

2.1. *Observation angle 90 degree*

The sketch of the experimental setup of the proposed geometry is shown in figure 1a. In the simulation model describing the electron beam, the PXR-target (crystal), a possible filter, and a view screen (scintillator), the screen was assumed to consist of a matrix of 10x10 μm pixel size, located at 10 mm distance from the target center under an observation angle of 90°.

To improve the resolution in this case it is proposed to use of thin metal filters, see e.g. figure 1b. For instance a 25 μm thick Ti foil (*K*-absorption edge of 5.1 keV) can be used to absorb the high intense but low-energetic PXR radiation which has a broad angular distribution and therefore will spoil the spatial resolution.

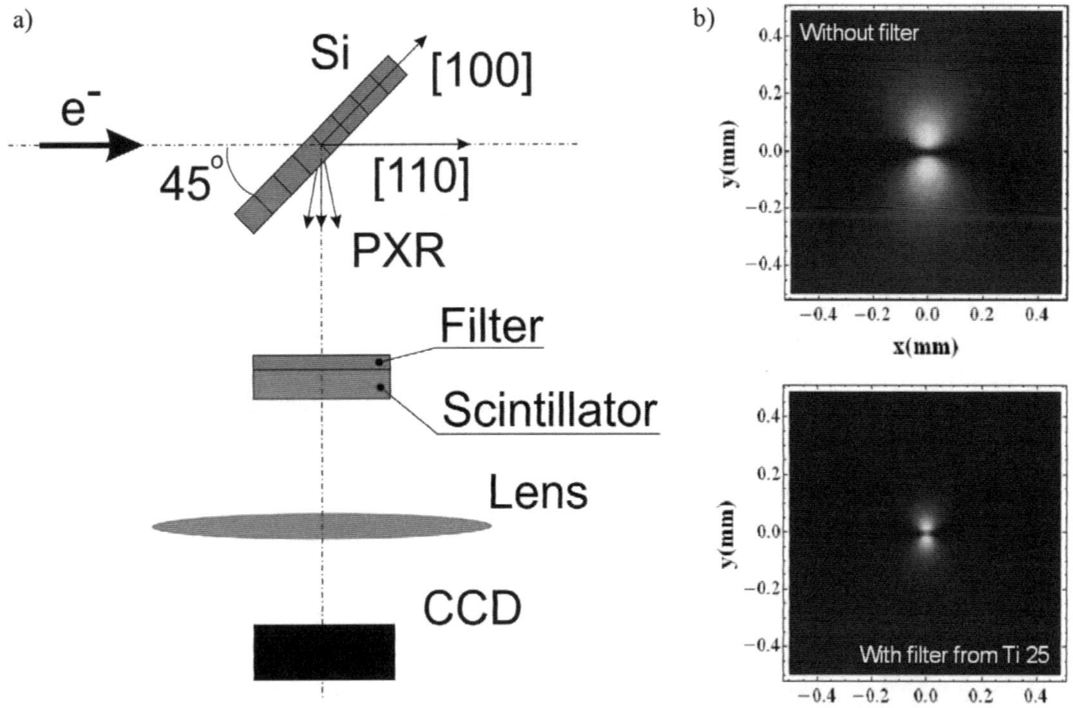

Figure 1. Experimental scheme (a); image of the PXR beam at the view screen from a single electron (b).

The intensity of the PXR main reflection (12 0 0) downstream this filter with a photon energy of 19.5 keV amounts 10^{-6} ph/e^- see figure 2. In addition this figure shows the spatial and the spectral distribution of PXR at the scintillator from a real electron beam with a transverse size of 35×50 μm, a beam size which seems reasonable for a proposed test experiment

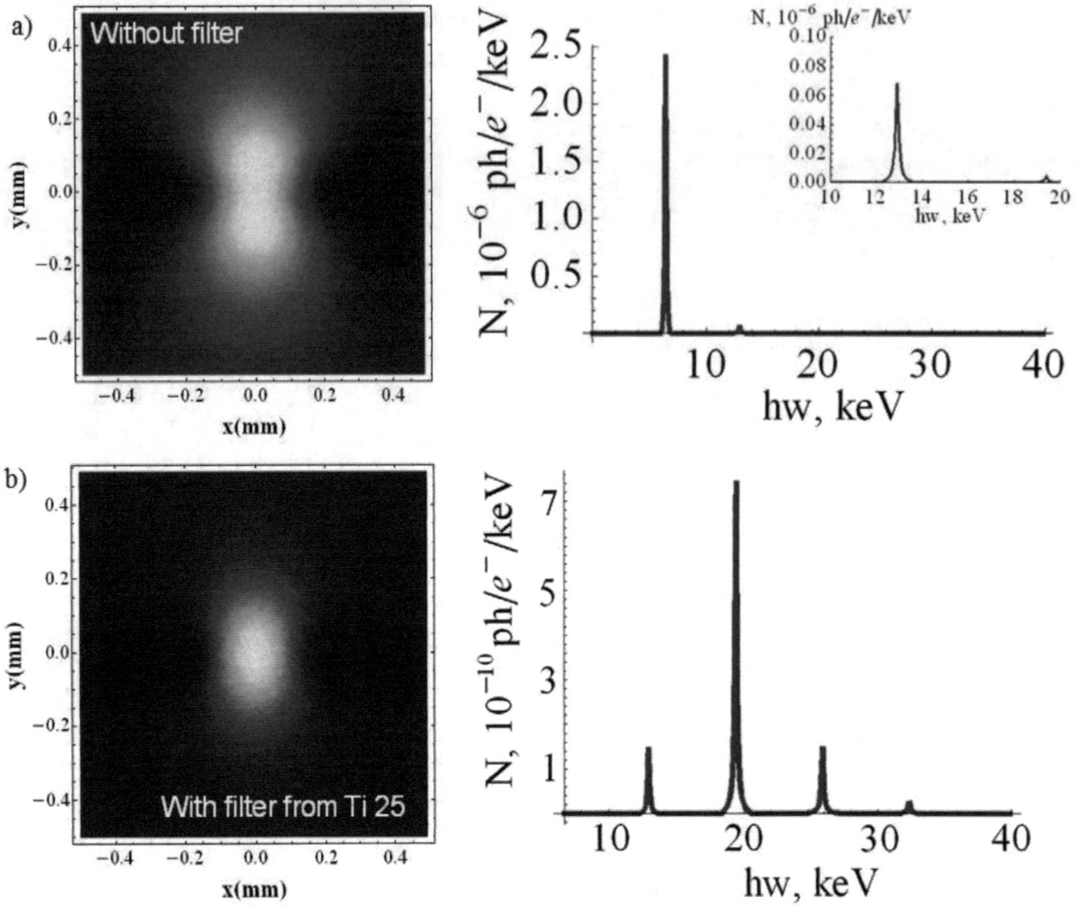

Figure 2. The image and spectrum of the PXR beam generated by a real electron beam of Gaussian distribution with $\sigma_x = 35$, $\sigma_y = 50$ μm: a) without filter; b) with filter consisting of a 25 μm thick Ti foil.

Figure 3 summarizes the simulation results for the obtained PXR spot sizes (FWHM) at the view screen as function of the electron beam cross-section. As can be seen, for small electron beam spots there is a significant broadening of the transverse resolution which can however be improved by using the proposed geometry with filter.

Journal of Physics: Conference Series **357** (2012) 012018 doi:10.1088/1742-6596/357/1/012018

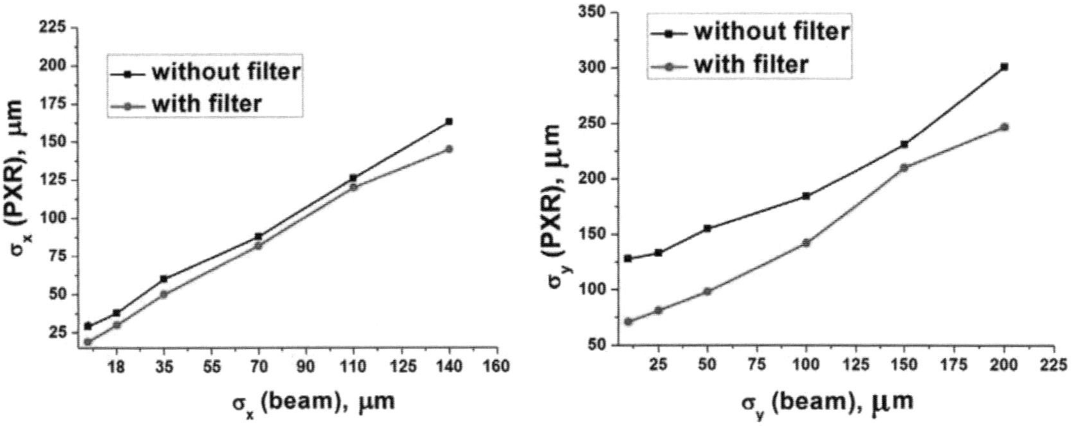

Figure 3. Dependency of the PXR spot size at the view screen (FWHM) as function of the electron beam size.

2.2. Observation angle 30 degree

When the observation angle is 30° (figure 4), the photon energy of the first reflex (4 0 0) is 17.7 keV. Such a high photon energy allows to obtain an X-ray image without filters with sufficient spatial resolution. The intensity of the PXR reflection in this geometry amounts 10^{-3} ph/e^- (figure 5), which in principle should be sufficient for beam imaging.

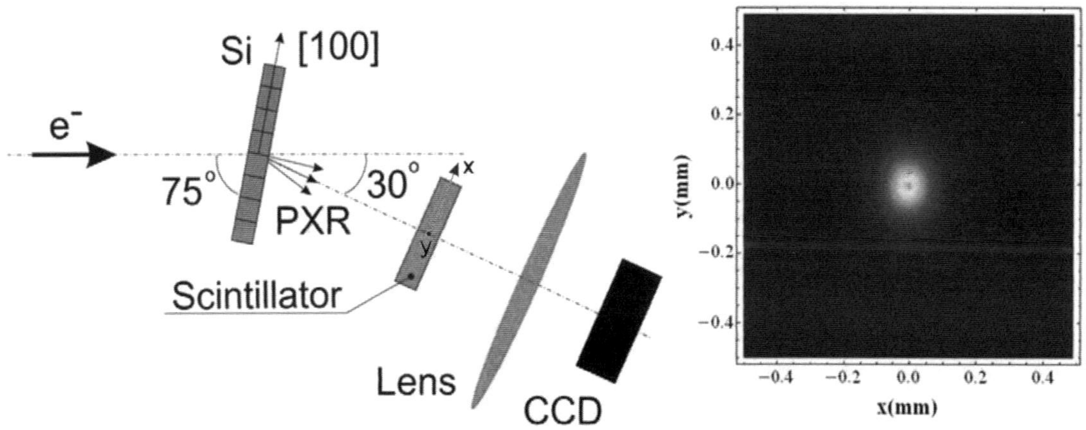

Figure 4. Experimental scheme (a); image of the PXR beam at the view screen from a single electron (b).

Journal of Physics: Conference Series **357** (2012) 012018 doi:10.1088/1742-6596/357/1/012018

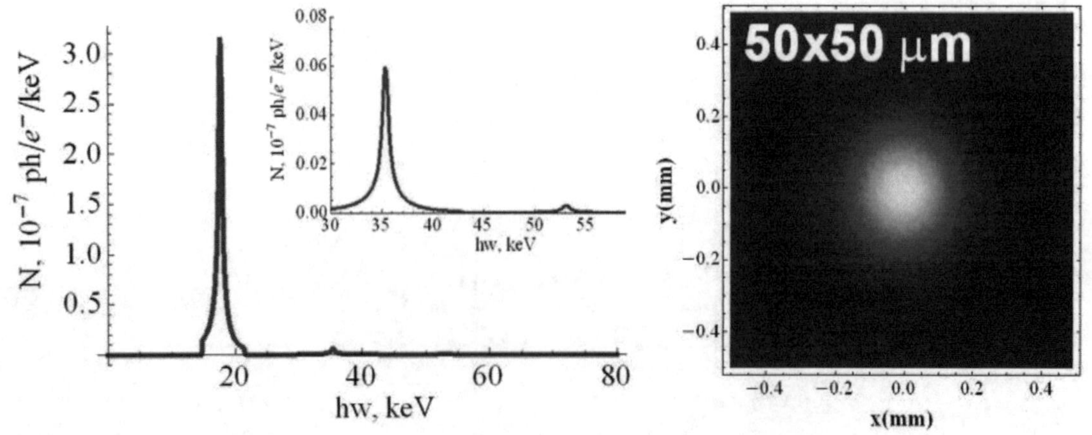

Figure 5. Spectrum and image of the PXR beam on the scintillator from real electron beam of Gaussian distribution with $\sigma_x = 50$, $\sigma_y = 50$ μm.

Figure 6 summarizes the simulation results for the obtained PXR spot sizes (FWHM) at the view screen as function of the electron beam cross-section. In this case, both transverse beam sizes were chosen to be equal. As can be seen, for this observation geometry the resolution broadening is drastically reduced. However, it is still too large to resolve electron beam spots in the order of a few microns.

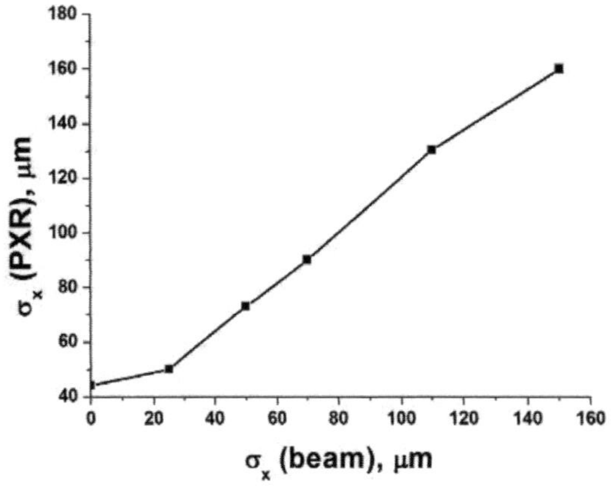

Figure 6. Dependency of the PXR spot size at the view screen (FWHM) as function of the electron beam size.

3. Conclusion

According to the simulation results it should be possible to measure transverse beam sizes in the order of $\sigma \geq 50$ μm with sufficient resolution for beam energies above 1 GeV and the observation geometry of 30 deg.

The main parameter that determines minimum resolution of this method is determined by the thickness of the crystal target that decreases with decreasing observation angle in Laue case (in Bragg case length of the path in the crystal increases), so it is necessary to carry out further optimization of the parameters in order to define a suitable geometry with an appropriate target crystal for a future test experiment.

This work was partly supported by Government of the Russian Federation "On the Federal Target Program" GK № 11.519.11.2030.

References

[1] Ter-Mikaelian M L 1972 *High Energy Electromagnetic Processes in Condensed Media* (New York: Wiley)
[2] Garibyan G M and Yan Shi 1971 *Zh. Éksp. Teor. Fiz.* **61** 930
[3] Baryshevsky V G and Feranchuk I D 1971 *Zh. Éksp. Teor. Fiz.* **61** 947
[4] Schagin A V and Maruyama X K 1997 *Accelerator-Based Atomic Physics: Technique and Applications*, ed S M Shafroth and J C Austin (New York: AIP Press) p 279
[5] Rullhusen P, Artru X and Dhez P 1998 *Novel Radiation Sources Using Relativistic Electrons* (Singapore: World Sci)
[6] Nitta H 1992 *Phys. Rev.* B **45** 7621
[7] Gogolev A S 2008 *Generation of parametric X-ray radiation is moderately relativistic charged particles: dis. Candidate. nat. mat. Science/* (Tomsk: TPU)

Journal of Physics: Conference Series **357** (2012) 012019 doi:10.1088/1742-6596/357/1/012019

Non-intercepting diagnostic for high brightness electron beams using Optical Diffraction Radiation Interference (ODRI)

A Cianchi[1], V. Balandin[2], M. Castellano[3], L. Catani[4], E. Chiadroni[3], G. Gatti[3], N. Golubeva[2], K. Honkavaara[2], G. Kube[2]

[1] INFN-Roma Tor Vergata & University of Rome "Tor Vergata", Via della Ricerca Scientifica, 1-00133 Rome, Italy
[2] DESY, Notkestrasse 85, 22607 Hamburg, Germany
[3] INFN-LNF, Via E. Fermi, 40 - 00044 Frascati, Rome, Italy
[4] INFN-Roma Tor Vergata, Via della Ricerca Scientifica, 1-00133 Rome, Italy

E-mail: `cianchi@roma2.infn.it`

Abstract. High-gain Free Electron Lasers and future Linear Colliders require development of modern electron linacs with high brightness beams. Conventional intercepting transverse electron beam diagnostics, e.g. based on Optical Transition Radiation (OTR), cannot tolerate such high power beams without remarkable mechanical damages on the diagnostics device. Optical Diffraction Radiation (ODR) is an excellent candidate for measurements of the transverse phase space parameters in a non-intercepting way. One of the main problems of this method is the low signal to noise ratio, mainly due to the unavoidable synchrotron radiation background. This problem can be overcome by using two slits on metallic foils, placed at a distance shorter than the radiation formation zone. In this case a nearly background-free ODR interference pattern is produced allowing the determination of the beam size and angular divergence. The accuracy on these parameters can be increased by exploiting both ODR polarization states, as well as different wavelengths. Here we report measurements of the ODR interference between two slits with different aperture sizes in a non-collinear geometry, carried out at FLASH (DESY, Germany). Our results demonstrate the unique potential of this technique to determine the beam parameters.

1. Introduction

The beam size measurement is the fundamental diagnostic in linear accelerators allowing both the appropriate matching along the machine as well as the emittance control. Unfortunately conventional beam size diagnostic are mainly based on intercepting devices. Such a device can be seriously damaged or even destroyed by high brightness beams. Optical Diffraction Radiation (ODR) is a good candidate to be used as non intercepting diagnostic for the determination of the beam size. The particles go through an hole in a metallic screen, whose dimensions are smaller than the radial extension of the particle electromagnetic field (EM) that is in the order of $\gamma\lambda$, where γ is the usual relativistic factor and λ the observed wavelength.

The physics of ODR is well known in literature [1]. The first suggestion of using ODR from a rectangular aperture as non-intercepting measurement of particle beam sizes is discussed in [2].

The first application of this technique to the particle beam diagnostics has been performed at ATF (KEK) ([3], [4], [5]).

Nevertheless two main problems remained: the low ODR signal-to-noise ratio, mainly due to the synchrotron radiation (SR) background [6] produced in the upstream magnetic elements and the ambiguity in the determination of the beam size due to an offset of the particle trajectory in the slit with respect to the geometrical center, according to Eq.6 in Ref [2]. The last issue can be mitigated by using a complementary diagnostic able to track the beam trajectory inside the aperture.

To overcome these limitations we came out with the idea of using a system of two slits with different aperture sizes, placed at a distance much shorter than the formation length. We called this system Optical Diffraction Radiation Interference (ODRI).

2. ODRI

An introduction and explanation of the ODRI theory can be found in [7]. Here we report just the fundamental details. The setup is illustrated in Fig. 1

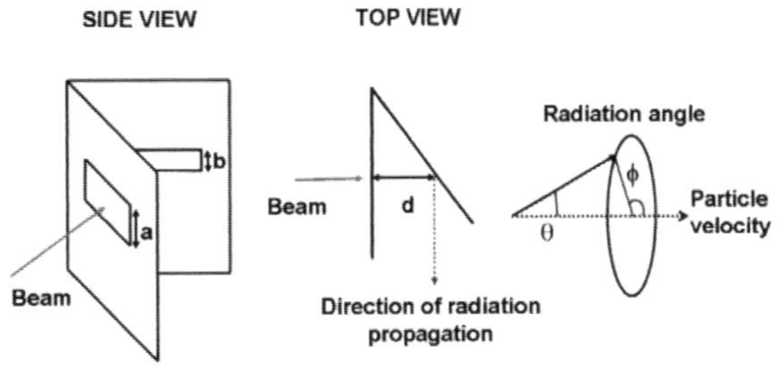

Figure 1. Sketch of the two-slits setup. Left: side view; Center: top view; Right: definition of the coordinate system. Dimension a is the aperture size of the first slit, b the size of the second slit, and d the distance between slits center.

We designed this apparatus for an electron energy in the range of 1 GeV, so the radiation formation length (L $\approx \gamma^2\lambda$) in the optical wavelength range is of the order of few meters. A 1 mm size slit is realized in a metallic screen, placed normal to the beam axis. A second slit, 0.5 mm wide, is opened by means of lithographic technique in a silicon aluminated wafer, placed at 45 degrees with respect to the beam axis. The distance between the centers of the two apertures is about 2 cm.

The forward diffraction radiation (FDR), emitted when the charges pass through the first aperture, interferes with the backward diffraction radiation (BDR), produced by the interaction of the EM field with the second screen. The first screen acts also as a mask for the incoming SR background. For this purpose the best choice could be represented by two identical slit apertures. However in this case, due to the small distance between slits, compared to the formation length, the FDR and BDR interference would cancel almost totally the intensity. So we used 1 mm and 0.5 mm slits, these values being a reasonable compromise between the need to shield from SR background and to prevent negligible total intensity.

As clearly explained in [7] the two slits must be slightly misaligned with respect to each other (tens of microns), in order to allow the discrimination between the beam transverse size σ and the angular divergence σ'. The simultaneous measurement of both σ and σ' paves the way to use this device in order to directly retrieve the beam emittance.

This setup is able to clearly estimate the emittance, the angular divergence and even the position of the beam inside the slit, as we already demonstrated in [7]. In this paper we further extend our study comparing the data obtained under the same beam condition but observing different wavelengths or polarization states. The results confirm the validity of our analysis and the approximations. These are: perfect metal model for the targets, negligible effect of the interferential filter bandwidth and of the energy spread (in the order of 0.1%), particle distribution gaussian both in position and momentum, far field approximation for the radiation fields.

3. Experimental Setup

The measurements have been carried out at the FLASH free-electron laser user facility at DESY (Hamburg) [8]. FLASH consists of an electron source to produce a high quality electron beam, followed by superconducting linac with TESLA-type accelerating modules, and an undulator section to produce FEL radiation (see Fig. 2).

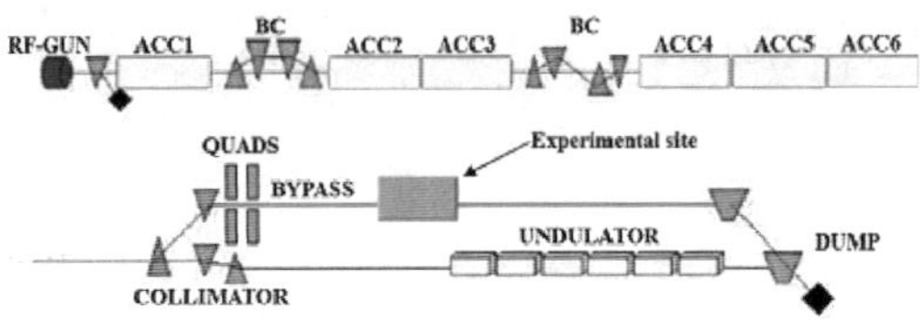

Figure 2. FLASH layout during our experiment (not to scale). The site of the experiment is indicated.

In addition there is a second electron beam line to by-pass the undulators. Our experimental station is placed in this line about 40 m away from the last bending dipole. The energy of the beam was around 1.2 GeV and the typical number of bunches per bunch train was 20. The bunch train repetition rate was fixed to 5 Hz. In our experiment, we used an electron bunch charge of about 0.4 nC. We integrated for 2 seconds, resulting in a total integrated charge of 80 nC. This charge is enough to damage a conventional aluminated OTR screen, so for the comparison with measurement obtained with OTR we run with a single pulse, 0.4 nC.

A typical normalized projected transverse rms emittance measured at the injector is around 2 mm mrad (1 nC uncompressed bunch). Some emittance growth may occur when the beam is transported to the by-pass line where our experiment is located.

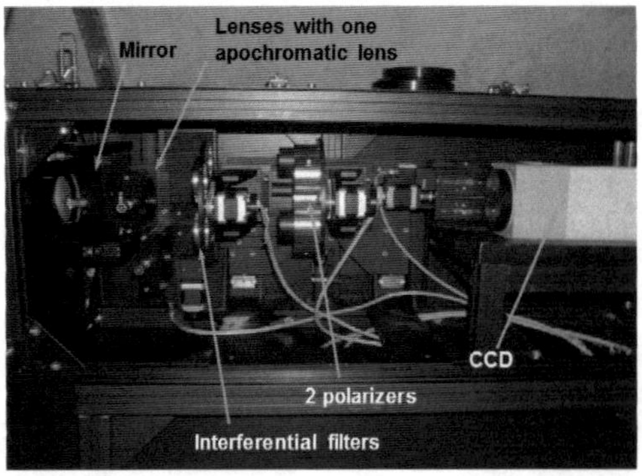

Figure 3. Optical system used in the measurements. The main components are illustrated in the image. All components are remotely controlled.

Radiation from the target is reflected by a mirror and transported through an optical system to the camera.

The main components of the optical setup are highlighted in Fig. 3. Both, an achromatic doublet to image the beam and an apochromatic lens, with custom design to reduce the influence of chromatic aberrations to obtain the ODRI angular distribution can be used. The lenses have different focal lengths, f=250 mm and f=500 mm, in order to focus on the same image plane. Two filter wheels can hold several narrow band interference filters and two Glan-Thomson polarizers to select vertical and horizontal polarization. The polarizer lengthens the optical path, thus increasing the focal length. However, this change can be corrected by slightly changing the longitudinal camera position. A cooled, high sensitivity, 16-bit CCD camera is used (Hamamatsu ORCA II-BT-512G model type C4742-98-26LAG2). The camera main features are the very high quantum efficiency in the whole visible spectrum, in particular at 800 nm, the negligible thermal noise, and the long exposure time.

4. New Results

The goal of these measurements was to check the validity of the model for different wavelengths and to investigate also the horizontal polarization of the radiation, i.e. the component parallel to the longest side of the slits.

The data analysis has been performed using the Cernlib fitting routine *Minuit* [9]. The angular distribution of the whole beam is obtained by summing up 5000 distributions produced by a single particle with different vertical positions within the slit and angular divergences, both Gaussian distributed. Since the resulting expression cannot be solved analytically, a Monte Carlo approach is used instead and the results are fed into the *Minuit* routine. Every time the fitting procedure changes the starting parameters, new distributions are generated through the Monte Carlo code.

4.1. Vertical Polarization

We compared the angular distributions obtained collecting the radiation at 800 nm and 550 nm in order to retrieve the values of σ_y and $\sigma_{y'}$. The results are shown in Fig. 4 and Fig. 5.

The retrieved parameters for 800 nm and 550 nm are respectively 81 μm and 76 μm for σ_y and 203 μrad and 193 μrad for $\sigma_{y'}$. Both values are well inside the 10% difference that we estimated is a good threshold for measurement reliability. From this comparison it is clear that the model is consistent and does not strongly depend on the wavelength.

At the same time, from the experimental point of view, in this range of energy and beam size, the longer wavelength seems to be preferable. Using a longer wavelength results in a wider angular distribution, increased fringes visibility and easier fit procedure. So, the wavelength should be chosen as long as possible, compatible with the detector efficiency, the slit aperture and the beam size at given energy.

4.2. Horizontal Polarization

We compared also the vertical polarization of the angular distribution at 800 nm with the horizontal polarization at the same wavelength in Fig. 6 and Fig. 7. For our knowledge these are the first ever measurements of the angular distribution of the horizontal polarization of ODR, even if in the case of ODRI. Some features are very different compared to the other polarization state. Here the central symmetry line does not carry any relevant information, while the interesting profiles are the lines separated by about $1/\gamma$ from the center, where the maximum of the intensity is located. From the profile of the bottom line we can retrieve the beam parameters and compare them with the results given by the other polarization. The parameter values for vertical and horizontal polarization are 77 μm and 70 μm respectively for the σ_y and 203 μrad and 211 μrad for $\sigma_{y'}$.

Journal of Physics: Conference Series **357** (2012) 012019 doi:10.1088/1742-6596/357/1/012019

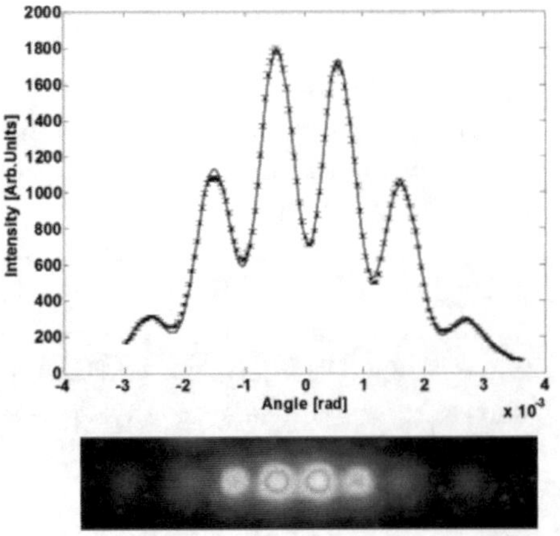

Figure 4. Top: angular distribution of the ODRI radiation at 800 nm with superimposed fit of data points. Bottom: image of the angular distribution on the detector. Fit results:σ_y=81 μm, $\sigma_{y'}$=203 μrad.

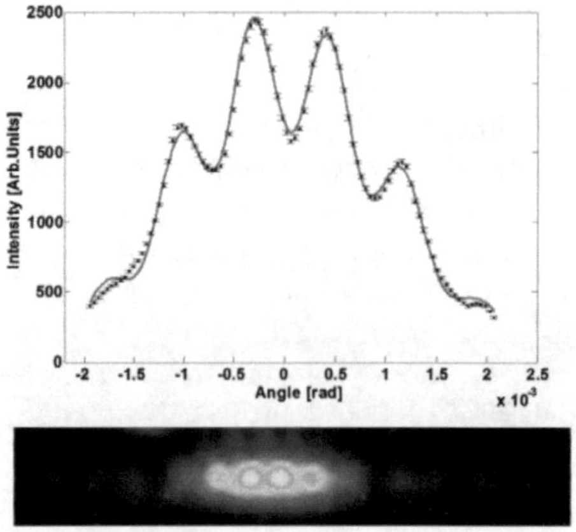

Figure 5. Top: angular distribution of the ODRI radiation at 550 nm with superimposed fit of data points. Bottom: image of the angular distribution on the detector. Fit results:σ_y=76 μm, $\sigma_{y'}$=193 μrad.

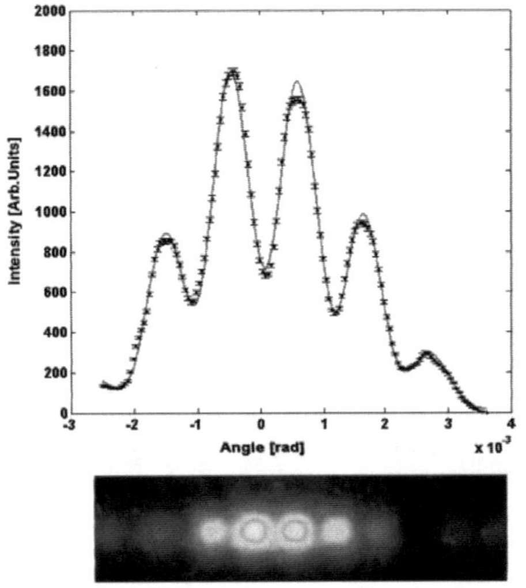

Figure 6. Top: angular distribution of the ODRI radiation at 800 nm with superimposed fit of data points. Bottom: image of the angular distribution on the detector. Fit results:σ_y=77 μm, $\sigma_{y'}$=203 μrad.

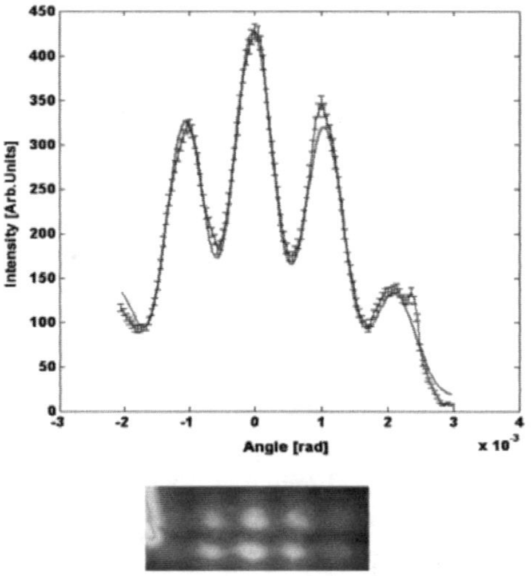

Figure 7. Top: angular distribution of the ODRI radiation at 800 nm with superimposed fit of data points. Bottom: image of the angular distribution on the detector. Fit results:σ_y=70 μm, $\sigma_{y'}$=211 μrad, $\sigma_{x'}$=122 μrad

Again these values are consistent with each other. From the horizontal polarization it is also possible to measure the angular divergence $\sigma_{x'}$=122 μrad, while the vertical polarization is not sensitive to this parameter. From the analysis of the two dimensional angular distribution an unexpected asymmetry between the upper and the lower part of the image is evident. A clear signature of unwanted background light is also visible in the left side of the image. Further investigations are needed in order to clearly identify this problem. At this point these behaviors can be explained only with some residual contribution of the SR background.

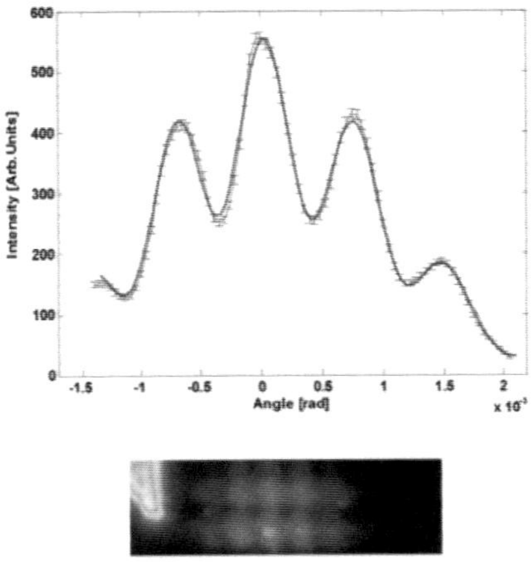

Figure 8. Top: angular distribution of the ODRI radiation at 550 nm with superimposed fit of data points. Bottom: image of the angular distribution on the detector. Fit results:σ_y=70 μm, $\sigma_{y'}$=160 μrad, $\sigma_{x'}$=123 μrad

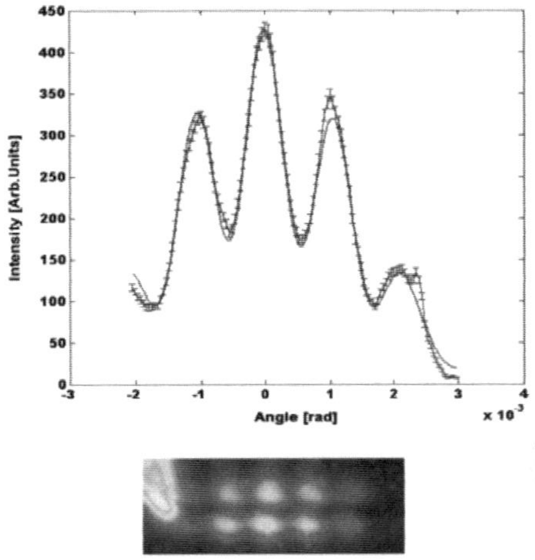

Figure 9. Top: angular distribution of the ODRI radiation at 800 nm with superimposed fit of data points. Bottom: image of the angular distribution on the detector. Fit results:σ_y=75 μm, $\sigma_{y'}$=211 μrad, $\sigma_{x'}$=122 μrad

As further benchmark of our model we compared also the angular distribution of the horizontal polarization at two different wavelengths in Fig. 8 and Fig. 9. The value obtained from the fit are 70 μm and 75 μm respectively for the case at 550 nm and 800 nm while the $\sigma_{y'}$ are 160 and 221 μrad, and the $\sigma_{x'}$ are 123 μrad and 122 μrad. From the images the unwanted background is also clearly visible on the left side.

To further check our results we compared the retrieved values of the beam dimension obtained by ODRI with conventional measurement with single pulse imaging an aluminated OTR screen. In all the cases the agreement is better than 10

5. Conclusion

The optical diffraction radiation interference (ODRI) can be successfully used as non intercepting beam size monitor. Moreover, the beam angular divergence can be retrieved as well from the analysis of the radiation angular distribution. In this paper we extend our analysis considering angular distributions produced at different wavelengths and with different polarization states. The comparison between different wavelengths shows a very good agreement for the vertical polarization component. Also the cross check between the two polarization states is totally satisfactory being the difference well below the 10 %. For the first time the horizontal polarization of ODR has been observed. An unwanted bright spot is visible in this case, probably originating from an unshielded background component. Further investigations are needed in order to understand the reason.

Acknowledgments
We thank DESY for the opportunity to perform our experiment at the FLASH facility. Special thanks are given to the FLASH operators for their help to setup the machine. We thank also Christian Wiebers for his important contribution to the mechanical construction of the experiment setup and Hans-Christian Schrder for the design of the apochromatic lens.

References
[1] Ter-Mikaelian 1972 *High Energy electromagnetic Process in Condensed Media* (New York, Wiley-Interscience)
[2] Castellano M 1997 Nucl. Instrum. Methods Phys. Res., Sect. A **394**, 275.
[3] Karataev P, Araki S, Hamatsu R, Hayano H, Muto T, Naumenko G, Potylitsyn A, Terunuma N and Urakawa J. 2004 Phys. Rev. Lett. **93** 244802.
[4] Karataev P, Araki S, Hamatsu R, Hayano H, Muto T, Naumenko G, Potylitsyn A, Terunuma N and Urakawa J 2005 Nucl. Instrum. Methods Phys. Res., Sect. B **394** 158
[5] Muto T, Araki S, Hamatsu R , Hayano H, Hirose T, Karataev P, Naumenko G, Potylitsyn A and Urakawa J 2003 Phys. Rev. Lett. **394** 104801
[6] Naumenko G A 2003 Nucl. Instrum. Methods Phys. Res., Sect. B **394**, 184
[7] Cianchi A, Castellano M Catani L, Chiadroni E, Honkavaara K. and Kube G. 2011 Physical Review Special Topics - accelerators and beams **14**, 102803
[8] Ackermann W et al. 2007 Nat. Photon. **1**, 336
[9] James F 1994 *Minuit reference manual* (Geneva, CERN)

Journal of Physics: Conference Series **357** (2012) 012020 doi:10.1088/1742-6596/357/1/012020

Experimental Research of the Diffraction and Vavilov-Cherenkov Radiation Generation in a Teflon Target

M Shevelev, G Naumenko, A. Potylitsyn and Yu Popov
Tomsk Polytechnic University, Lenina str. 2, Tomsk, 634050, Russia

E-mail: naumenko@tpu.ru

Abstract. Geometry of Vavilov-Cherenkov (VChR) radiation when an electron moves close to a dielectric target is in analogy to diffraction radiation (DR) geometry. In this case we may expect DR generation from the upstream face of the target besides that VChR. The joint observation of these booth types of radiation is very interesting from the pseudo-photon viewpoint, which is applicable for relativistic electrons. Unexpected results obtained in our experiment insist on reflection about nature both DR and VChR. The experiment was performed on the relativistic electron beam of the microtron of Tomsk Polytechnic University.

1. About problem

The nature of Vavilov-Cherenkov radiation (VChR) in dielectric targets, as well as the transition radiation, traditionally is considered as a dynamical polarization of the target material with macroscopic permittivity ε by an electron electromagnetic field, followed by the propagation of this excitation in a target with velocity $v = c/n$, where c is the speed of light in vacuum, n is a refractive index [1]. In traditional representation, VChR appears when a particle moves in a medium with a velocity higher than the phase velocity of an electromagnetic wave in this medium. However, for high-energy electrons with the Lorenz factor $\gamma \gg 1$, the characteristic transverse size of an electron field $\gamma\lambda$ can be macroscopic (here, $\lambda = 2\pi c/\omega$ is the wavelength). Under these conditions, VChR can appear when the electron moves near the target [1–3] (see figure 1). Therewith the radiation angular distribution is satisfied to the well-known condition $\cos\theta_{ch} = 1/n\beta$, where β is the electron velocity in units of c.

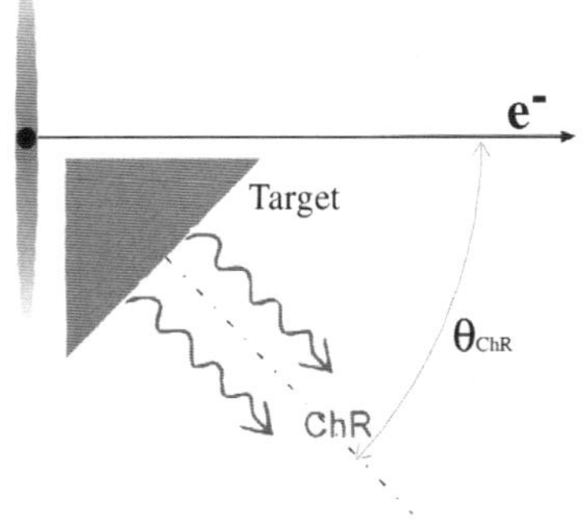

Figure 1. Geometry of VChR emission.

Published under licence by IOP Publishing Ltd

By analogy with transition radiation, the appearance of diffraction radiation (DR) should be expected due to the interaction of the field of the electron with the edge of the dielectric target; this radiation propagates inside the target in an angular cone that is determined by the velocity of the particle and the refractive index along the direction of electron motion (see figure 2).

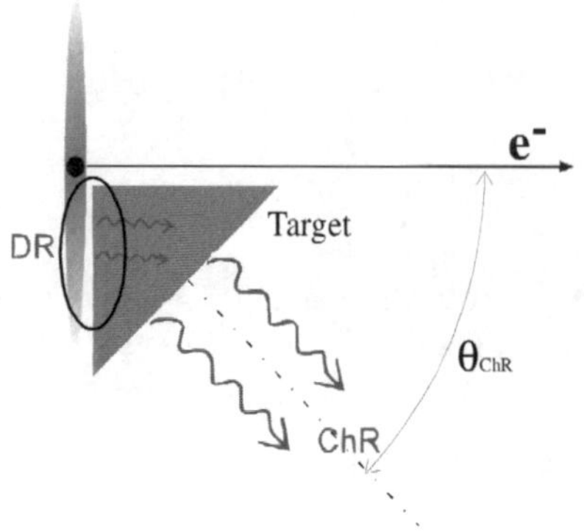

Figure 2. Geometry of VChR and DR simultaneously emission.

On the other hand, the pseudo-photons viewpoint may be also used for analyzes of an interaction of the relativistic electron electromagnetic field with matter. The "pseudo-photon" method proposed by Fermi [4] and developed by Williams [5] is widely used for theoretical studies of electromagnetic processes (see for instance [6] and [7]). According to this approach, the field of a charged particle may be replaced by a field of photons, which in this case are called pseudo-photons (in [8] is used the term "virtual quanta"; this term should be differ from the term "virtual quanta" in quantum mechanics). This approach provides a good accuracy for the ultra-relativistic particles when the particle velocity is close to the light velocity ($v_p \approx c$) (see [6]) and when the longitudinal electric field component of the particle is negligible. In this case the particle field has the same properties as the field of real photons. Using the pseudo-photons viewpoint, the DR in a dielectric target, when an electron moves close to the target, may be simply explained. When pseudo-photons of electron field cross the upstream face of a target, they continue to propagate in a target material like usual photons with the velocity $v = c/n$, taking into account the refraction on the upstream face. However, as for a large n the pseudo-photons velocity is less then electron velocity, pseudo-photons lag behind the electron, and the electron becomes semi-bare.

There is the question, how can we explain the VChR using the pseudo-photons viewpoint? On the first view it seems impossible. This misunderstanding puts in doubt the possibility of application of this viewpoint. The experiment investigations presented in this paper are devoted to resolve this problem.

2. Experimental set-up and procedure

The experiment was carried out in the extracted electron beam of the Physical and Technical Institute of Tomsk Polytechnic University microtron with parameters presented in Table 1.

The experiment was performed on the extracted electron beam of the microtron. The beam is extracted from the vacuum chamber through a 50 μm thick beryllium foil. The beam parameters are listed in Table 1.

Table 1. Electron beam parameters.

Electron energy	6.1 MeV ($\gamma=12$)	Bunch period	380 psec
Train duration	$\tau \approx 4\mu$sec	Bunch population	$N_e = 6\cdot10^8$
Bunches in a train	$n_b \approx 1.6\cdot10^4$	Bunch length	$\sigma \approx 1.9$~2.4 mm

For the radiation measurements the room temperature detector DP20M (figure 3) was used, with parameters described in [9]. Its main elements are a low-threshold diode, a broadband antenna and a preamplifier.

Figure 3. Detector DP20M.

The detector efficiency in the wavelength region $\lambda = 3\sim16$ mm is estimated to be constant to *a* ±15 % accuracy. The detector sensitivity is 0.3 V/mWatt. A wave-guide with a cutoff λ_{cut}=17 mm was used to cut the long-wave background of the accelerator RF system. The high frequency limit of the wavelength interval is defined by the bunch form-factor. This limit (λ_{min} = 9 mm) was measured using discrete wave filters [10] and a grating spectrometer. The coherent radiation intensity for $\lambda > 9$ mm) is by 8 orders larger than the incoherent one and has a power level =1 Watt per steradian. It means that one can investigate coherent radiation in this wavelength range without difficulty.

To exclude the prewave zone effect [11] a parabolic telescope (see figure 4) was used for the angular distribution measurement. This method was suggested and tested in [9] and gives the same angular distribution as in the far field zone ($R \gg \gamma^2\lambda$).

The scheme of experimental set-up is shown in figure 4. Teflon target in the form of a triangular prism with dimensions 175×175×74 mm was used in experiment. The target was placed at a distance of 220 mm from the extraction window; the distance from the target to the electron beam (impact parameter) was 20 mm. The refractive index for Teflon target was measured using the method described in [11] and was equal to $n = 1.41 \pm 0.1$ in the wavelength range from 10 to 20 mm.

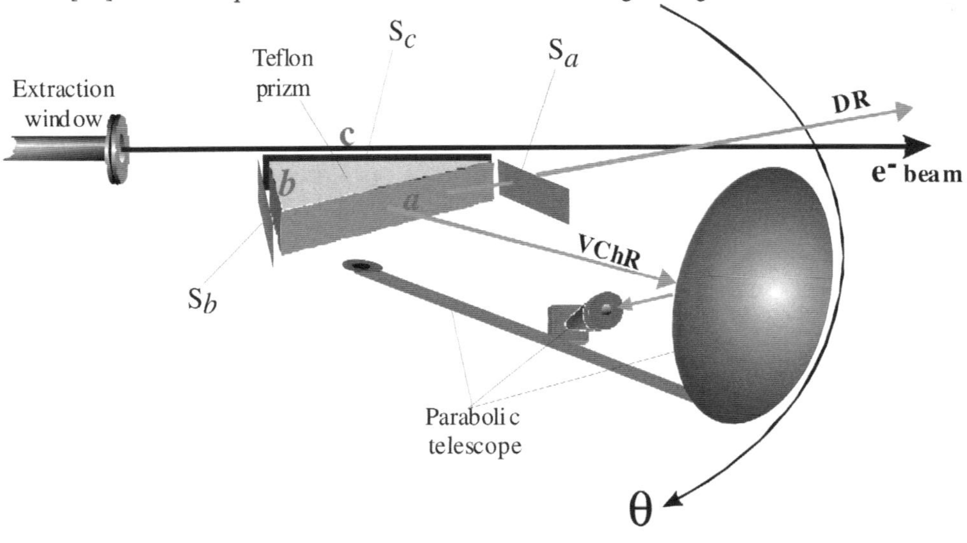

Figure 4. Scheme of VChR and DR simultaneously measurement. S_a, S_b and S_c are the removable conductive screens. *a*, *b*, *c* are the vertical faces of the target.

In experimental analysis, if necessary were used alternately conductive screens S_a, S_b and S_c (see figure 4). Screens S_b and S_c were set close to the face of the prism.

3. Measurements

Measurements of the coherent radiation intensity on the angle θ were performed with a step 0.5. In figure 5 circles shows the measured radiation intensity on the angle without screens S_a, S_b and S_c. The angle $\theta = 0^\circ$ corresponds to the direction of the electron beam. The peak at the angle $\theta = 44^\circ$ corresponds to VChR.

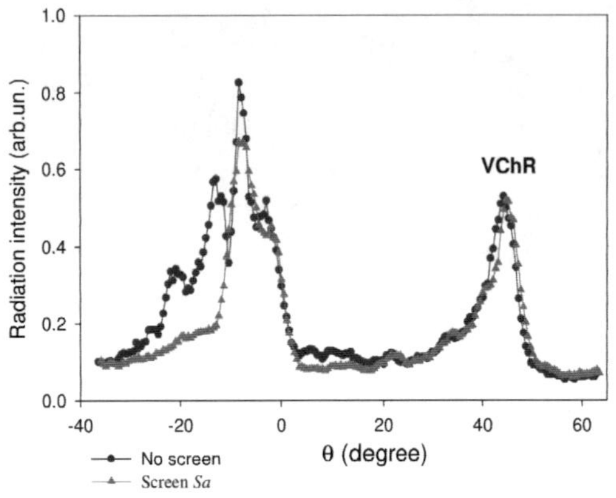

Figure 5. Measured angular distributions including screening of the face *a*.

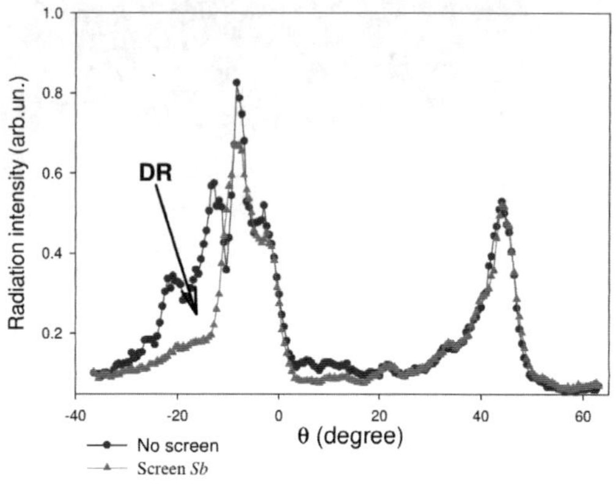

Figure 6. Measured angular distributions including screening of the face *b*.

For the second step, the screen S_a was installed. The radiation angular distribution for this geometry is shown in figure 5 by triangular points. It is clear, that the difference between these distributions (at the angle $\theta \approx -20^\circ$) is a radiation from the downstream face *a* of the target. The peak at the angle $\theta = -8^\circ$ corresponds to a backward transition radiation from the parabolic mirror.

To clarify the nature of the radiation at the angle $\theta \approx -20^\circ$, the screen S_a was removed and screen S_b was installed. The radiation angular distribution for this geometry is shown in figure 6 by triangular points. The curve showed in this figure by circles, is the angular distribution without screens. One may see, that figures 5 and 6 are almost the same. Taking into account the shadowing effect from screen S_b (see [13]), we can conclude, that the radiation at the angle $\theta \approx -20^\circ$ is a DR from the upstream face *b* of the target, which propagates inside the target and is refracted on the downstream face *a*.

One should be called attention to very important fact: the screen S_b screens the DR but does not exert influence on the VChR (peak at angle $\theta = 44^\circ$ in figure 6). Below we will turn back to this problem.

For more detail analyzes of the nature of radiation we had removed the screen S_b and installed screen S_c (see figure 4). The measured radiation angular distribution is shown in figure 7 (triangular points). For comparison the angular distribution without screens is shown by circles.

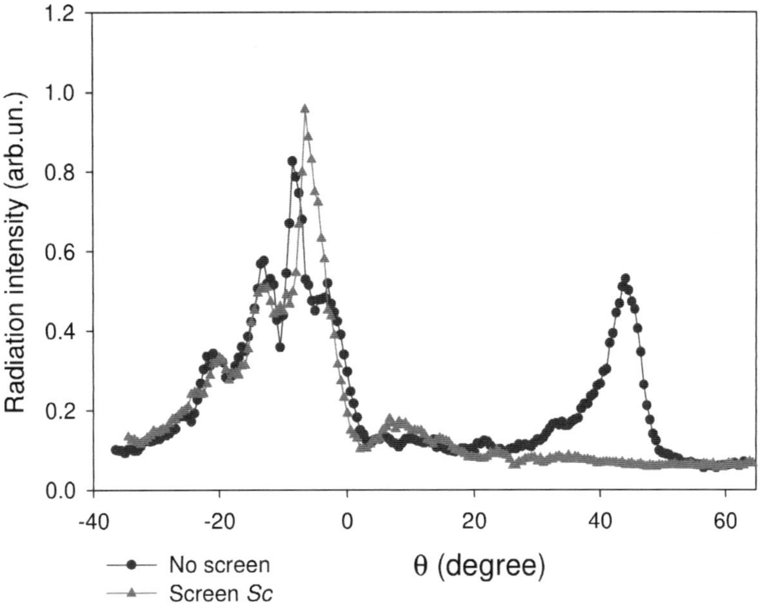

Figure 7. Measured angular distributions including screening of the face *c*.

One may see in figure 7, that in this geometry the VChR disappears. However the screen S_c does not exert influence on the DR.

4. Discussion

Let us make the detail analysis of obtained experimental results. First of all the experiment results show the simultaneous generation of DR and VChR inside a target, and in doing so the DR is generated from the upstream face of the target, because the screen S_c does not exert influence on the DR, but the screening of the upstream face *b* (figure 4) suppresses a DR generation (according to [14] the surface currents are not induced on a downstream surface of a thick conductive screen, therefore radiation is not emitted from this screen in forward direction). This deduction is in agreement with both traditional and pseudo-photon viewpoints. It is clear that the electron field area up to radius about $\gamma\lambda$ affects on a DR generation.

Another generation mechanism we see for VChR. The screening by screen S_b does not exert influence on the VChR. Ie VChR is not a result of bulk dielectric polarization. We can conclude from result, shown in figure 7, that VChR is generated from the face *c* of the prism, which is parallel to the electron beam trajectory. According to the pseudo-photons viewpoint, when pseudo-photons of electron field cross the upstream face of a target, they continue propagate in a target material like usual photons with the velocity $v = c/n$. For a large n the pseudo-photons velocity is less then electron velocity and pseudo-photons lag behind the electron, causing DR. In so doing, the electron becomes semi-bare and a shadow appears directly downstream to the face *b* (in figure 4) of the prism (see [13]). According to [15] this state is unstable and the electron field is recovered to the usual state at the distance about $\gamma^2\lambda$. In the recovery process, the recovered electron field is refracted immediately inside the prism and also lags behind the electron. This process is continuous throughout the length of the target along the face *c*. In so doing, the refracted pseudo-photons become the sources of VChR for a direction, which is satisfied to the Vavilov-Cherenkov condition.

Finally we can conclude that macroscopic Maxwell equations may be used (in principle) for calculating the radiation emitted by relativistic electrons passing through or near material targets. Nevertheless phenomenological concepts like *pseudo-photons* and *semi-bare electron* sometimes are more useful for an intuitive understanding of the main features of radiation mechanisms. Our experimental analysis of radiation from a dielectric prism, when electrons move close to some face of the prism, shows that the VChR can be explained in frame of pseudo-photon viewpoint together with the *semi-bare electron recovery* concept of Feĭnberg [15]. In principle, as it was shown above, the *semi-bare electron* concept follows from the pseudo-photon viewpoint, but the interpretation of the electron field recovery process, needs the additional justification.

Acknowledgment
This work was partly supported by the warrant-order 1.226.08 of the Ministry of Education and Science of the Russian Federation and by the Grant RFBR 11-02-91177-GFEN_a.

References

[1] Bolotovskii B M 1961 *Sov. Phys. Usp.* **4** 781
[2] Takahashi T, Oyamada M, Kondo *et al.* 2000 *Phys. Rev.* **62** 8606
[3] Potylitsyn A P, Popov Yu A, Sukhikh L G *et al. J. Phys.: Conf. Ser.* **236** 012025
[4] Fermi E 1924 *Zh. Phyz* **29** 315
[5] Williams E J 1935 *Kgl. Danske Videnskab. Selskab. Mat.-fys. Medd.* **13** 4
[6] Ter-Mikaelian N L 1972 *High-Energy Electromagnetic Processes in Condensed Media* (Wiley, New York)
[7] Ter-Mikaelian N L 1969 *The Influence of the Medium on High Energy Processes at High Energies* (Publishing House of Academy of Science of Armenia, Yerevan)
[8] Jackson J D 1998 *Classical Electrodynamics, 3rd ed.* (J. Willey&Sons, New-York)
[9] Kalinin B N, Naumenko G A, Potylitsyn A P *et al.* 2006 *JETP Letters* **84** 3 110-114.
[10] Hanke K 19.04.1996 *CLIC note* **298**
[11] Verzilov V A 2000 *Phys. Lett. A* **273** 135
[12] Shevelev M V, Naumenko G A, Potylitsyn A P 2009 *Izvestiya Vuzov, Fizika* 52 11/2 261
[13] Naumenko G, Artru X, Potylitsyn A *et.al.* 2010 *Journal of Physics: Conference Series* **236** 012004
[14] Naumenko G, Potylitsyn A, Popov Yu et al. 2010 *Journal of Physics: Conference Series* **236** 012024.
[15] Feĭnberg E L 1979 *Sov Phys Usp* **22** 6 479

Investigation of Coherent Diffraction Radiation from a Dual Target system at CTF3.

K. Lekomtsev (1), **G. Boorman** (1), **R. Corsini** (2), **P. Karataev** (1) **and T. Lefevre** (2)

(1) John Adams Institute at Royal Holloway University of London, Egham, Surrey, TW20 0EX, UK.
(2) CERN CH-1211, Geneva 23, Switzerland.

E-mail: Konstantin.Lekomtsev.2009@live.rhul.ac.uk

Abstract. A Coherent Diffraction Radiation (CDR) experiment was built at the CLIC Test Facility 3 at CERN. Two silicon wafers are positioned on one side of the beam and the CDR radiation originating from them is translated towards the Michelson Interferometer. We have demonstrated that the first target cuts off the backgrounds, generated by the beam from upstream of the experimental setup. The results of the theoretical and experimental investigations of the CDR from the two targets and a work towards longitudinal diagnostics of sub-picosecond electron bunches is presented. The radiation spatial distribution measurements were performed and compared with the theory. A capability to obtain interferometric measurements of CDR at the experimental setup was proven. Bunch shape instabilities at CTF3 were investigated, as well.

1. Introduction

A novel scheme for the drive beam generation has been proposed for the future Compact Linear Collider (CLIC), in which a long bunch train with a low bunch repetition frequency will be accelerated with a low RF frequency and, then, recombined to increase the bunch repetition frequency from 1.5 GHz to 12 GHz. The optimisation and monitoring of the longitudinal charge distribution in the bunch is crucial for an optimal performance of the CLIC drive beam as well as for maximization of the luminosity at the CLIC interaction point [1].

Diffraction Radiation (DR) is a physical phenomenon occurring when a charged particle moves in the vicinity of a conducting screen. The coulomb field of the particle interacts with the screen and surface currents are produced, which become a source of the DR. The first theoretical approach to the DR was developed in 1950s, however, the first experimental investigations were performed only in 1995 by Shibata et al. [2]. Using an electron beam with the energy of 150 Mev, the DR was generated by a circular aperture in millimetre and submillimetre wavelength regions.

The DR generated from conducting screens has been experimentally investigated in the last 15 years [3, 4, 5] and it was proven to be very promising tool for the longitudinal [6, 7] and transverse [8, 9] diagnostics of electron beams. The main advantage of DR compared to transition radiation, for instance, that it does not affect the beam. Other important property of this physical phenomenon is instantaneous emission, which makes it possible to perform single shot, time resolved measurements. Experiment-wise, one is interested in observation of the Coherent

Diffraction Radiation (CDR), which is N times more intensive than incoherent radiation, where N is the number of particles in the beam. The coherence of the effect is achieved when electrons in a bunch radiate in phase, i.e. the wavelength of radiation is comparable or larger than the bunch length.

At the CLIC Test Facility 3 we have built an experimental setup based on the CDR from a dual target system. Two silicon wafers, coated with aluminium, are positioned on one side of the beam and the radiation, generated from them, is transferred towards the Michelson interferometer. The upstream target serves as a barrier for the Coherent Synchrotron Radiation (CSR) backgrounds coming from upstream of the experimental setup (see figure 1). The ultimate goal of the experiment is to reconstruct longitudinal parameters of the bunch from the spectral measurements obtained at the Michelson interferometer. In the current configuration of the setup ultra-fast Schottky Barrier Diode (SBD) detectors are used for the spectral measurements. The intensity of the CDR is defined by the following equation:

$$S(\omega) = S_e(\omega) N^2 F(\omega), \tag{1}$$

where $S(\omega)$ is the measured spectrum, $S_e(\omega)$ is the single electron spectrum which has to be predicted precisely , N is the number of electrons in the bunch, $F(\omega)$ (the goal of the measurements) is the bunch form factor which contains information about the longitudinal distribution of electrons in the bunch.

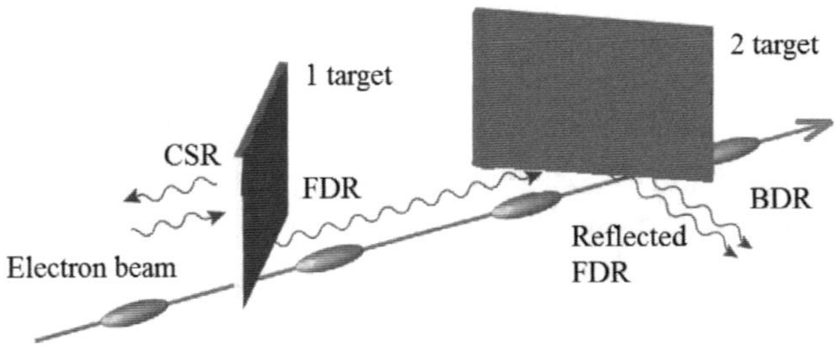

Figure 1. Schematic layout of the Dual Target System at CTF3. FDR - Forward Diffraction Radiation; BDR - Backward Diffraction Radiation; CSR - Coherent Synchrotron Radiation.

In this paper we concentrate on the investigation of the CDR from a dual target configuration. A theoretical model, based on the Classical Diffraction Radiation theory, was developed to investigate a geometry of the DR spatial distribution generated by a single electron while interacting with the targets. The CDR spatial distributions were obtained using the two SBD detectors and the measurements were compared with the theory. In order to understand the background contribution in the experimental data better, the dedicated measurement of the CSR reflection from the downstream target was performed. A capability to obtain interferometric measurements of CDR at the experimental setup was proven. Bunch shape instabilities at CTF3 were also investigated.

2. Calculation of Coherent Diffraction Radiation from the dual target system

A detailed description of the theoretical model was presented in [10]. In this section the model and the key dependences, used for the comparison with our experiment, are shown. For the calculations we shall use a classical theory of Diffraction Radiation, based on Huygens principle of plane wave diffraction. In reality the classical DR theory describes backward DR only. However, for a metallic foil and millimetre wavelengths we can use an ideal conductor approximation. In this case backward DR (BDR) characteristics coincide with forward DR (FDR) ones. A particle field is introduced as a superposition of its pseudo-photons and when they are scattered off a target surface, they are converted into the real ones and propagate either in the direction of a specular reflection (BDR) or along the particle trajectory (FDR).

Consider an electron moving along the z-axis. Each point of the target surface can be represented as an elementary source of the radiation. Two polarization components of the DR can be represented as a superposition of the waves from all elementary sources at a distance **r** from the target [11]:

$$E_r^{DR} = -\frac{1}{4\pi^2}\iint E'(x_r, y_r)\frac{e^{ik|\mathbf{r}|}}{r}dx_r dy_r. \tag{2}$$

$E'(x_r, y_r)$ is the amplitude of the radiation source positioned on the target surface, (x_r, y_r) are the coordinates of the radiation source on the target surface, r is the distance from the radiation source at the target to the observation point and k is the wavenumber.

The process of the CDR emission is considered in two main steps: firstly, propagation of the FDR from the first target towards the second one, diffraction at the second target and propagation towards the observation plane, and secondly, propagation of the BDR from the second target towards the observation plane (see figure 1). The DR spatial distribution in general form can be written as:

$$\frac{d^2W^{DR}}{d\omega d\Omega} = 4\pi^2k^2a^2\left(|E_x^{DR}|^2 + |E_y^{DR}|^2\right), \tag{3}$$

where E_x^{DR} and E_y^{DR} are the horizontal and vertical polarisation components of E_r^{DR}, respectively.

Once the two radiation components are obtained, we can derive the DR distribution from the two targets. We observe interference between the BDR from the second target and the FDR from the first target, therefore the following formula for the radiation spatial distribution is applied:

$$\begin{aligned}\frac{d^2W_r^{DR}}{d\omega d\Omega} &= 4\pi^2k^2a^2\left[\left(ReE_{r_1} - Re\left[E_{r_2}\exp\left(\frac{ikd}{\beta}\right)\right]\right)^2\right. \\ &\left.+\left(ImE_{r_1} - Im\left[E_{r_2}\exp\left(\frac{ikd}{\beta}\right)\right]\right)^2\right],\end{aligned} \tag{4}$$

where E_{r_1} is the FDR from the first target diffracted from the second one and E_{r_2} is the BDR from the second target, r_1, r_2 are the surfaces of the targets, $\beta = v/c$ is the speed of an electric charge in terms of the speed of light, the $\exp(ikd/\beta)$ defines the phase delay due to the particle moving from the first target to the second one, a is the distance from the second target to the observation plane and d is the distance between the targets.

Understanding of the radiation distribution geometry is crucially important for the optimisation of the experimental setup. In the current configuration the two targets are positioned on one side of the beam, therefore the fields of the DR interfere with each other. The CDR distributions as functions of the second target impact parameter and the horizontal coordinate at the observation plane were calculated for the two main configurations of the experiment.In the first configuration the impact parameter of the first (upstream) target is at 30 mm from the beam and the second target impact parameter is variable in the region of 10 to 30 mm. In the second configuration the first target impact parameter is 10mm and , as in the first configuration, the second target impact parameter is variable in the region of 10 to 30 mm.

In figures 2 and 3 the vertical polarisation component of the CDR spatial distribution is presented for the two main configurations of the setup, the distributions were calculated for the parameters shown in table 1. In figure 2 the radiation distribution was calculated for the first configuration of the targets as a function of the second target rotation and impact parameter. In figure 3 the CDR distribution was calculated for the second configuration of the targets as a function of the second target rotation and impact parameter. Zero degree rotation in the figures corresponds to the mirror reflection from the downstream target. The distributions have a single mode nature and the radiation intensity decreases while the second target is gradually lifted up, i.e. h_2 increases. In figure 3 the maximum intensity is suppressed by approximately 50% due to the destructive interference between the targets.

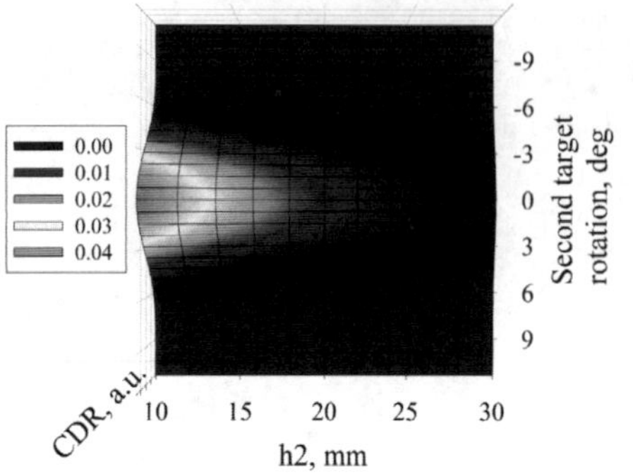

Figure 2. Vertical polarisation of the CDR distribution for the first configuration of the targets. The first target impact parameter $h_1 = 30$ mm.

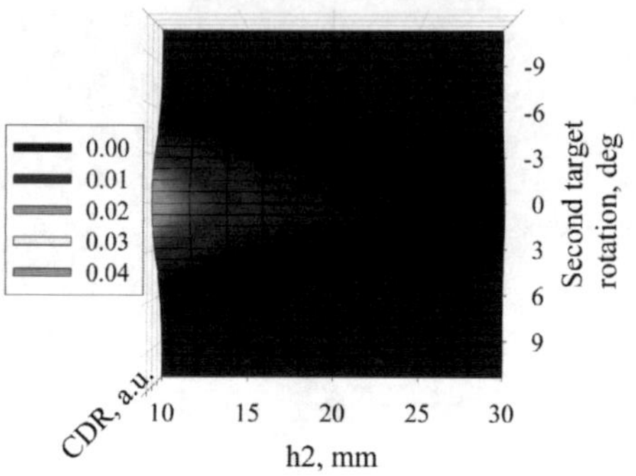

Figure 3. Vertical polarisation of the CDR distribution for the second configuration of the targets. The first target impact parameter $h_1 = 10$ mm.

3. Experimental investigation of the CDR from a Dual Target system

3.1. Experimental setup

The experimental setup is located at the Combiner Ring Measurements (CRM) line of the CLIC Test Facility 3 (CTF3). The CRM line is a straight section deviated from the first quarter of the Combiner Ring (figure 4). Once the bending magnet, located at the entrance to the CRM line, is switched off, the beam is directed straight into the beam dump passing the dual target system and the CDR from the targets can be observed. If the bending magnet is switched on, the CSR, generated upstream of the experimental setup, can be reflected by the downstream target and measured by the same detection system.

Two six-way crosses are installed in the CRM line, they contain two targets, which are positioned above the beam at 45deg with respect to the beam propagation direction. The targets are silicon wafers coated with aluminium. They are mounted on the Ultra High Vacuum (UHV) manipulators, which provide the vertical translation and rotation of the second target and a vertical translation of the first target.

The radiation, originated from the targets or reflected from them, is transferred through a CVD diamond window with a viewing diameter of 30 mm. Diamond exhibits a broadband transparency in the far-infrared and millimetre wavelength range. The CVD window has a thickness of 0.5 mm, which is smaller than or comparable to the observation wavelength and, therefore, the viewport absorption and the distortion of the transmitted radiation spectrum due to multiple reflections are minimized. The radiation, originated from the targets, is translated vertically by a periscope towards the floor to avoid X-ray backgrounds from the horizontal particle beam plane. On the optical table a Michelson interferometer is installed (see figure 5).

The polariser (P1) transmits the vertical polarisation component towards the interferometer and reflects the horizontal polarisation component toward the second detector, positioned

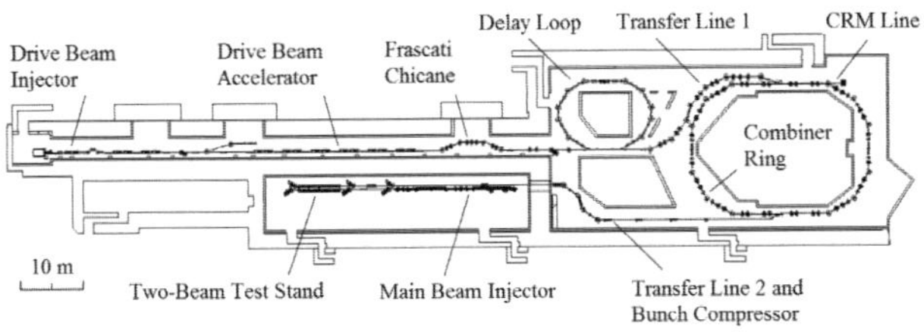

Figure 4. Schematic layout of the CLIC Test Facility 3.

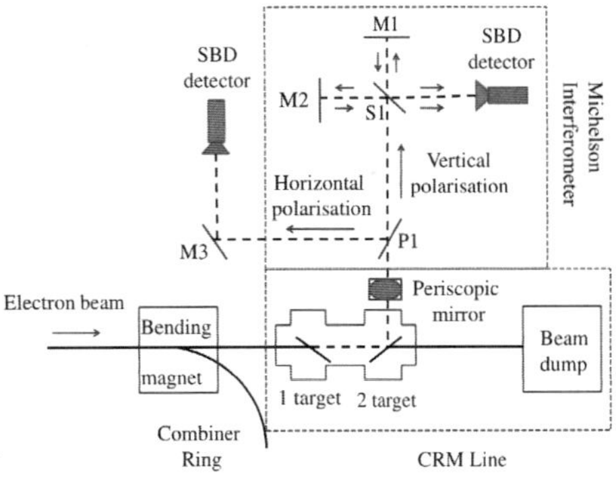

Figure 5. Schematic layout of the CDR experiment.

Table 1. CTF3 and CDR experiment parameters.

Parameter	Value	Unit
Beam energy (γ)	235	
Bunch charge	2.3	nC
Bunch spacing frequency	3	GHz
Target dimensions (projected)	40x40	mm
Observation wavelength (λ)	5	mm
First target impact parameter (h_1) (upper position)	30	mm
First target impact parameter (lower position)	10	mm
Second target impact parameter (h_2)	10	mm
Distance between the targets (d)	0.25	m
Distance from the second target to the observation plane (a)	2	m

beside the interferometer and used for bunch shape instability studies. The parameters of the experimental setup and CTF3 are presented in table 1.

In the experiment the ultra fast Schottky barrier diode (SBD) detectors with a response time less than 1ns are used to measure power in the narrow frequency ranges. The detectors are sensitive in 50 - 75 GHz (DXP15) and 60 - 90 GHz (DXP12). They are polarisation sensitive and only one detector can be used at a time for the interferometric measurements.

Data acquisition is performed using a 10-bit Acqiris DC282 digitiser. The digitiser can provide four channel sampling at up to 2 GS/s, or dual and single-channel sampling at 4 Gs/s and 8 Gs/s, respectively. In the current configuration 4 Gs/s sampling for two channels is used. An external trigger provides a precise synchronisation with the electron beam. The data is read out for every single bunch train and transferred to the digitiser through high quality RF cables.

3.2. Measurements of the CDR spatial distribution

In the experimental setup the radiation is transferred through a periscope mirrors towards the optical table where it is split into two polarisation components, the vertical component is translated towards the beam splitter and further on, when recombined, towards the detector in the interferometer. In order to understand the spatial geometry of the radiation detected in the

Journal of Physics: Conference Series **357** (2012) 012021 doi:10.1088/1742-6596/357/1/012021

interferometer, the measurements of the CDR spatial distributions as functions of the second target rotation angle and the impact parameter were performed for the two main configurations of the experimental setup. For the first configuration the first (upstream) target is located 27 mm away from the beam, the second target position is in the range of the discrete points of 7 to 25 mm away from the beam. Later on, the scan is normalized by the current. The similar scan is performed for the second configuration of the setup when the first target is 7 mm away from the beam.

In figure 6 the CDR distribution for the first configuration, measured by the detector sensitive in 50-75 GHz, is presented. The second target travelling range in the horizontal plane is from 7 to 23 mm. It has a single mode shape. The distortions caused by the backgrounds coming from the upstream of the experimental setup are seen in the scan. Figure 8 demonstrates the radiation distribution for the same configuration, but measured using the SBD detector sensitive in 60-90 GHz spectral range. The second target travelling range in the horizontal plane is from 7 to 25 mm. When compared with the distribution in figure 6, the latter scan shows a smoother radiation distribution, most likely due to the milder background conditions at the time of the measurement.

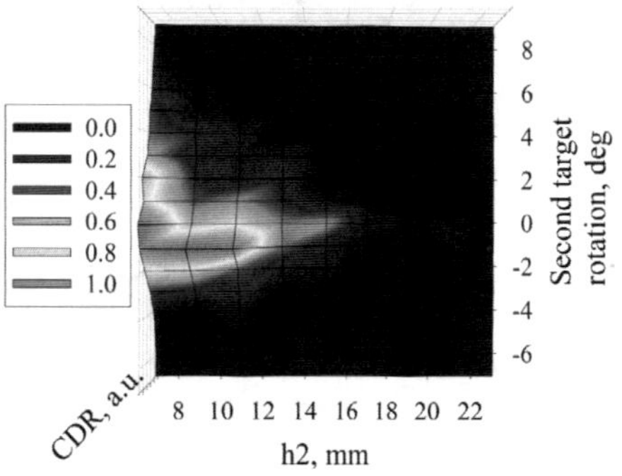

Figure 6. Vertical polarisation of the CDR distribution measured by DXP15 detector. The first target impact parameter $h_1 = 27$ mm.

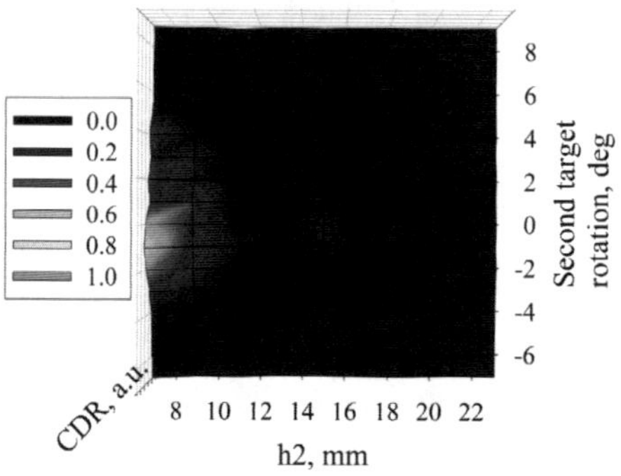

Figure 7. Vertical polarisation of the CDR distribution measured by DXP15 detector. The first target impact parameter $h_1 = 7$ mm.

In figures 7 and 9 the first target is positioned 7 mm away from the beam, consequently the intensity of the radiation decreases as a result of destructive interference between the two targets. In this configuration the first target also serves as an obstacle for the coherent backgrounds coming from the upstream.

The distributions for the first and the second target configurations qualitatively agree with the theoretical calculations: figures 6 and 8 correspond to the theoretically calculated distribution in figure 2; figures 7 and 9 can be compared with figure 3. The theoretical CDR distributions were calculated for the wavelength $\lambda = 5$mm. The detectors in use sensitive and demonstrate a flat frequency response in the following wavelength regions: DXP15(4-6 mm) and DXP12(3.33 - 5 mm). In the region of 3.33 - 6 mm the CDR spatial distribution does not change considerably and, therefore, the radiation distribution calculated for λ=5 mm can be compared with the experimental results. All four experimental scans demonstrate a general agreement with the theory: they have a single mode nature, the maximum intensity of the radiation decreases when the first (upstream) target is positioned closer to the beam and the intensity of the CDR gradually decreases while the second target impact parameter increases. However, in figure 6 one may see the asymmetry in the distribution, as compared with the experimental result in figure 8 and theoretically calculated CDR distribution in figure 2, which is yet to be understood.

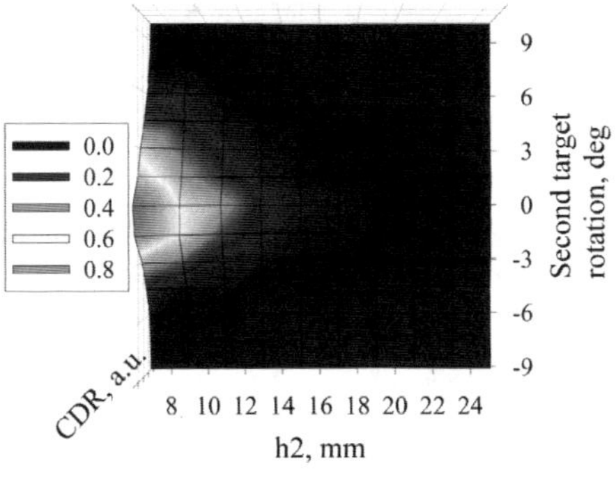

Figure 8. Vertical polarisation of the CDR distribution measured by DXP12 detector. The first target impact parameter $h_1 = 27$ mm.

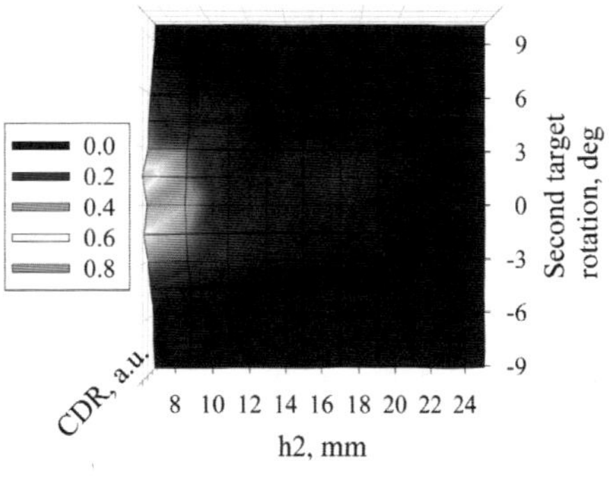

Figure 9. Vertical polarisation of the CDR distribution measured by DXP12 detector. The first target impact parameter $h_1 = 7$ mm.

4. Coherent Synchrotron Radiation backgrounds

Coherent backgrounds coming from the upstream of the experimental setup significantly complicate the experimental data interpretation. A dedicated measurement of the Coherent Synchrotron Radiation (CSR) reflection from the downstream target was performed. At a time of the measurement the bending magnet at the entrance to the CRM line was switched on, so the beam circulated in the Combiner Ring and generated CSR, which propagated into the CRM line.

The configuration of the setup makes it possible to measure the reflection of the CSR from the second target. By rotating it one can measure the intensity of the radiation as a function of the rotation angle. When the upstream target is at its highest position, it is far from the center of the six-way cross and does not block the CSR. However, when it is gradually lowered down a portion of the oncoming radiation is cut off and not reflected towards the detector.

In figure 10 the four rotation scans of the CSR reflection are presented. The red curve shows the CSR reflected from the second target when the first target is at its highest position and the second target in the centre of the six-way cross. The green curve corresponds to the configuration of the setup when both targets are at the centre of a six way cross and the maximum CSR suppression should be achieved. It is clearly seen that intensity suppression with the factor 5 achieved by positioning the first target in the centre of the beam pipe, however, some intensity is still observed for this configuration. This might be due to the residual reflections in the coupled six-way crosses. The residual part of the CSR background can be reduced by putting the targets closer together.

5. Interferometric measurements

The Michelson interferometer is used to perform spectral measurements of the CDR, originated from the targets. The measurements are performed in the following way: a detector in the interferometer registers vertical polarisation component of the radiation; the DAQ system reads out a signal from the detector for each train of bunches; an interferometric scan is performed by moving the mirror M1 (figure. 5) on the translation stage; three to ten signal readouts per position of the M1 are performed to collect statistics. A length of the interferometric scan can

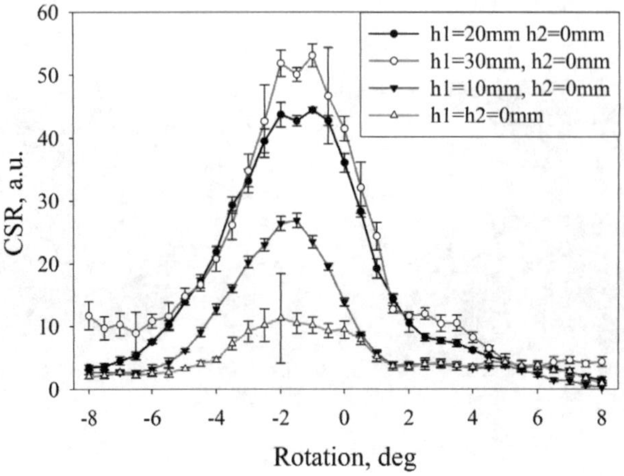

Figure 10. The rotation scans of the CSR reflection from the second target.

vary and it was limited to the maximum travelling range of the translation stage (100mm for UTS100 stage, which was used in the measurements).

Each point of the interferogram is obtained by calculating an average integrated intensity of the CDR radiation over a number of shots, obtained for each position of the moveable mirror in the Michelson interferometer using the DXP15 detector. The integration for each signal is performed over a narrow slice of the pulse (approximately 25 ns), this was done because the bunch length variation throughout the train is rather severe. A sample interferogram, obtained while both targets are at the same distance from the beam, is presented in figure 11. A sample signal of the detected radiation with the marked integration region is presented in figure 12.

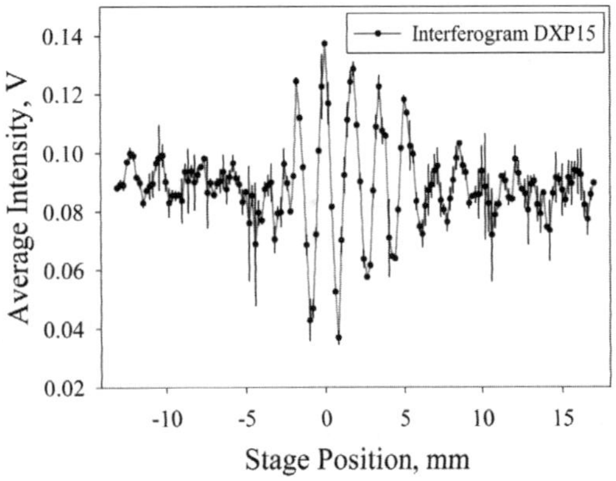

Figure 11. A sample interferogram obtained for the dual target system: $h_1 = h_2 = 7$ mm.

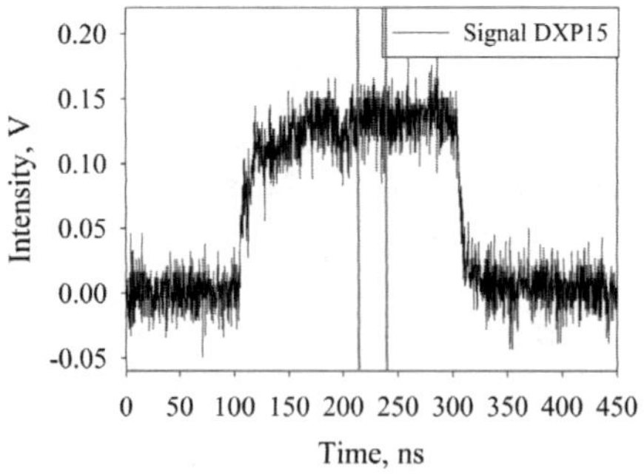

Figure 12. A sample signal from the DXP15 detector.

The measurement was performed using DXP15 detector when both target were at 7 mm from the beam; the stage traveling range for the measurement was $x = 30$ mm which resulted in the path difference of $z = 2x = 60$ mm and the corresponding signal delay in the interferometer of $T = z/c = 200$ ps. This yields the maximum spectral resolution of 5 GHz. The resulting spectrum is shown in figure 13, where the red line marks a low frequency cut-off, defined by the detector. The DXP15 detector is sensitive in 50-75 GHz, which gives five spectral data points in this frequency region. Due to a very narrow frequency region where the data were obtained, this measurement has a very limited applicability for a bunch profile reconstruction.

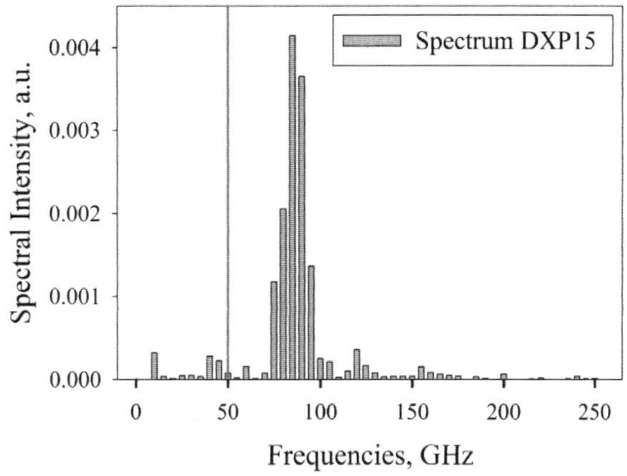

Figure 13. Spectrum obtained from the interferogram in figure 11.

A bunch length stability along the train is a key to successful spectral measurements. When the bunch length vary along the pulse, the CDR is generated at different wavelengths from different parts of the pulse, therefore complicating the result interpretation. Additional problematic issue is instability of a bunch shape. At CTF3 bunch length manipulations are performed before the beam is inserted into the recombination rings and after it is extracted from them. In the linac section of CTF3 the bunch length is in the region of 1 to 7 ps, then the beam is directed through the stretching chicane where the bunches are stretched up to 15 to 20 ps to achieve an effective recombination in the Delay Loop and the Combiner Ring. Later on, the bunches are shortened to 1-2 ps to achieve an effective RF transfer in the Power Extraction and Transfer Structures (figure 4). When the beam goes through the Frascati chicane the bunches can become distorted and micro-bunching structures can occur, which results in distortion of the signals, obtained using the high frequency SBD detectors.

Overall, the experimental setup was proved to be capable of taking interferometric data despite significant bunch shape instabilities and the RF drifts in the machine. An immediate solution of a bunch shape variation problem could be installation of the low frequency detectors, which would be less sensitive to the changes of a bunch shape.

6. Conclusions

In this paper we have presented the results on the investigation of the CDR from the dual target system at CTF3. The theoretical model, based on the Classical Diffraction Radiation theory, was developed to theoretically calculate the radiation spatial distributions. The experimental spatial distributions were measured at the CDR setup and demonstrated a general agreement with the theory. Additional studies of the CSR contribution into the measurements were performed as well. Installation of the second target made it possible to effectively cut off a dominant part of the backgrounds coming from the upstream of the experimental setup.

The ultimate goal of the experiment is to reconstruct longitudinal parameters of the bunch from the interferometric measurements obtained at the Michelson interferometer. The experimental results showed that the interferometric measurements can be performed with the new two-target configuration. However, limited bandwidth of the detectors in use limits applicability of the spectral measurements for the longitudinal bunch profile reconstruction.

Bunch length and shape instabilities and their influence on the measurements were discussed. A clear understanding of the problematic issues and the hardware constraints was achieved. The ways towards the further development of the experimental setup were identified: firstly, usage of low frequency detectors with a broader bandwidth to suppress the influence of a bunch shape variation and, secondly, as a more long term-solution, shot-by-shot measurements and installation of a grating spectrometer.

References

[1] Geschonke G et al. 2002 *CTF3 Design report 2002-08* (CERN/PS)
[2] Shibata Y et. al. 1995 *Physical Review* E **52** 6787
[3] Vnukov I E, Kalinin B N, Naumenko G A, Padalko D V, Potylitsyn A P and Chefonov O V 1998 *JETP Letters* **67** 802
[4] Urakawa J et. al. 2001 *Nuclear Instruments and Methods* A **472** 309
[5] Muto T et. al. 2003 *Physical Review Letters* **90** 104801-1
[6] Feng B et. al. 2001 *Nuclear Instruments and Methods* A **475** 492
[7] Castellano M et. al. 2001 *Physical Reiew* E **63** 056501-1
[8] Karataev P et. al. 2004 *Physical Review Letters* **93** 244802
[9] Karataev P et. al. 2005 *Nuclear Instruments and Methods* B **227** 158
[10] Lekomtsev K et. al. 2010 *Il Nuovo Cimento* 34C **N4** 261-268
[11] Ter-Mikaelyan M L 1972 *High Energy Electromagnetic Processes in Condensed Media* (New York: Wiley-Interscience)

Journal of Physics: Conference Series **357** (2012) 012022 doi:10.1088/1742-6596/357/1/012022

PIC code KARAT simulation of different types of polarization radiation generated by relativistic electron beam

K P Artyomov[1,2], V V Ryzhov[1,2], G A Naumenko[3], M V Shevelev[3]

[1] Institute of High Current Electronics SB RAS, Tomsk, Russia
[2] Tomsk Polytechnic University, Tomsk, Russia
[3] Nuclear Physics Institute at Tomsk Polytechnic University, Tomsk, Russia

E-mail: artyomov@to.hcei.tsc.ru

Abstract. Different types of polarization radiation generated by a relativistic electron beam are simulated using fully electromagnetic particle-in-cell (PIC) code KARAT. The simulation results for diffraction radiation, transition radiation, Smith-Purcell radiation and Vavilov-Cherenkov radiation are in a good agreement with experimental data and analytical models. Modern PIC simulation is a good tool to check and predict experimental results.

1. Introduction

Modern scientific installations are very complex devices. Numerical simulation is a way of reducing the development time of installation and production costs. Simulation allows comprehensive analysis of the experimental conditions, investigate many parameters in a very wide range and choose their optimal values.

Particle-in-cell simulation enables you to visualize charged particles interactions with the targets, and also generation, propagation and interference of the radiation fields. The particle-in-cell method allows one to follow the dynamics of charged particles taking into account own fields of particles and external electromagnetic fields for any given geometry.

Using PIC simulations in problems devoted to polarization radiation generation allows to analyze the radiation fields formation in the near field, taking into account the form factor of the targets and spectral composition of the generated radiations.

In this paper the following problems of the polarization radiation generation by a relativistic electron beam will be considered:

- Smith-Purcell radiation. Generation of the resonant diffraction radiation from inclined gratings for bunch length measurements.
- Transition radiation. Coherent backward transition radiation from a bunched relativistic electron beam.
- Diffraction radiation. Generation of the diffraction radiation in two serial conductive targets.
- Vavilov-Cherenkov radiation. Coherent Vavilov-Cherenkov radiation from a bunched electron beam passing near a dielectric target.

All calculations were performed using a desktop PC. Duration of some calculations did not exceed several days.

2. Particle-in-cell code KARAT

KARAT is a fully electromagnetic code based on the particle-in-cell (PIC) method [1]. It is primarily aimed to the solution of non-stationary electrodynamic problems having complicated geometry and involving dynamics of relativistic electrons and non-relativistic ions. In particular, KARAT is well suited to the simulation of high-current electron devices such as vircators, free-electron lasers, gyrotrons, etc. It is also appropriate for the modeling of physical phenomena in laboratory and space plasmas.

There are three versions of the code: one, two and three spatial dimension problems. In all cases three electromagnetic field components and three momentum components are taken into account.

A finite difference scheme with overstepping on the rectangular shearing grid is used to solve Maxwell's equations. The electric field is corrected with the help of Boris' method [2]. The scheme's unique feature is a special grid shear that provides fulfillment of the boundary conditions without any extrapolation.

In addition, the code has several features useful for modeling phenomena associated with charged particle beam devices. One may adjust the code for potential approximation of electromagnetic problems.

The boundaries consist of perfectly matched layers that absorbing outgoing waves from the interior of a computational region without reflecting them back into the interior.

The KARAT code has been compared to analytic solutions and applied successfully for many physical problems. The simulation results are in good agreement with experimental data.

3. PIC code KARAT simulations

3.1. Generation of the resonant diffraction radiation from inclined gratings for the bunch length measurement

In [3] the resonant diffraction radiation from an inclined grating and its application to measure the length of bunches of charged particles was analytically studied. It is well known that the Smith-Purcell radiation (SPR) obeys the dispersion relation, because of this SPR yield depends on the polar angle. Moreover, the coherence condition can occur under certain angles for the corresponding wavelength, which makes it possible to use this effect for the bunch length measurements.

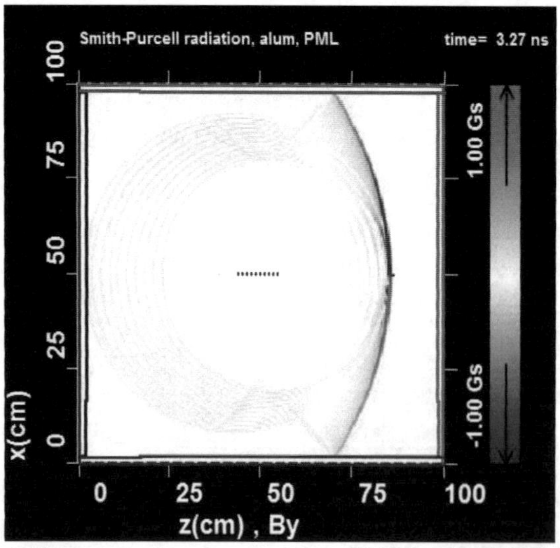

Figure 1. Contour map of B_y component of the electromagnetic field.

The mechanism for bunch length diagnostics based on measuring the radiation intensity by two detectors versus grating inclination angle was experimentally investigated. Both detectors are placed in fixed positions on different sides relative to the beam trajectory.

It was demonstrated that a variation of the grating inclination angle leads to a wavelength shift of the emitted radiation at fixed observation point. The wavelength can easily be tuned by changing the grating orientation with respect to the beam axis.

Figure 1 represents a variant of the modeled geometry. For clarity, grating is 10 times zoomed. The electron beam is injected from the left boundary along the grating. The figure shows the B_y component of the electromagnetic field. It is clearly seen the transition radiation field, which accompanies the electron beam. Transition radiation is generated when the beam injected into the calculation region.

Also in figure you can see the Smith-Purcell radiation. SPR is a spherical wave diverging from the grating. Registration of radiation carried out simultaneously by two detectors placed on both sides of the grating.

Parameters used in simulation: electron beam energy – 100 MeV, bunch length – 3 ps, number of electrons per bunch – 10^8.

Simulation was done using a planar two-dimensional x-z version of the KARAT code.

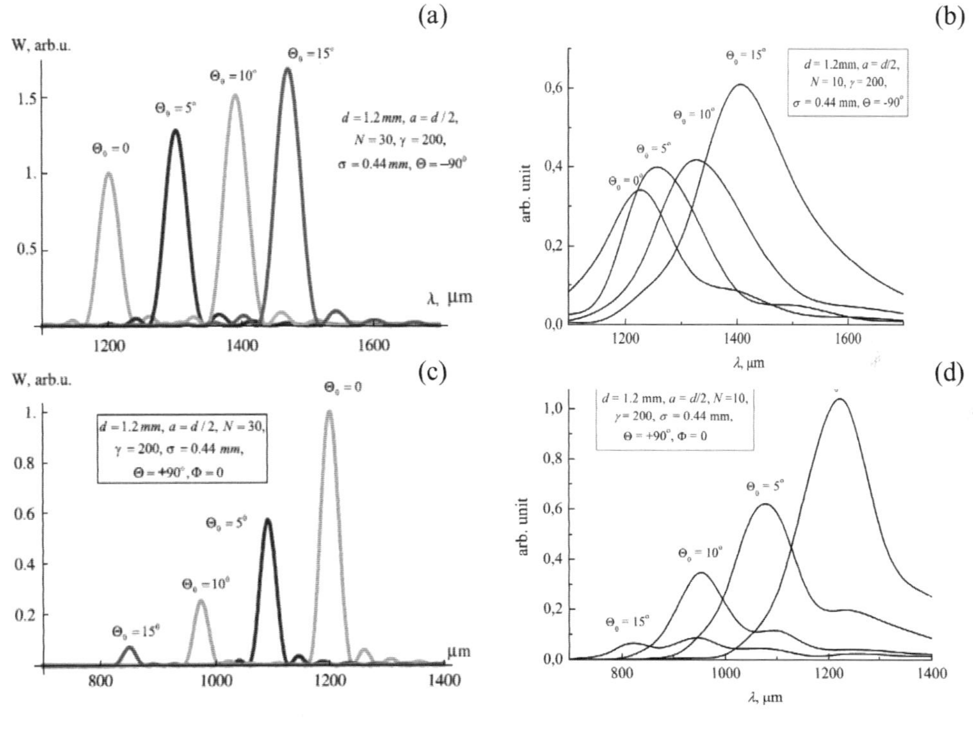

Figure 2. Spectrum of CRDR for different grating inclination angles: (a) and (c) – theoretically calculated, (b) and (d) – simulated by KARAT code.

Comparison of the theoretical calculations and KARAT simulation is shown on figure 2. There is good agreement in position and intensity of spectral lines of the coherent resonant diffraction radiation (CRDR) depending on the inclination angle for both detectors.

The large width of spectral lines in the simulation is caused by a small distance between the grating and detectors.

3.2. Coherent backward transition radiation from a bunched relativistic electron beam

The calculations were performed for the beam parameters, target size and geometry similar to that used in [4]. Actually, the modeling geometry was the same as in the previous problem, but diffraction grating on the path of the electron beam was replaced by a conductive foil. The simulation results were compared with analytical data on the backward transition radiation at a wavelength of 15 mm and the experimental data for the range of 11–17 mm.

Parameters used in simulation: electron beam energy – 6.1 MeV, bunch length – 3 mm, number of electrons per bunch – 10^8.

Simulation was done using a planar two-dimensional x-z version of the KARAT code.

Registration of transition radiation was carried out using a point-detector array placed at equal distance from the intersection of electron beam and the target near the specular angle.

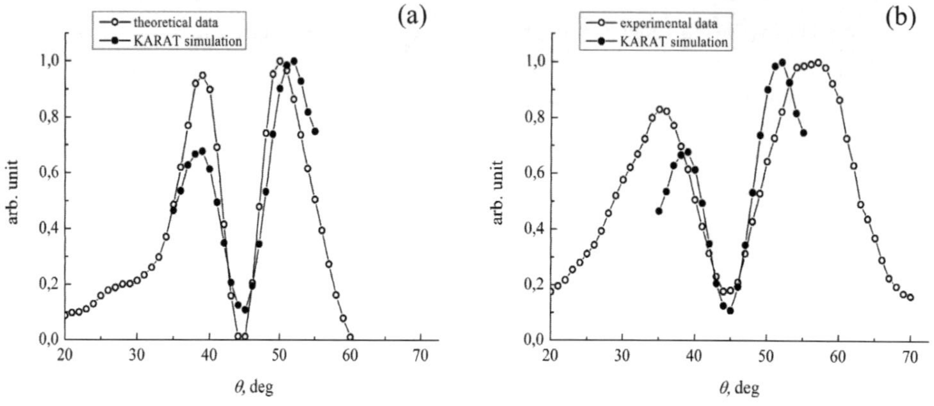

Figure 3. BTR orientation dependences obtained theoretically, experimentally and simulated by KARAT code.

The figure 3(a) shows a comparison of the orientation dependences of the backward transition radiation (BTR) obtained in the KARAT simulation and calculated theoretically. One can see a good agreement between the two peaks. The difference in the height of the peaks can be explained by a much broader spectrum of the radiation being registered in the simulation.

The figure 3(b) shows a comparison of the orientation dependences obtained in simulation and experiment using the "parabolic telescope" with detector DP21-M in the wavelength range 11–17 mm [4].

There is an acceptable agreement. One can see the asymmetry of the two peak characteristic of backward transition radiation. The differences may be interpreted by the difference in real and model electron bunch profile. Various distances from the target to the detector in the experiment and simulation can also affect the results of the measurements.

3.3. Generation of the diffraction radiation in two serial conductive targets

In [5] the generation of diffraction radiation in two serial conductive targets was experimentally investigated. It was found that the interaction of an electron beam with the first target has a significant effect on the interaction with the second target, and depends on the distance between the targets. With increase of the distance between targets the amplitude of the radiation angular distribution from the second target was also increased.

The dimensions of the model were up to 1000 × 1500 mm. Inside modeling region two conducting targets were placed at an angle of 45° to the electron beam trajectory. The beam and targets parameters were corresponded to the experimental conditions. Registration of the backward diffraction radiation

(BDR) from the second target was carried out at the maximum possible distance from the front edge of the target to minimize the effect of pre-wave zone.

Parameters used in simulation: electron beam energy – 6.1 MeV, bunch length – 3 mm, number of electrons per bunch – 10^8.

Simulation was done using a planar two-dimensional x-z version of the KARAT code.

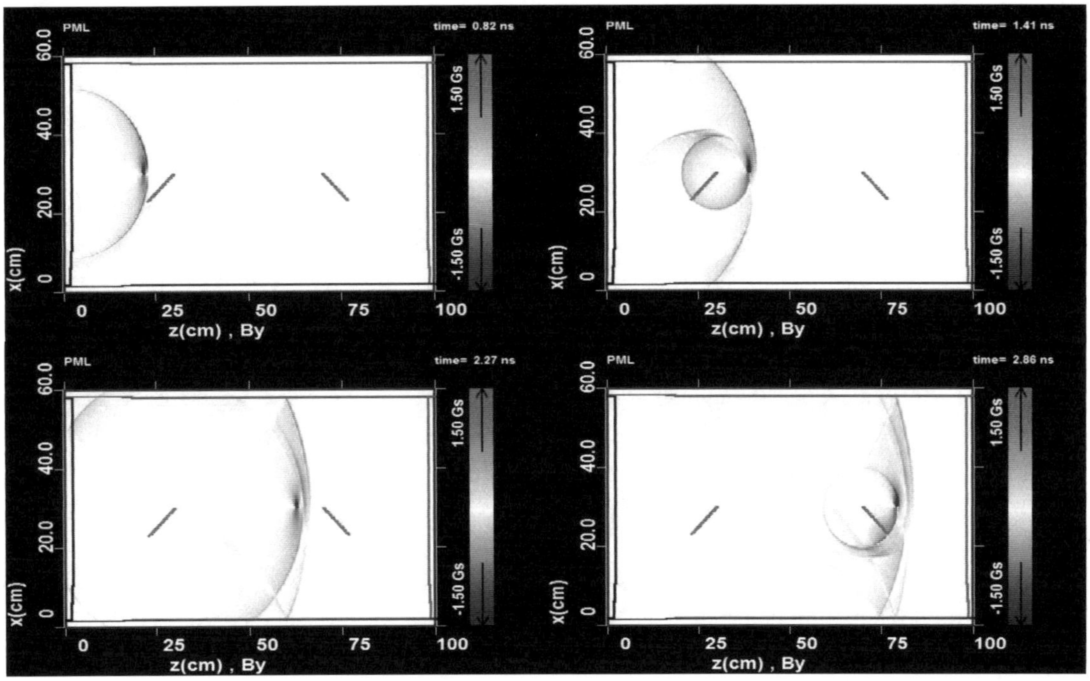

Figure 4. Contour map of B_y component of the electromagnetic field at a different stage of simulation.

The figure 4 shows a demonstration model. It is aimed to show the fields of transition and diffraction radiation generated by an electron beam. Low electrons energy of 1 MeV has been selected for the effective spatial separation of the beam field and radiation fields. Because of that radiation field is well ahead of the beam at distances less than the distance between the targets. This allows clearly see the process of forming a diffraction radiation fields.

The figures show the contour map of the B_y component of the electromagnetic field allowing to analyze the simulated processes in the most convenient way. When the beam was injected into the computational region transition radiation was necessarily also generated, but it does not affect the final result.

It is clearly seen that both targets are the sources of the forward (FDR) and backward (BDR) diffraction radiation. They are of the different direction of the B_y component of the electromagnetic field. Moreover at the second target there is an interference of the FDR from the first target and BDR from the second target in the radiation registration direction (straight down). In the case of relativistic energy, such as 6 MeV, the interference will be more significant.

With increasing distance between the targets negative contribution of the FDR from the first target was reduced and the amplitude of the radiation from the second target was increased.

The figure 5 shows the dependences of the radiation intensity on the observation angle for different distances between targets simulated by KARAT code (a) and measured experimentally (b).

There is good agreement between the amplitude of the BDR angular distribution depending on the distance between the targets and the position of the maxima of distributions. The simulation showed that under experimental conditions there is the interference of the diffraction radiation from the first and second target which has a significant affect on the resulting angular distribution.

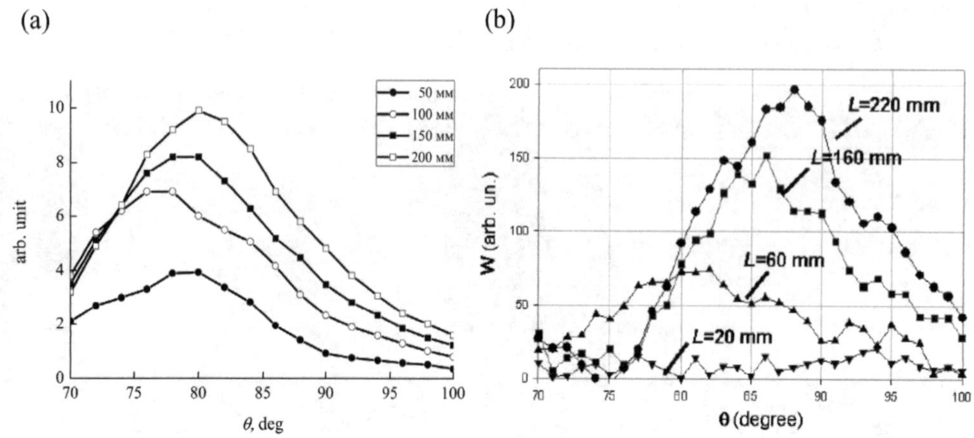

Figure 5. Dependences of the radiation intensity on the observation angle for different distance between targets simulated by KARAT code (a) and measured experimentally (b).

The discrepancy between the position of the maxima of distribution and the specular angle (90 degrees) caused by the difference in the phenomena of electromagnetic diffraction and diffraction radiation generated by an electron beam.

Simulation gives results similar to the experimental one and following the dynamics of radiation generation. In addition, simulation enables to visualize the sources of radiation fields and estimate their relative influence.

3.4. Vavilov-Cherenkov radiation. Coherent Vavilov-Cherenkov radiation from a bunched electron beam passing near a target

In [6] the method of non-invasive bunch length diagnostic based on an analysis of the spectra of coherent Vavilov-Cerenkov THz range was considered. The paraffin and teflon prisms were used as the Cherenkov targets. The electron beam was injected along the large face of the prism. Figure 6 shows the experimental setup. Simulation was carried out in accordance with the experiment.

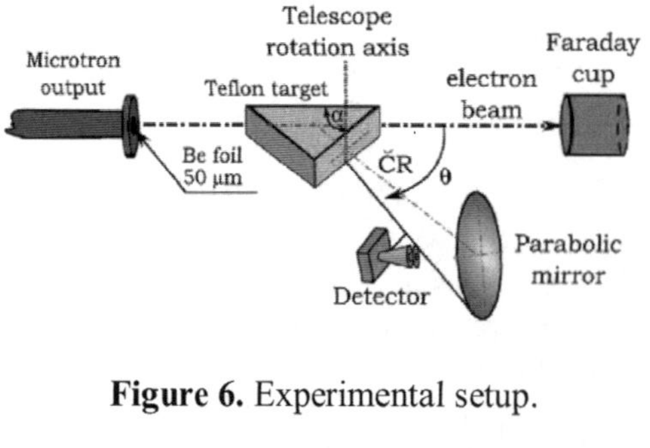

Figure 6. Experimental setup.

Measurements of the angular characteristics were carried out at the maximum distance to reduce the effect of the pre-wave zone. Modeling was performed for prisms with dielectric permittivity $\varepsilon = 1.988$ ($\alpha = 45°$) and $\varepsilon = 2.22$ ($\alpha = 40°$).

Parameters used in simulation: electron beam energy – 6.1 MeV, bunch length – 3 mm, number of electrons per bunch – 10^8.

Simulation was done using a planar two-dimensional x-z version of the KARAT code.

The figure 7 shows a demonstration model. It was aimed to show the formation of the Vavilov-Cherenkov radiation (VCR) in the dielectric prism. One can clearly see the front of VCR, and also re-reflection radiation inside the prism.

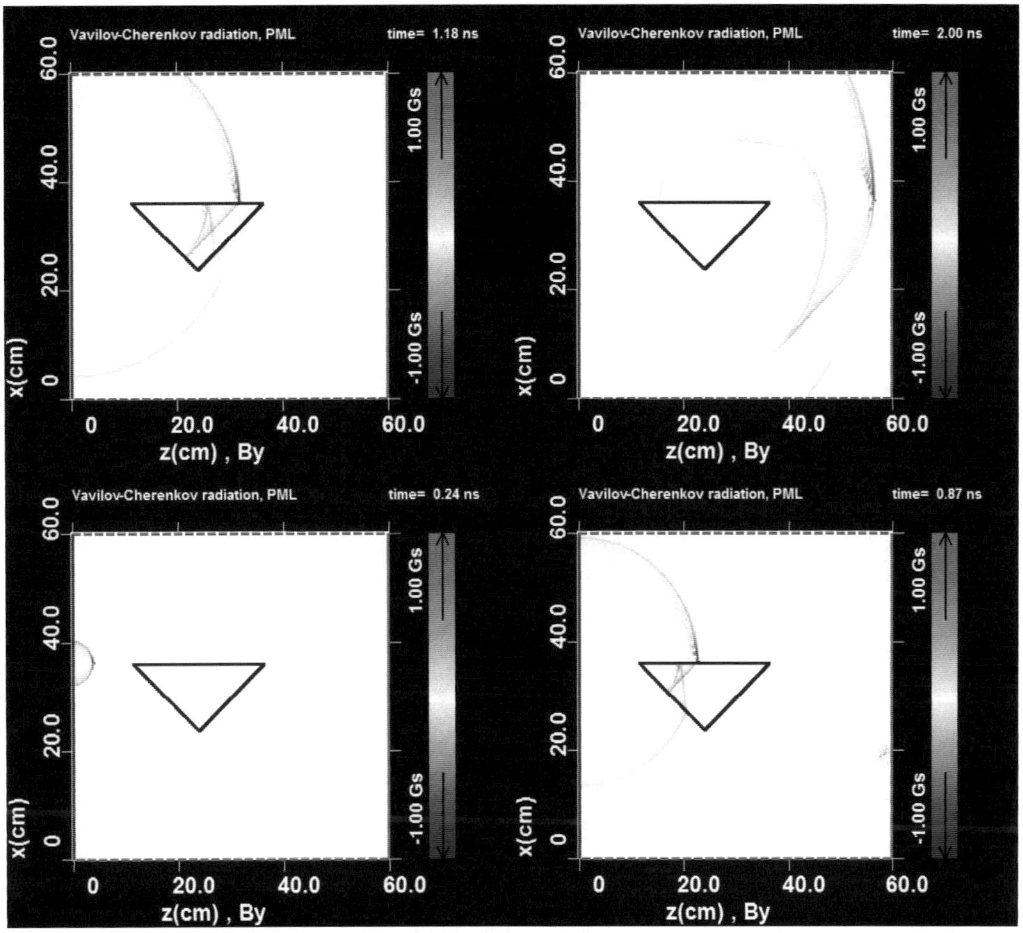

Figure 7. Contour map of B_y component of the electromagnetic field at a different stage of simulation.

The radiation output from the prism had minimal losses because the angle of prism was chosen to be equal to the Cherenkov radiation propagation angle.

There is a good agreement in the position of the maxima of the angular distributions on both figures 8 and 9. In addition, the distributions obtained in the KARAT simulation agree very well with the experimental data.

KARAT simulation shows VCR generation in a prism, a reflection of the generated field from the internal faces of prism and the contribution of reflected pulses in the total signal. The simulation takes into account a wide spectral range of radiation and the form-factor of the electron beam, which may influence the form of the angular distribution of the VCR. These effects may be particularly important in the case of materials with frequency dispersion.

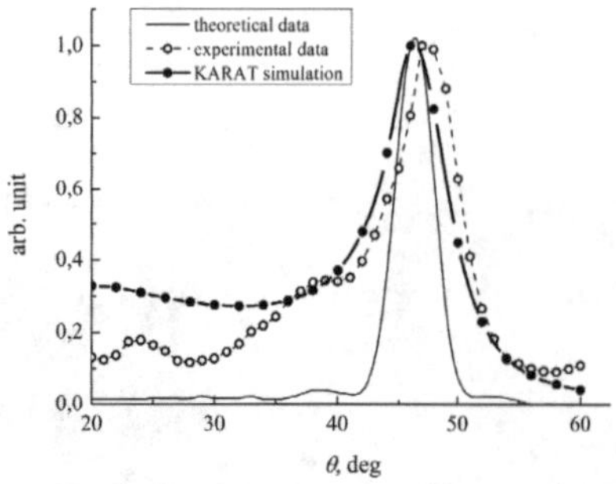

Figure 8. The angular distributions of coherent VCR for paraffin target with ε = 2.22 (α = 40°).

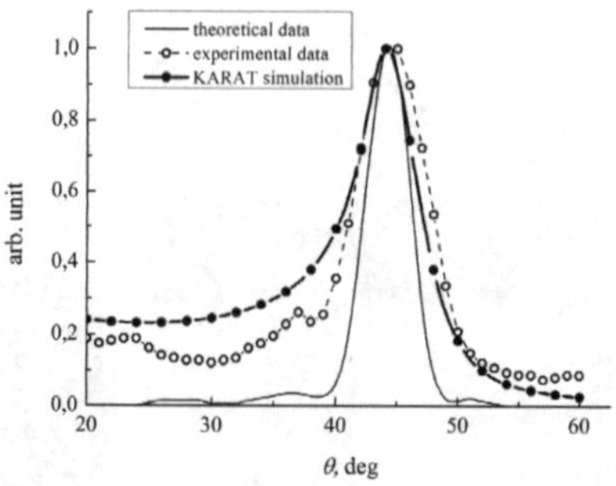

Figure 9. The angular distributions of coherent VCR for teflon target with ε = 1.988 (α = 45°).

4. Conclusion

The agreement of the KARAT simulation code results for the Smith-Purcell radiation, transition radiation, diffraction radiation and Vavilov-Cherenkov radiation with theoretical and experimental data was shown.

Particle-in-cell simulation allows analyzing the interaction of the electron beam with a targets and generation of various types of the radiation as well as optimizing experimental setup parameters.

Modern PIC simulation is a good tool for cross-check and prediction of the experimental results. In some cases, numerical simulation can partially or completely replace the real experiment, or assist in setting up the installation and detecting apparatus.

Acknowledgment

The work was partly supported by the grant of Ministry of Education and Science of the Russian Federation “Advancement of the Scientific Potential of High Education” No.0.81.2011.

References

[1] Tarakanov V P 1992 *User's Manual for Code KARAT* (Springfield: BRA)

[2] Birdsall C K, Langdon A B 1985 *Plasma Physics via Computer Simulation* (McGraw-Hill Book C.)

[3] Potylitsyn A P, Karlovets D V, Kube G 2008 Resonant diffraction radiation from inclined grating and bunch length measurements *Nucl. Instr. and Meth.* B **266** 3781–3788

[4] Naumenko G A, Aleinik A N, Aryshev A S, Kalinin B N, Potylitsyn A P, Saruev G A, Sharafutdinov A F 2005 Coherent transition and diffraction radiation from a bunched 6.1 MeV electron beam *Nucl. Instr. and Meth.* B **227** 70–77

[5] Naumenko G, Artru X, Potylitsyn A, Popov Yu, Sukhikh L “Shadowing” of the electromagnetic field of a relativistic electron arXiv:0901.2630v1

[6] Potylitsyn A P, Gogolev S Yu, Karlovets D V, Naumenko G A, Popov Yu A, Shevelev M V, Sukhikh L G 2010 Coherent Cherenkov radiation from a short bunch passing near a target and possibility of a bunch length diagnostics *Proceedings of IPAC'10 (Kyoto, Japan)* pp 1074–1076

Journal of Physics: Conference Series **357** (2012) 012023 doi:10.1088/1742-6596/357/1/012023

Sub-millimeter Bunch Length Non-invasive Diagnostic Based on the Diffraction and Cherenkov Radiation

M Shevelev[1*], H Deng[2], A Potylitsyn[1], G Naumenko[1], J Zhang[2], Sh Lu[2], S Gogolev[1], D Shkitov[1]

[1] Tomsk Polytechnic University, Lenin avenue, 30, Tomsk, 634059, Russia
[2] Shanghai Institute of Applied Physics, Chinese Academy of Science, Shanghai 201800, China

E-mail: s062014@yandex.ru

Abstract. A layout for the investigation the coherent Cherenkov radiation from a dielectric target with a large spectral dispersion and the coherent diffraction radiation from a conducting screen as a tool for non-invasive longitudinal electron beam profile diagnostics are proposed for the 20~30MeV Linac at Shanghai Institute of Applied Physics (SINAP). In this paper the status of the joint experiment and future plans are presented.

1. Introduction

Non-invasive methods of the sub-millimeter bunch length diagnostics are very important nowadays for such facilities as free electron lasers with typical bunch lengths of some hundreds of femtoseconds. One of the ways to measure bunch length is based on coherent radiation that is generated at wavelengths comparable to, or longer than, the bunch length, when all electrons in the bunch irradiate more or less in phase. The intensity of coherent radiation is proportional to the square of the bunch population. The spectral distribution of the coherent radiation contains the information about the electron distribution in a bunch. Coherent Diffraction Radiation (CDR) and Coherent Cherenkov Radiation (CChR) are suggested as the mechanisms for coherent radiation generation due to their non-invasive nature.

2. Experimental setup

The future experiments will be performed at the facilities of the femtosecond accelerator in the SINAP. It generates a multi-bunch electron beam with 68 pC per bunch and 20–30 MeV energy electron beam. The main beam parameters are summarized in table 1.

Journal of Physics: Conference Series **357** (2012) 012023 doi:10.1088/1742-6596/357/1/012023

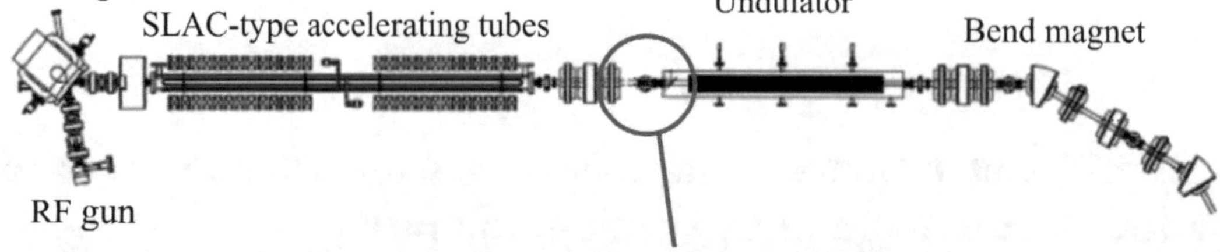

Figure 1. SINAP linac schematic layout.

To register a THz radiation the Golay Cell Type OAD-7 is proposed. This is the room temperature detector what makes it very attractive to use for accelerator physics and other application. It has excellent sensitivity and flat frequency response over a wide wavelength range. The detector parameters are presented in table 2.

Table 1. Electron beam parameters of SINAP linac facility

Electron energy	20–30 MeV
Beam charge	0.068 nC
Normalized emittance	~ 10π mm·mrad
Macro-bunch repetition	3.125~50 Hz
Micro-bunch repetition	2856 MHz
Macro-bunch duration	2–3 μs
Micro-bunch duration (FWHM)	~ 250 fs

The linac consists of S-band thermionic RF gun, an alpha-magnet and the SLAC-type accelerating tube (figure 1). After pre-acceleration in the RF gun the electron beam is compressed by the alpha-magnet from few tens of picoseconds to few hundreds of femtoseconds. Then the electron beam is accelerated up to 20–30 MeV by the SLAC-type tube.

Table 2. Detector parameters.

Optimal operating wavelength rage	15 ÷ 1000 μm
Response rate	28 ms
Maximum signal input power	10 μW
System noise output @ 20 Hz	6.2 $\mu V \cdot Hz^{-1/2}$
System optical responsivity @ 20 Hz	29.7 kV/W
Detector optical NEP (20 Hz)	208 $pW \cdot Hz^{-1/2}$
Aperture diameter	6 mm
Window material	HDPE

3. CChR for bunch length measurement

In conventional case Cherenkov Radiation (ChR) is emitted when the charge particle travels inside the medium with velocity that exceeds the speed of light in this transparent medium. The radiation cone in this case is defined by the following condition:

$$\cos\theta = \frac{1}{n\beta},$$

where θ is the radiation angle, β is the electron velocity in the speed of light units, n is the refracted index.

ChR may also appear while a particle travels in the vacuum in a vicinity of the medium due to the fact that ChR is generated due to dynamic polarization of the medium by the charged particle's electromagnetic field [1–3]. For the high-energy particles, the Lorentz-boosted electromagnetic field has the transverse dimension of about $\gamma\lambda$ (γ is the particle Lorentz-factor, λ is the radiation wavelength).

If a dielectric medium has a frequency dispersion, the different wavelength would be radiated under different angles. One may change the complicated spectral measurements by more convenient angular ones in order to obtain bunch form-factor and the bunch length.

Target made from CsI will have large frequency dispersion in THz region [4]. Manufacturing such a Cherenkov target as prism one may obtain wide angular distribution that may be measured with good accuracy. The dependence of the CsI refractive index on the frequency from [4] is presented in figure 2.

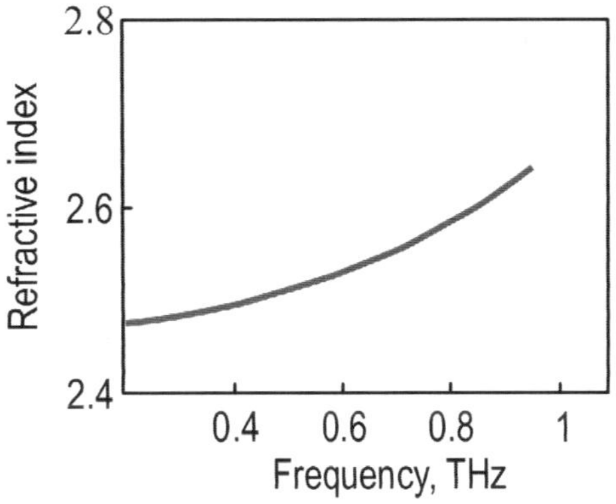

Figure 2. Refractive index dependence on frequency for CsI.

To calculate the target geometry for arrangement of experiment it is needed to use the frequency dependence, the Cherenkov criterion and the Snell's law. One may write $\frac{d\eta}{dn} = \frac{d\eta}{df} \times \frac{df}{dn}$, where η is the observation angle, f is the frequency and n is the refracted index. To determine the target geometry, the maximum reasonable value of $\frac{d\eta}{dn}$ should be used. The calculated $\frac{d\eta}{dn}$ dependence on the angle of prism using the data from figure 2 is shown in figure 4. Thus, for the experimental target shown in figure 3 the optimal prism angle α is equal to 45 degree.

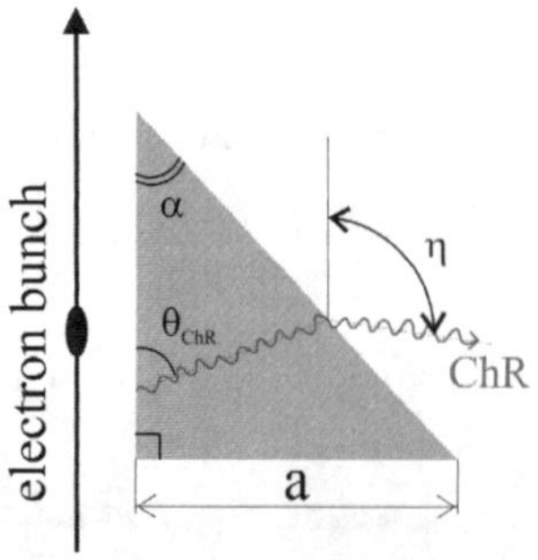

Figure 3. Experiment scheme.

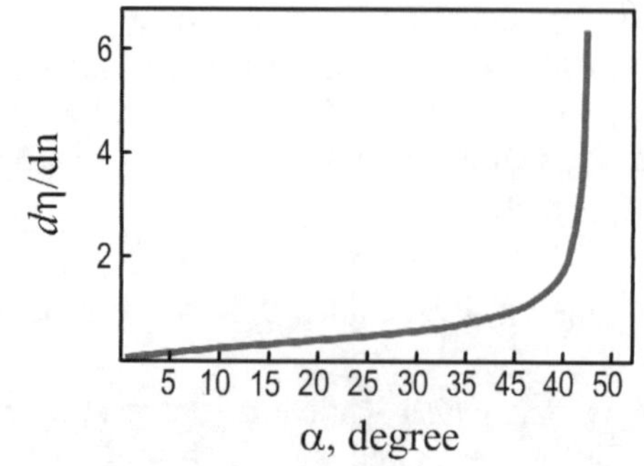

Figure 4. Calculated $\frac{d\eta}{dn}$ dependence on the angle α of Cherenkov prism.

There are many brands of the same dielectric material, which can differ from each other in a number characteristics (e.g. content of admixture, fabrication method etc.), consequently, different brands may have different properties including the refractive index. For joint experiments three different dielectric targets were manufactured. Therefore we are going to measure the dielectric targets spectral dispersion in the THz region prior to start experimental research. The materials and dimensions of the targets are listed in table 3.

Table 3. Sample dimensions and geomentry

Material	Dimension a , mm (figure 3)	Height, mm	Angle α , degree (figure 3)
Pure CsI	40	40	45
Impure CsI (Tl)	40	40	45
PbF2	30	30	45

The theoretical angular distribution was simulated for parameters of the SINAP linac and characteristic of the CsI Cherenkov target using the universal model describing polarization radiation mechanism [5] (figure 5).

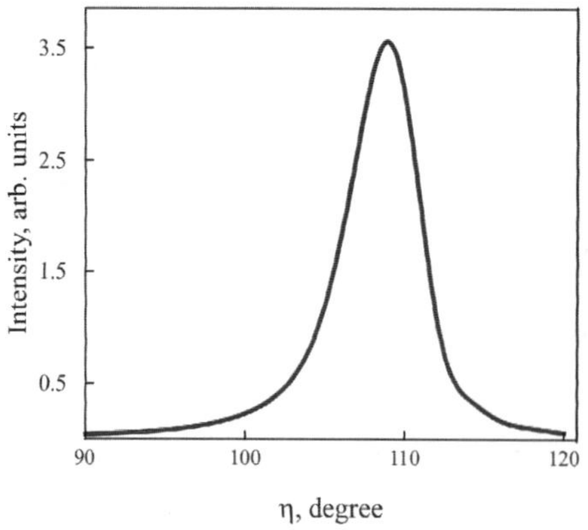

Figure 5. Calculated angular distribution of CChR from the CsI target for electron bunch energy 20 MeV, bunch length 75 μm and $\lambda = 0.04 \div 10$ mm.

Journal of Physics: Conference Series **357** (2012) 012023 doi:10.1088/1742-6596/357/1/012023

4. CDR for bunch measurement

Diffraction Radiation (DR) appears when a relativistic charged particle moves in the vicinity of a medium (or target) with some impact parameter comparable to or smaller than the charged particle field area. The particle field polarizes the target atom thereby induced currents changing in time, which in turn give rise to DR.

Bunch length may be reconstructed from a coherent radiation spectrum. The interferometers, e.g. Michelson, Mach-Zender, Fabry-Perot, Martin-Puplett, which are used for this purpose are very cumbersome and expensive. The design of simple interferometer consisting of two CDR aluminium targets from both side of beam is suggested (figure 6). Moving one of these targets it is possible to measure some kind of interferogram without any additional spectrometer equipments.

The theoretical curves of interferograms measured by detector placed at distance 153.2 mm from CRD target were simulated using the developed DR model [6] (figure 7).

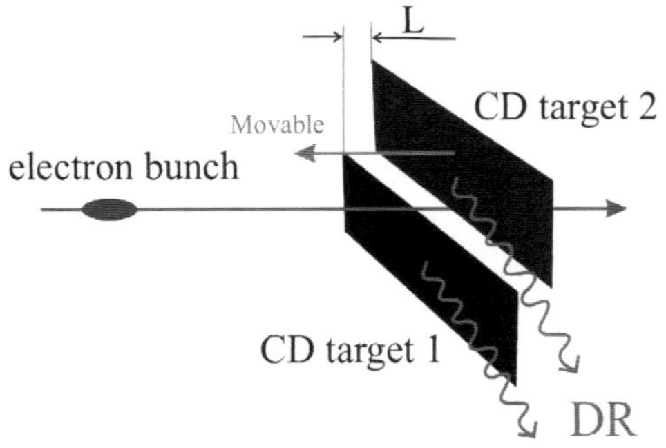

Figure 6. Scheme of CDR interferometer.

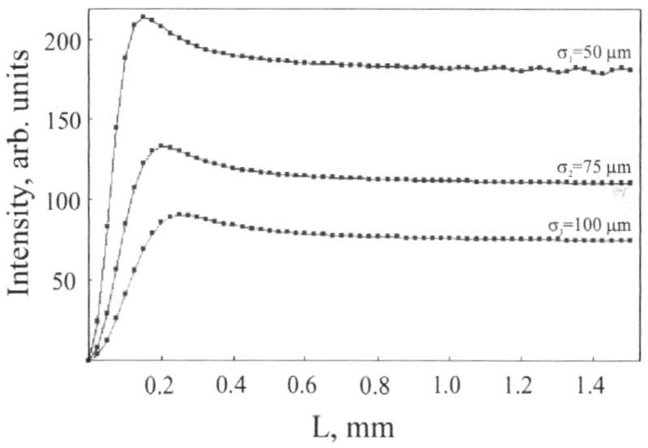

Figure 7. Theoretical calculated interferograms for electron bunch energy 20 MeV, $\lambda = 0.04 \div 2$ mm and different values of bunch length.

Main part of information about the bunch length is grouped in the first oscillation of the inerferogram. Therefore we may try to build a criterion for the bunch length estimation using this part of interferogram. If we define the interferogram function as $S(L)$, then the simplest criterion may be represented using the dimension theory as $\sigma_z = f(x)$, where $x = \left(S(L) \Big/ \frac{dS(L)}{d(L)} \right)\Bigg|_{L=L_{\max}}$, $L_{\max}$ is the value of variable L in the point of maximal $\frac{dS(L)}{dL}$, σ_z is the bunch length in Gaussian approximation of the longitudinal distribution of electrons in bunch, the function $f(x)$ is yet undefined.

For definition of this function we calculated the dependence $x(\sigma_z)$ from theoretical curves of interferogram in Gaussian approximation of the longitudinal distribution of electrons. Finally with a good accuracy was obtained $\sigma_z = x$.

5. Conclusion and future plan

In conclusion, it is worth pointing out that CChR generated by the ultra-relativistic electron bunch passing near a target with frequency dispersion and CDR generated by a DR target consisting of the two planar shifting halves seem to be a promising non-invasive mechanisms to estimate the length of bunch without additional spectrometers. The next step is to carry out experiments at the Shanghai linac facilities during May 2012.

Acknowledgement
This work was supported by joint Russian-Chinese grant NSFC (11111120065) and RFBR No 11-02-91177-GFEN_a and the Russian Ministry of Education and Science under a contract No P790.

Reference

[1] Bolotovskii B M 1961 *Sov. Phys.Usp.* **4** 781
[2] Takahashi T, Shibata Y, Oyamada M, Ishi K, Ikezawa M, Kondo Y 2000 *Phys. Rev.* E **62** 8606
[3] Potylitsyn A, Popov Yu, Shukhikh L et al. *J. Phys.: Conf. Ser.* 2010 **236** 012025
[4] Meijer A, Pijpers J, Nienhuys H, Bonn M, van der Zande W 2008 *J. Opt. A;Pure Appl. Opt.* **10** 095303
[5] Karlovets D, Potylitsyn A 2009 *JETP Letters* **90**(5) 368
[6] Potylitsyn A et al. 2010 *Diffraction Radiatation from Relativistic Particles* (Berlin: Springler)

Journal of Physics: Conference Series **357** (2012) 012024 doi:10.1088/1742-6596/357/1/012024

An event generator for crystal source Application of the CLIC positron baseline

O Dadoun[1], G Le Meur[1], F Touze[1], A Variola[1], X Artru[2], R Chehab[2], M Chevallier[2], V Strakhovenko[3]

[1] LAL, IN2P3-CNRS and Université de Paris Sud, 91898 Orsay Cedex, France
[2] IPN-Lyon, IN2P3-CNRS and Université Claude Bernard, 69622 Villeurbanne, France
[3] BINP, Av. Lavrentyeva, 11, 630090 Novosibirsk, Russia

E-mail: `dadoun@lal.in2p3.fr`

Abstract. A high energy electron beam propagating at glancing angles along a mono crystal axis experience a collective scattering from many crystal atoms and may hence, be channeled; it performs an oscillating trajectory and emits a large number of photons. This property is used to obtain an intense beam of photons which generate a large number of e^+e^- pairs in an amorphous target. The radiation in crystals is handled by dedicated programs as FOT while the interaction between charged particles and amorphous material is well described in Monte Carlo codes such as Geant4 or EGS. Thus a Monte Carlo code associating the crystal program FOT with Geant4 has been worked out and a presentation is given for this new program named G4Fot. An application to the future positron baseline of the Compact Linear Collider (CLIC) using a tungsten crystal oriented on its $\langle 111 \rangle$ axis and an amorphous material, the so called hybrid source is also presented.

1. Introduction

A fast charged particle interacting with the atoms of a material produces radiation. In amorphous media the contributions from the interactions with the individual atoms as bremsstrahlung and pair production are added incoherently. Those interactions are well described by the Bethe-Heitler mechanism which is included in various codes as `EGS` [1], `Fluka` [2] or `Geant4` [3]. If the particle is entering in a crystal with glancing incidence along the axes or planes, instead of being incoherently scattered by the atoms, it will be submitted to correlated deflections from the atoms of the rows or planes ; the particles trajectory can then be calculated averaging the crystal potential along the crystalline direction with two consequences : a larger radiated energy with a softer radiation spectrum than with the bremsstrahlung and consequently a much larger number of photons. The radiated energy becomes higher in such process than with the bremsstrahlung above certain incident energy value, as for example 1 GeV for the tungsten crystal [4]. The photon spectrum produced by 5 GeV energy electrons beam impinging with normal incidence on an amorphous and a crystal tungsten target oriented on its $\langle 111 \rangle$ axis is presented figure 1. Both targets have the same thickness of 1.4 mm. In this figure the vertical scale is $E \times dN/dE$, where E is the photon energy. In an amorphous target, the bremsstrahlung spectrum has a $1/E$ behaviour and exhibits on the figure an almost constant shape. On the other hand, in the crystal an enhancement of soft photons production is observed. Moreover the difference of the photon yield, defined as the number of photons produced at the exit of the

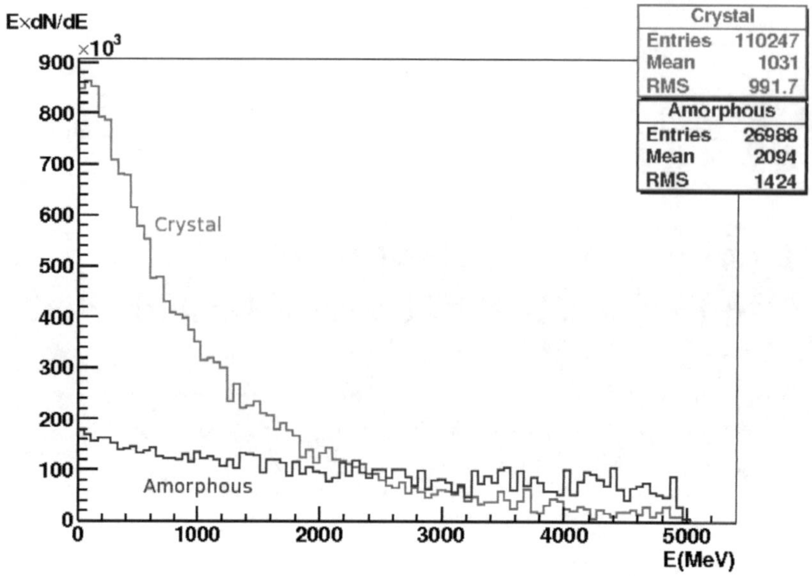

Figure 1. Emerging photons from 6000 electrons at 5 GeV impinging on 1.4 mm of tungsten crystal oriented on its $\langle 111 \rangle$ axis and amorphous target. Normal incidence of the electrons with respect to the targets was considered.

crystal divided by the number of incident electrons, between crystal and amorphous material is around 4.5. The larger number of photons and their softer energy are interesting for using such process to generate large numbers of e^+e^- pairs for positron source applications. This method has been proposed some years ago in the framework of future lepton colliders [5].

The simulations worked out to describe this method have used a dedicated code `Fot` [6] calculating the crystal effects, written by X. Artru from IPN-Lyon and associated to `EGS` to derive the positron production. In order to carry out much more intensive simulations the crystal code `Fot` has been associated to `Geant3` code by A. Jejcic [7]. The combined codes have been used to analyze the results of the *Proof of Principle* experiment at Orsay [8]. That combined program was precise enough but rather slow in term of computing point of view. In order to develop very intensive simulations for the preparation and analysis of the CERN experiment WA103, another crystal program simulation, more rapid, written by V.M. Strakhovenko from BINP (call VMS simulation later on) has been used and also associated to `Geant3`. The discrepancies between the two combined codes were within 10%. With the goal of using `Fot` as an event generator for `Geant4` and for adding new functionalities, the `Fot` program written in `Fortran` has been converted into `C++`. The new combined program, called `G4Fot` has already provided some results which are presented hereafter.

2. Simulations

The radiation emitted from the interaction between a fast electron and the the individual atoms of an amorphous media is called incoherent bremsstrahlung (IB). In crystalline targets and when the electron trajectory is close enough to a major crystallographic axis the amplitudes of the bremsstrahlung emission at certain wavelengths may interfere constructiveley. This will result in an enhancement of the intensity at this wavelength compared to ordinary incoherent emission. This type of radiation is called coherent bremsstrahlung (CB). Moreover if particles enter with glancing angle to axis they may be trapped in "channels" performing stable periodical trajectories along atomis rows or planes. This radiation is then called channeling radiation (CHR) or Kumakhov radiation. Those physics processes have been implemented in the `Fot` simulation. The basic principles of this simulation will now be presented. For more details the reader can refer to the original X. Artru's article [6].

2.1. Fot simulation

Taking advantage of the stronger axis potential with respect to the planar one, the incident electrons are crossing the crystal longitudinal axis z at glancing angles. They will feel a continuous potential, equal for the transverse coordinate $r_\perp$ to :

$$\overline{V}(r_\perp) = \langle\langle V(r_\perp, z)\rangle_t\rangle_z \ .$$

Where $V(r_\perp, z)$ is the atom potential at the location of the projectile and $\langle\ \rangle_t$ means average over t ; This implies conservation of the transverse energy, defined by :

$$E_\perp = \frac{1}{2}E\psi^2 - e\overline{V}(r_\perp), \tag{1}$$

where ψ is the particle angle to the axis. For $E_\perp = 0$ an angle can be defined as the critical angle for channeling radiation. Derived from the relation 1, this critical angle or Lindhard angle is defined by :

$$\psi_c = \sqrt{\frac{2eV_0}{E}}. \tag{2}$$

If the incident angle is somewhat larger than the critical angle, the electrons may cross planes or strings at regular intervals and interferences are occurring providing CB with shorter wavelengths than with channeling. Electrons can also develop ordinary bremsstrahlung when they cross the crystal with random orientation. In order to evaluate the radiation rate, `Fot` calculates the electron trajectories in the crystal and the subsequent photon emission probabilities. This program is using the semi-classical Baier-Katkov formula for radiation in a non-uniform field, taking into account the multiple scattering [9].

Two mains parameters could be defined in the simulation : the maximal transverse angle $\psi_\perp^{max}$ and the maximal transverse energy $E_\perp^{max}$. Typically $\psi_\perp^{max} = 10^{-2}$ rad and $E_\perp^{max} = 100$ keV, decreasing those cut off will increase the computing time simulation. The transverse angle and the transverse energy of an electron must then be less than the maximums values to be trapped in the crystal channel. More precisely three different cases can occur during the electron propagation in the crystal :

(i) $\psi >> \psi_\perp^{max}$ and/or $E_\perp >> E_\perp^{max}$, only IB radiation occurred ;
(ii) $\psi > \psi_\perp^{max}$ and $E_\perp > E_\perp^{max}$, CB plus IB radiation ;
(iii) $\psi < \psi_\perp^{max}$ and $E_\perp < E_\perp^{max}$, CHR+CB and IB.

A radiated photon could then, if is energy is above $2 \times m_e$, where m_e is electron mass, produce in the crystal an e^+e^- pair. Besides the incoherent pair creation which is similar to what happens in an amorphous medium, two other kinds are expected. The coherent pair creation which is associated to the coherent bremsstrahlung and the pair production in strong fields which appears at very high energy. For the tungsten crystal at room temperature the associated threshold is for the latter around 22 GeV and go down to 13 GeV at $T = 77$ K [10]. For the cases considered in this paper which concern incident electron energies up to 10 GeV, the only process assumed will be the incoherent pair production well described by standard simulation of particle matter interaction such as `Geant4`.

The crystal temperature should also be taken into account in the axis potential. The temperature increase in the crystal due to energy deposition leads to thermal vibration with, as a consequence, the lowering of the available channeling potential.

Journal of Physics: Conference Series **357** (2012) 012024 doi:10.1088/1742-6596/357/1/012024

2.2. G4Fot simulation

Since Fot simulation do not take into account the e^+e^- pair creation, this process will be simulated using Geant4. Towards this goal Fot have been converted into C++ (Fot++), the common Fortran becomes C++ classes and the original algorithm is kept[1].

For an event, ie one incident electron impinging on a crystal target, the phase space of all radiated gamma (from IB, CB and CHR processes) simulated by the Fot++ are stored in memory. When the electron exits the target the simulation stops. The created photons in the target are then simulated by Geant4 and the material is considered as a pure amorphous target (figure 2).

In the case where the electron is dechanneled before exiting the crystal, it is not anymore simulated by Fot++. The dechanneled electron is then also simulated by Geant4 for the rest of is propagation in the material since it could radiate IB. If their energies are enough (typically in

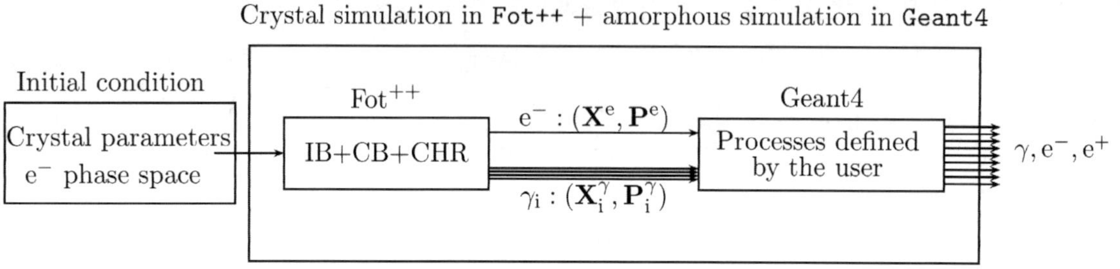

Figure 2. When an e^- impinges a crystal target Fot++ simulate all the produced photons from IB, CHB and CHR processes. Since the simulation can not take into account the pair creation process, photons phase space are keep in memory and simulated in an amorphous material using in Geant4. In the case where the electron is dechanneled before exiting the crystal, he s then also simulated by Geant4 for the rest of is propagation in the material considered as an amorphous one.

the GeV range) and the angle small enough the e^+e^- pairs created could also radiate in their turn photons from CB or CHR processes. Those second generation photons are not taken into account in G4Fot .

2.3. G4Fot benchmark

The photon flux produced in the crystal depends on the energy, the divergence of the electron beam and the crystal thickness. In the context of the positron source using the hybrid scheme for the future linear collider, a normal incidence to the $\langle 111 \rangle$ plane of the crystal has been considered. To limit the energy density deposition in the crystal the energy of the impinging electron taken below 10 GeV. Moreover, to have a reasonable number of positron produce, the energy should be greater than 3 GeV. According to the beam energy, different radiator thicknesses have been chosen, they correspond to some optimum values [7]. For instance at 10 GeV the crystal thickness is 1 mm and increases up to 1.6 mm for 3 GeV.

As already mentioned, the e^+e^- created in the crystal could in their turn be rechanneled and emits again some CB and/or CHR. This second generation photons is not taken into account in the simulation. Then, in principle, it underestimates somehow the number of produced photons. Since those photons populates the lower part of the energy spectrum compared to pure bremsstrahlung (see figure 1), the photons mean energy is slightly overestimated. This

[1] It is important to mention that the first goal was to obtain the same results with the two codes. It is why some classes, historically described in the Fortran code, have been implemented while the Standard Template Library from C++ could be used.

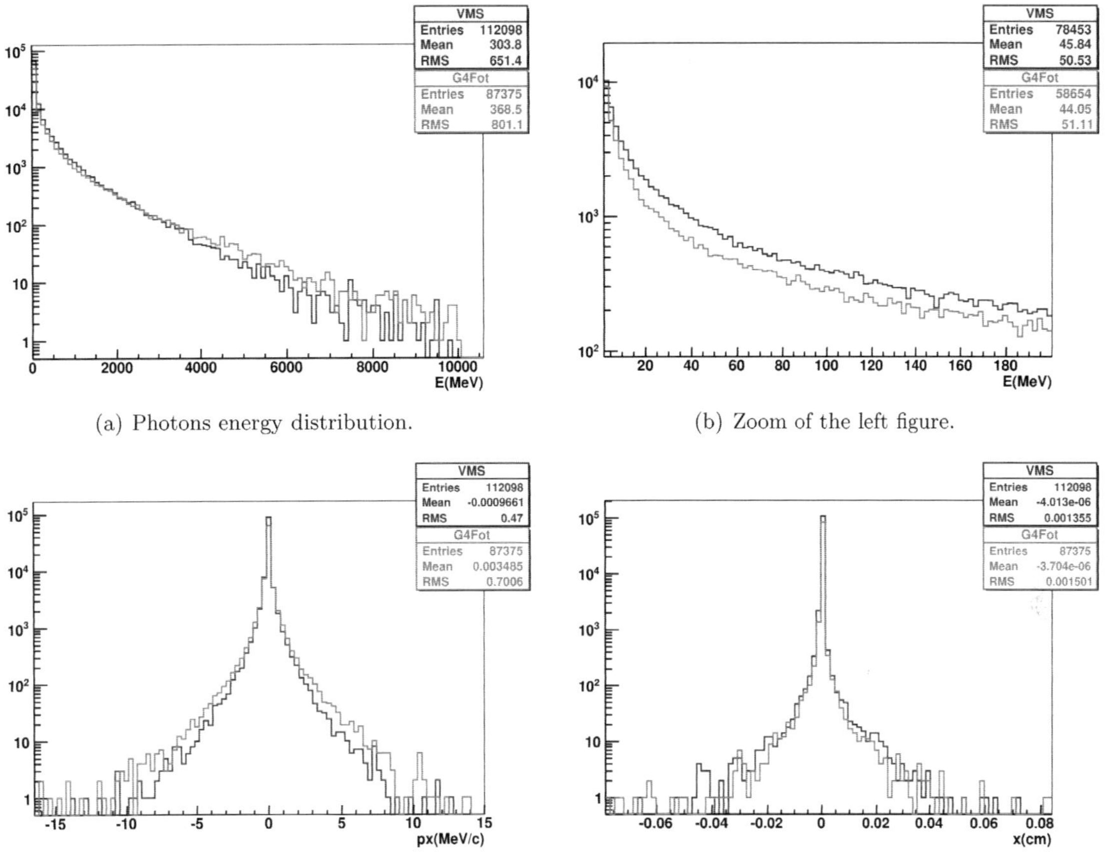

(a) Photons energy distribution.

(b) Zoom of the left figure.

(c) Photon transverse momentum px.

(d) Photons transverse x position.

Figure 3. A 10 GeV electron beam energy impinge 1 mm of crystal target thickness. Figures above presents the photons phase space at the exit of the crystal, produced by `G4Fot` and `VMS` simulations. The Simulations was done using 5000 incident electrons.

could be observed on figure 3(a) and figure 3(b) where a 10 GeV electron beam impinges on 1 mm crystal tungsten target. The other consequence is the lack of created pairs in the crystal as shown as the statistics boxes figure 4(a) and figure 4(c)).

Systematics comparison of the particles phase space at the exit of the crystal have been performed between both simulations using different incident electron beam energies. The comparison has shown a discrepancy between 10% and 20% depending on the incident electron beam energy considered.

3. CLIC positron baseline

The CLIC study considers the hybrid source using channeling as the baseline for unpolarized positron production. It uses a 5 GeV electron beam impinging on a tungsten crystal target of 1.4 mm. With the crystal oriented on its $\langle 111 \rangle$ axis the resulting intense photon beam is then impinging on a 10 mm amorphous tungsten target producing positrons by e^+e^- pair creation. Downstream the amorphous target, a capture section based on an adiabatic matching device followed by a 2 GHz Pre-Injector Linac focuses and accelerates the positron beam up to around 200 MeV (figure 5). The separation between the crystal and the amorphous is then used to bend

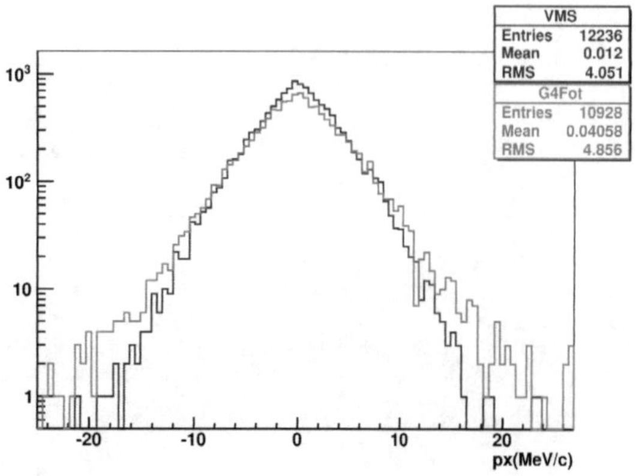

(a) Electrons transverse momentum px.

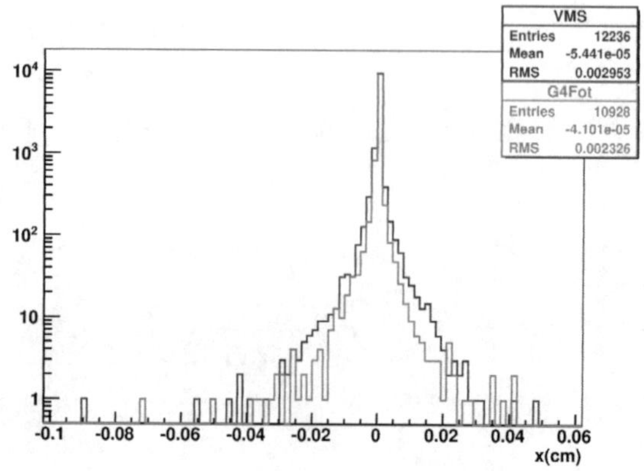

(b) Electrons transverse position x.

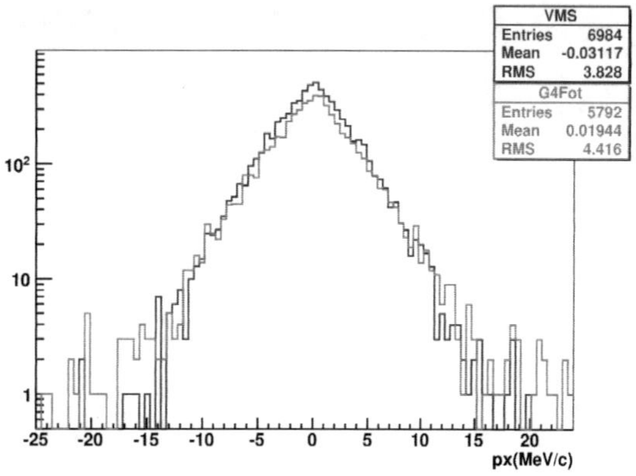

(c) Positrons transverse momentum px.

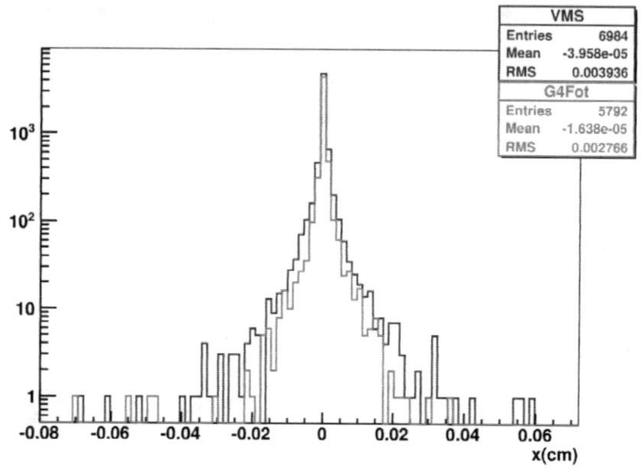

(d) Positrons transverse position x.

Figure 4. A 10 GeV electron beam energy impinge 1 mm of crystal target thickness. Figures above presents electrons and positrons phase space at the exit of the crystal, produced by `G4Fot` and `VMS` simulations. The Simulations was done using 5000 incident electrons.

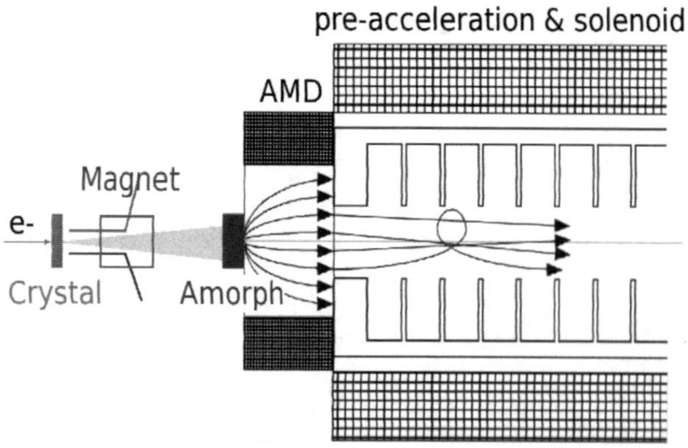

Figure 5. CLIC positron generator (adapted from Figure 4 of [20]).

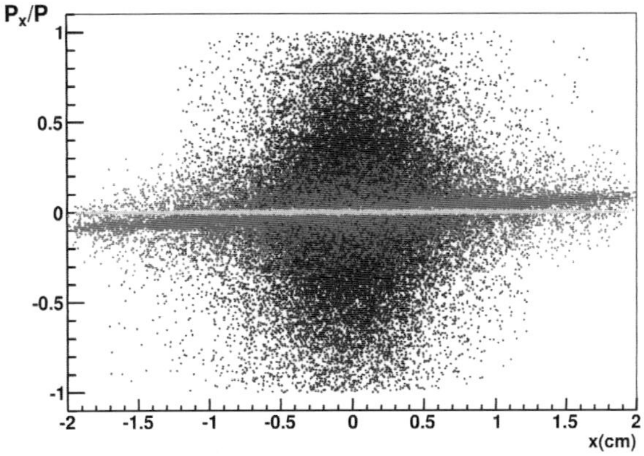

Figure 6. Positrons emittance at the exit of the amorphous target, the AMD and the Pre-Injector Linac at 200 MeV.

the charged particles produced by the crystal radiator [17]. This limits the energy despostion in the tungsten amorphous target. After the amorphous target, the AMD transform the positron phase space after the target into a larger dimension and smaller momentum spread which is easier to transport in the Pre-Injector Linac (figure 6). At the exit of the Pre-Injector Linac a positron yield of 0.8 is obtained. This yield do not fulfill the CLIC requirements. However, recent studies suggest that the incident electron beam intensity should be increased to obtain the requested number of e^+ captured into the Pre-Damping Ring. Nevertheless an increase of the electron beam intensity of 25% to 35% is still acceptable to keep in the energy deposition limitation.

4. Conclusion

The CLIC positron baseline using a crystal and an amorphous tungsten target have been presented. This very promising method, specially to produce the high intensity positron beam requested for future linear collider (a scheme for ILC can be found here [21]), can now be simulated using one simulation named `G4Fot`. This simulation described in this paper have been compared to an another simulation and for the energy considered namely between 3 and 10 GeV an agreement between 10 and 20 % have been found.

References

[1] Hirayama H, NamitoY, Bielajew A F, Wilderman S J and Nelson W R 2005 "The EGS5 Code System" SLAC-R-730 (2005) and KEK Report 2005-8 (2005) Nelson W R , Hirayama H and Rogers D W O 1985 "The EGS4 Code System", SLAC 265, SLAC, December 1985

[2] Fasso A , Ferrari A, Ranft J , and Sala P R, "FLUKA : a multi-particle transport code", CERN-2005-10 (2005), INFN/TC 05/11, SLAC-R-773

[3] Agostinelli S *et al.* 2003 Geant4 - a simulation toolkit *Nucl. Instr. and Meth.* A **206** 250

[4] Baier V N, Katkov V M, Strakhovenko V M 1986 *Phys. Stat. Solidi* B **133** 583

[5] Chehab R *et al.* 1989 Study of a positron source generated by photons from ultrarelativistic channeled particles (Proc. PAC , Chicago)

[6] Artru X 1990 A simulation code for channeling radiation by ultrarelativitic electrons or positrons *Nucl. Instr. and Meth.* B **48** 278

[7] Artru X *et al.* 1994 Positron source using channelling in a tungsten crystal *Nucl. Instr. and Meth.* A **344** 443-454

[8] Artru X *et al.* 1996 Axial channelling of relativistic electrons in crystals as a source for positron production *Nucl. Instr. and Meth.* B **119** 246-252

[9] Baier V N, Katkov V M 1968 **26** 854 and Baier V N, Katkov V M 1969 *Soviet Physics JETP* **28** 807

[10] BaierV N, Katkov V M and Strakhovenko V M 1989 Interaction of high energy electrons and photons with crystals *Ups. Friz. Nauk* **159** 455-491

[11] Artru X & Chevallier M 1989 Channeling radiation : Theory, semi-classical simulations, Radiation Effects and Defects in Solids : Incorporating Plasma Science and Plasma Technology, 1029-4953; 1994 **130** 415-432

[12] *An experimental study of the radiation arising due to* 10 GeV *electrons and positrons passing through single crystal* (in russian) 1989 ed Tsyganov E N *Fiz. Elem. Chast. Atom. Yadra* **20** 5-50

[13] Kimball J C and Cue N 1985 *Phys. Rep.* **125** 69

[14] Artru X *et al.* 1993 Observation of channeling and blocking effect in pair creation in Ge crystal *Physics Letters* B **313** (1993) 483-490

[15] Takahashi T 2010 Status of Hybrid target R&D at KEK-Linac, LCWS 2010

[16] Dadoun O *et al.* 2009 Study of an hybrid positron source using channeling for CLIC, CLIC Note 808

[17] O. Dadoun et *al.* 2010 The Baseline Positron Production and Capture Scheme for CLIC (IPAC 2010)

[18] Chehab R *et al.* 2002 *Phys. Lett.* B **525** 41

[19] Okuno H *et al.* 2003 *Nucl. Instr. and Meth.* B **201** 259

[20] Kamitani T, Rinolfi L 2007 Positron production for CLIC, CLIC Note 465

[21] Xu C *et al.* 2011 A positron source using an axially oriented crystal associated to a granular amorphous converter, submitted to Chinese Physics C (2011).

Journal of Physics: Conference Series **357** (2012) 012025 doi:10.1088/1742-6596/357/1/012025

Planar positron channeling radiation generated in an acoustic superlattice

B Azadegan[1], L Sh Grigoryan[2] and W Wagner[3]

[1] Sabzevar Tarbiat Moallem University - P.O. 397, Sabzevar, Iran

[2] NAS Institute of Applied Problems in Physics - 25 Hr. Nersessian str., 0014 Yerevan, Armenia

[3] Helmholz-Zentrum Dresden-Rossendorf - P.O.Box 510119, 01314 Dresden, Germany

E-mail: w.wagner@hzdr.de

Abstract. Planar positron channeling in an acoustic superlattice and stimulation of the emission of channeling radiation at resonance of the particle motion with the acoustic field excited in a $PbTiO_3$ single crystal is considered in the framework of classical mechanics and electrodynamics. Based on computed particle trajectories, spectral-angular distributions of channeling radiation influenced by resonant ultrasound have been simulated. The presented method explains the influence of ultrasonic waves on the intensity of channeling radiation in a rather simple and descriptive manner.

1. Introduction

Channeling radiation (CR), which has been explained as a self-contained source of X- and γ-rays in 1976 [1], is a well investigated phenomenon today (see, *e.g.*, [2, 3, 4, 5, 6, 7]). At not too high energies (up to several tens of MeV) of the channeled light charged particles, CR emission has to be considered as a quantum process, whereus at highly relativistic energies classical mechanics and electrodynamics may be utilized for an adequate description of CR [8]. Since the decisive criterion is the number of bound states within the continuous transverse potential [9], which governs channeling, the situation is rather different for positrons and electrons at planar or axial channeling, respectively. Besides that, this number strongly depends upon the particle energy, atomic number and interatomic distances of the crystal as well [8, 10].

Furthermore, since channeling and associated CR emission is naturally controlled by the continuous potential of the considered crystallographic axis or plane of a single crystal, the idea to manipulate CR emission by means of dedicated modulation of the crystal lattice has been worked out already in the 1980s [11, 12]. Very first model calculations predicted that the intensity of CR may be increased if the oscillatory motion of the channeled particles becomes resonant with ultrasonic waves (US) excited in the crystal [13, 14]. It was suggested to apply piezoelectric single crystals in which US may be excited utilizing external electromagnetic RF-fields. Forthwith first measurements of electron CR on quartz, $LiNbO_3$ and CdS crystals were carried out at ultrarelativistic energies where the CR spectra show a smooth behaviour [15].

A consistent quantum theory of electron and positron CR at impact of US has been developed in a series of works published after year 2000 [16, 17, 18]. Based on it, systematic measurements of planar CR on quartz single crystals have been carried out at medium electron energies (14 - 32

MeV) as an important prerequisite for envisaged experimental investigations [19, 20]. Since the frequency of resonant US strongly depends upon the transition energy between bound channeling states, measurements where single CR lines are resolved were categorically necessary.

Eventually, the effect of US on planar electron CR could be verified experimentally for two different crystallographic planes of quartz [21, 22, 23]. It should be mentioned that, in agreement with preceding simulations, resonant enhancement of the CR intensity has been observed at hypersound frequencies of about 12 GHz.

In the framework of quantum theory, the mechanism of CR stimulation by resonant US has been explained by the dispersive behaviour of the eigenstates of channeled particles in the continuous planar potential [24]. Within an acoustic superlattice of certain US wavelength, the longitudinal momentum of channeled particles, which occupy corresponding transverse states, is no more an integral of motion. Hence, conservation of total energy directly leads to a quantum uncertainty [24] which appears in an intermixture of dedicated channeling states due to a resonant increase of the dispersion of transverse energy. The induced modification of the occupation of bound states finally influences the CR intensity.

At particle energies, where the classical treatment of channeling is valid (see, *e.g.*, [25]), a description of this resonance effect is still missing. This paper for the first time offers a descriptive classical explanation of the influence of resonant US on CR. Exemplarily and for simplicity, we consider planar positron channeling in a polyatomic $PbTiO_3$ crystal. Since the piezoelectric constants of this material are much larger than those of quartz, generation of US by external electromagnetic fields should principally be more effective.

In what follows we briefly consider theoretical basics which have been applied for detailed computing of particle trajectories in an acoustic superlattice. Based on these data, spectral-angular distributions of CR have been calculated by means of classical electrodynamics. The results presented provide a simple picture how resonant US influence channeling and CR emission.

2. Theory

Let us consider a relativistic charged particle planar channeled in a piezoelectric single crystal, and assume that longitudinal US are excited in the crystal along the direction of channeling, z. In this case, the quasi-static, continuous and periodic transverse potential of the channeling plane, $U(x)$, is modulated by another periodic function, $\Lambda(z)$. Since, for a relativistic particle, $z \approx ct$ (c−velocity of light), the potential $V(x,z)$ is time-dependent, and the usual procedure applied to calculate the continuous planar potential, namely averaging of the atomic potential in space (over y and z) and time (over thermal atomic vibrations), has to be modified. As shown in [26], $V(x,z)$ may approximately be written as the product $U(x)\Lambda(z)$ where

$$\Lambda(z) \cong 1 + 2 \sum_{m=1,2,...} (-1)^m J_m(mk_s a^*) \cos[m(k_s z + \varphi)] \tag{1}$$

does not yet depend on the crystal parameters (except, of course, the velocity of US in the crystal, u_s), J_m denotes the Bessel function, $k_s = \omega_s/u_s$ is the wave number of US, $a^* = a\,\sin(\omega_s t)$ their amplitude and ω_s the frequency. The (undisturbed) periodic planar potential in the most common form reads

$$U(x) = \sum_{n=1,2,...} \nu_n \exp(ingx) \qquad n = ... -1, 0, 1, 2, ... \tag{2}$$

where $\vec{g}$ is the reciprocal lattice vector normal to the crystallographic plane considered, and the coefficents of this Fourier series are given by

$$\nu_n = \frac{2\pi a_0 Z_1 e^2}{V_c} \sum_j \exp[M_j(\vec{g})] \cdot \exp(-i\vec{g}\cdot\vec{r_j}) \sum_{i=1}^{4} a_i \exp[-\frac{1}{4}(\frac{b_i}{4\pi^2})(ng)^2] \tag{3}$$

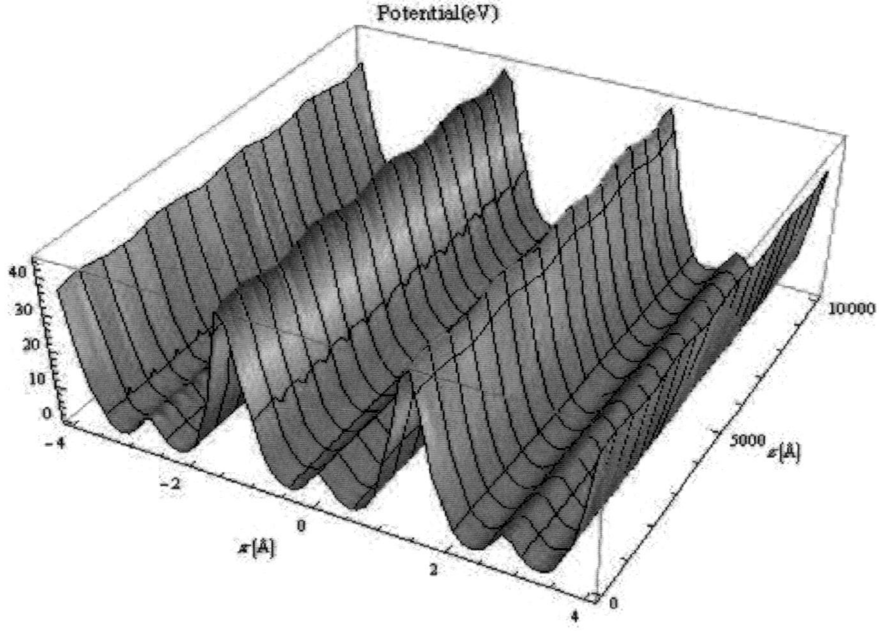

Figure 1. The (110) planar potential of a $PbTiO_3$ single crystal modulated by ultrasound of $k_s = 1.282\,\text{Å}^{-1}$.

(a_0, Z_1, e and V_c are the Bohr radius, atomic number, electron charge and the volume of unit cell, respectively, $\vec{r_j}$ is the radius vector to the lattice site of atom j in the unit cell, a_i and b_i are parameters defined in [27]). The Debye-Waller factor $M_j(\vec{g}) = g^2\langle u_j^2\rangle/2$ takes the one-dimensional mean-squared thermal vibration amplitude $\langle u_j^2\rangle$ of atom j into account.

The continuous potential, modulated by US, of the (110) plane of a $PbTiO_3$ single crystal, calculated numerically for positron channeling, is shown in figure 1. Due to the perovskite structure of $PbTiO_3$, the undisturbed (110) potential is by no means a harmonic one (as usually assumed for analytic calculations of positron channeling, *e.g.*, in [13]). Therefore, bound channeling states, which would result from a quantum calculation, are not equidistant. Accordingly, from a classical calculation one would expect a CR spectrum peaked at several frequencies. This situation partly resembles planar electron channeling where, as exploited in our former works [21, 22, 23], the US frequency had to be resonant to some CR transition for observing the effect of CR intensity enhancement. The classical description of CR generation is based on the calculation of the trajectory of the channeled particle (here $\vec{r}(t) = \vec{x}(t) + \vec{z}(t)$, $\vec{y}(t) \doteq 0$) obtained by solving the equations of motion

$$\gamma m\ddot{x}(t) = -\frac{\partial V(x,z)}{\partial x} \qquad \gamma m\ddot{z}(t) = -\frac{\partial V(x,z)}{\partial z} \tag{4}$$

(γm–relativistic particle mass, γ–Lorentz factor) at given initial conditions, *i.e.*, the coordinates of the point of incidence of the particle into the crystal and the projections of its momentum for an angle of incidence, Θ_0, with respect to the channeling plane

$$x(0) = x_0 \quad z(0) = z_0 \quad p_x(0) = p\sin(\Theta_0) \quad p_z(0) = p\cos(\Theta_0). \tag{5}$$

According to classical electrodynamics, the energy of CR radiated into a solid angle $d\Omega$ within

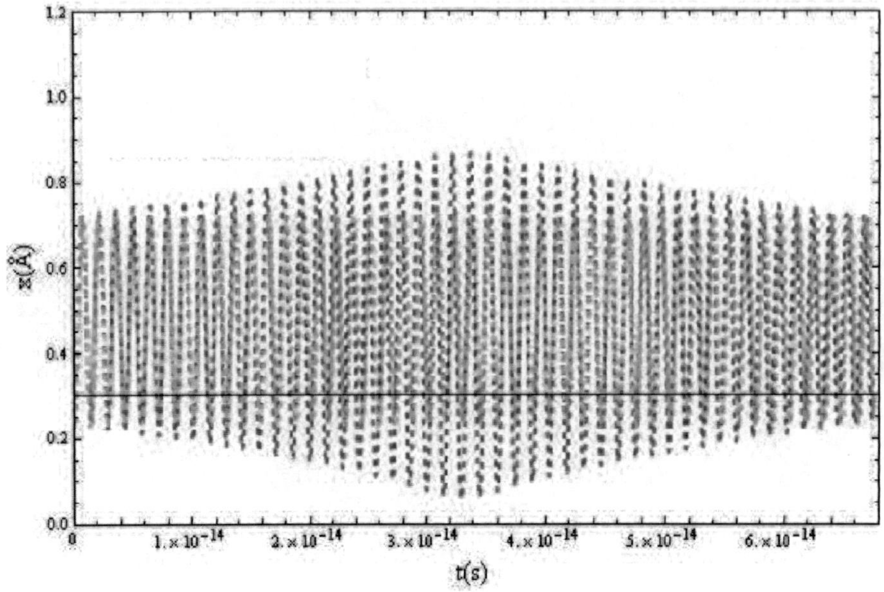

Figure 2. Trajectories $x(t)$ of positrons channeled along the (110) plane of a $PbTiO_3$ crystal at impact of resonant and non-resonant ultrasonic waves (see text).

a frequency interval $(\omega, \omega + d\omega)$ is given by

$$\frac{d^2E}{d\omega d\Omega} = \frac{e^2}{4\pi^2 c}| \int_0^\tau \exp[i(\omega t - \vec{k}\cdot\vec{r})]\frac{\vec{n}\times((\vec{n}-\vec{\beta})\times\dot{\vec{\beta}})}{(1-\vec{\beta}\cdot\vec{n})^2}dt|^2 \tag{6}$$

where $\vec{\beta} = \dot{\vec{r}}(t)/c$ means the particle velocity, $\vec{n}$ defines the direction of photon emission, $\vec{k} = \omega\vec{n}/c$ denotes the wave vector of CR, and τ is the time-of-flight of the particle through the crystal [4].

3. Trajectory calculations

In what follows we exemplarily demonstrate the output of a trajectory calculation for positrons of energy 40 MeV channeled along the (110) plane of an ultrasonic superlattice created in a $PbTiO_3$ single crystal. The equations of motion (4) have been solved numerically for the initial conditions (5): $x_0 = 0.235$ Å, $\Theta_0 = 0$. The period of transverse oscillations of a channeled particle amounts to $T = 1.408 \times 10^{-15}$ s. As shown in figures 2 and 3, the influence of US with a wave number of $k_s = 1.488 \times 10^{-3}$ Å^{-1} appears as an increase of the amplitude of transverse oscillations, $x(t)$, and transverse velocity, $\beta_x(t)$. Simultaneously, one observes a (coherent) decrease of the longitudinal velocity, $\beta_z(t)$ (see figure 4). This means that US, which are resonant to the oscillatory motion of the channeled particle [13, 18], force a transfer of longitudinal momentum to transverse one. Trajectories and velocities not influenced by US are also shown in figures 2 - 4. They depict the results of corresponding calculations but for wave numbers different from the resonant one by ± 5% ($k_s = 1.571 \times 10^{-3}$ Å^{-1} and $k_s = 1.406 \times 10^{-3}$ Å^{-1}). This finding again underlines the prediction of theory that the influence of US on CR emission is selective.

It is worth to remind here that the impact of resonant US in a quantum treatment of channeling appeared as a large increase of the dispersion of corresponding eigenvalues. Hence, the quantum uncertainty explored in [24] also reflects a resonant coupling of longitudinal and transverse momentum of the channeled particle. Since the effect is discrete, its verification for planar

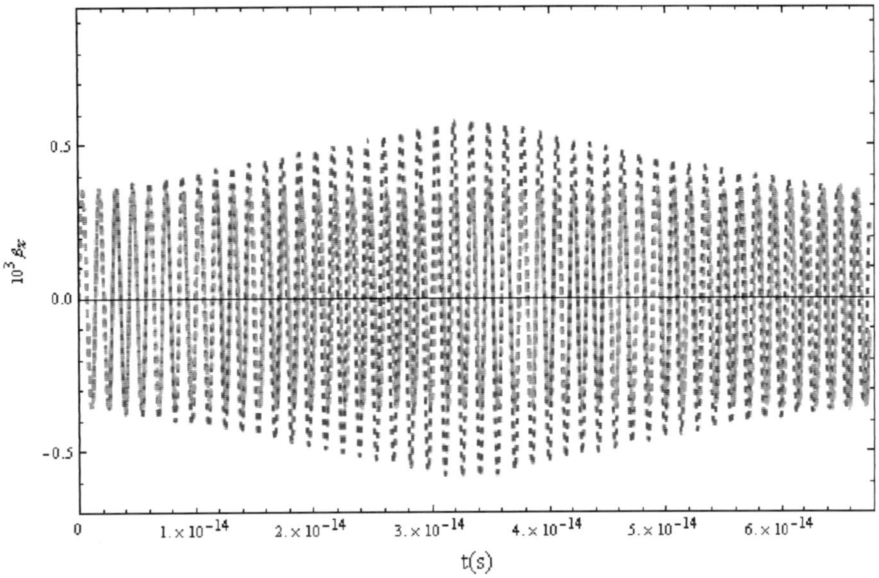

Figure 3. Transverse velocity $\beta_x(t) = \dot{x}/c$ of positrons channeled along the (110) plane of a $PbTiO_3$ crystal at impact of resonant and non-resonant ultrasonic waves (see text).

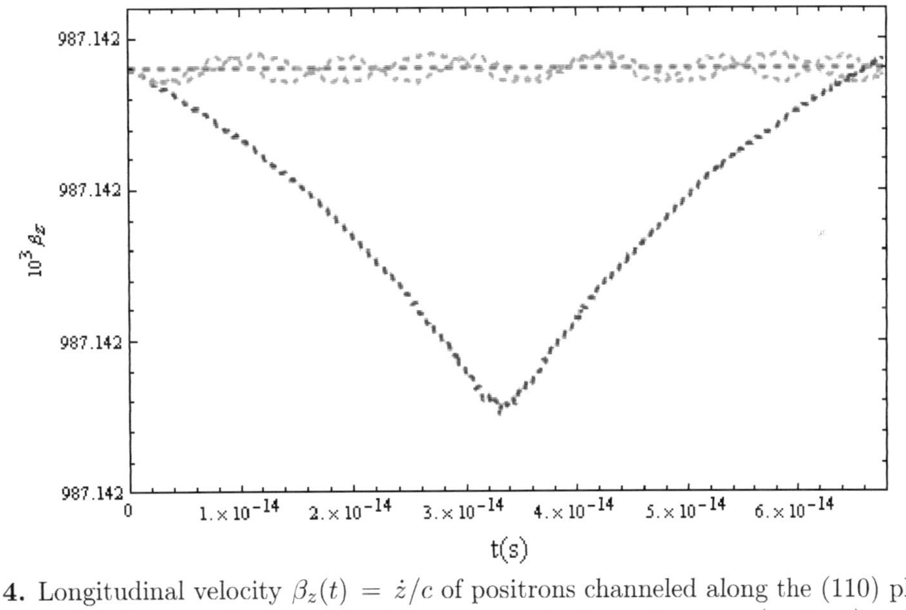

Figure 4. Longitudinal velocity $\beta_z(t) = \dot{z}/c$ of positrons channeled along the (110) plane of a $PbTiO_3$ crystal at impact of resonant and non-resonant ultrasonic waves (see text).

electron channeling [21, 22, 23] succeeded only for selected CR transitions at appropriate tuning of the US frequency.

4. Stimulation of CR emission by ultrasound

The radiation emitted by channeled positrons moving through the $PbTiO_3$ crystal on oscillating trajectories as found in Sect. 3 has been calculated according to (6). The corresponding spectral-angular distributions of forward emitted CR are shown in figure 5. At impact of US resonant with

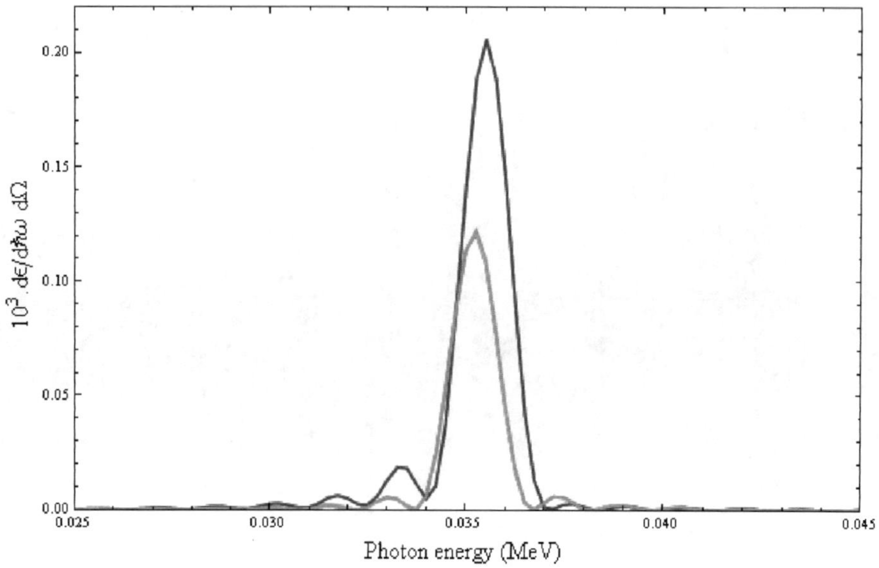

Figure 5. Spectral-angular distributions of forward emitted CR for positrons of energy 40 MeV impinging with $\Theta_0 = 0$ at $x_0 = 0.235\,\mathring{A}$ and channeled along the (110) plane of a $10\,\mu$m thick $PbTiO_3$ crystal at impact of resonant (larger amplitude) and non-resonant ultrasonic waves.

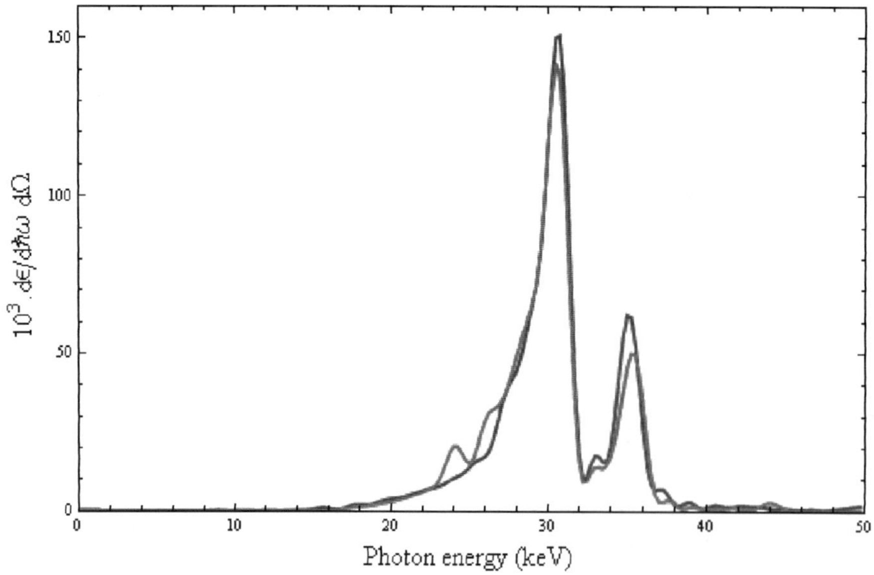

Figure 6. Spectral-angular distributions of forward emitted CR for positrons of energy 40 MeV channeled along the (110) plane of a $10\,\mu$m thick $PbTiO_3$ crystal at impact of US with wave number $k_s = 1.488 \times 10^{-3}\,\mathring{A}^{-1}$ (larger amplitude) compared to the CR spectrum without US (smaller amplitude).

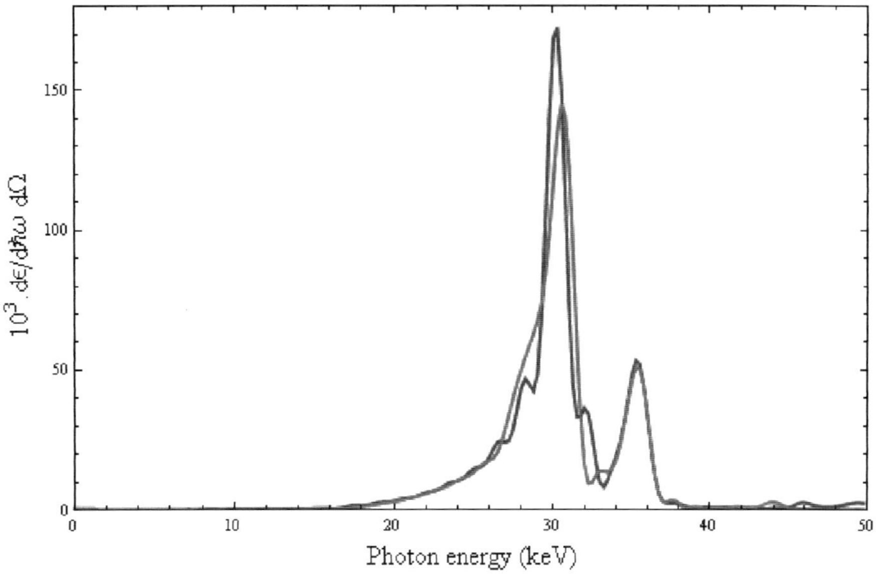

Figure 7. Spectral-angular distributions of forward emitted CR for positrons of energy 40 MeV channeled along the (110) plane of a 10 μm thick $PbTiO_3$ crystal at impact of US with wave number $k_s = 1.282 \times 10^{-3}\,\AA^{-1}$ (larger amplitude) compared to the CR spectrum without US (smaller amplitude).

the oscillatory motion of channeled particles one observes an enhancement of the CR intensity (upper curve in figure 5) at a photon energy of $E_{CR} = 2\gamma^2\,\hbar 2\pi/T$. Non-resonant US does not influence CR emission (lower curve in figure 5).

Naturally, variation of the initial conditions results in varied trajectories. Therefore, averaging over relevant initial parameters is necessary to simulate the residual CR spectrum. For a parallel positron beam ($\Theta_0 = 0$) one would observe CR spectra as shown in figure 6, where the upper spectrum reveals the influence of US with wave number $k_s = 1.488 \times 10^{-3}\,\AA^{-1}$, and the lower one depicts CR emission without US. The spectra are characterised by two maxima (CR "lines"). From comparison with figure 5 one may conclude that the relative influence of US with given k_s on the CR intensity is larger for the smaller CR peak at about 35 keV. A second example calculated for $k_s = 1.282 \times 10^{-3}\,\AA^{-1}$ is shown in figure 7. In this case, US mainly stimulate the intensity of the dominant first CR peak at about 30 keV.

5. Summary

This work presents first classical trajectory calculations for channeling in an ultrasonic superlattice. As may easily be realised, resonant ultrasound influences the transverse amplitude of the oscillatory motion as well as the transverse and longitudinal velocities of the channeled particle. Its longitudinal momentum is no more a conserved quantity. A resonant ultrasonic field changes the CR intensity as has been demonstrated by simulations of CR spectra for planar positron channeling in a $PbTiO_3$ single crystal. The possible magnitude of CR stimulation is, of course, subject to experimental investigations.

Due to the unharmonicity of the continuous potential of the crystallographic (110) plane of $PbTiO_3$, features found for positron channeling resemble planar electron channeling. Up to now, stimulation of CR emission by ultrasonic waves has only been proved for electron channeling in piezoelectric quartz crystals. Suitable positron beams were not available. However, efforts to modify the positron beam line of the Beam Test Facility at INFN LNF Frascati, Italy, are reported in [28] recently. The verification of the observed effect of CR intensity enhancement by resonant ultrasonic waves on other crystals and/or beams of relativistic charged particles is, therefore, of principal interest.

References

[1] Kumakhov M A 1976 *Phys. Lett.* **57** 17

[2] Sàenz M A and Überall H 1985 *Coherent Radiation Sources* (Berlin-Heidelberg: Springer)

[3] Kumakhov M A and Komarov F F 1989 *Radiation from Charged Particles in Solids* (New York: AIP)

[4] Kumakhov M A and Weddel R 1991 *Radiation of Relativistic Light Particles during Interaction with Single Crystals* (Heidelberg: Spektrum)

[5] Akhiezer A I and Shul'ga N F 1996 *High-Energy Electrodynamics in Matter* (Amsterdam: Gordon and Breach)

[6] Rullhusen P, Artru X and Dhez P 1998 *Novel Radiation Sources Using Relativistic Electrons* (Singapore: World Scientific)

[7] Baier V N, Katkov V M and Strakhovenko V M 1998 *Electromagnetic Processes at High Energy in Oriented Single Crystals* (Singapore: World Scientific)

[8] Andersen J U, Andersen S K and Augustyniak W M 1977 *K. Dan. Vidensk. Selsk. Mat. Fys. Medd.* **39** Nr. 10

[9] Lindhard J 1964 *Phys. Lett.* **12** 126

[10] Chadderton L T 1973 in: *Channeling* ed Morgan D V (London: Wiley) p. 287

[11] Baryshevsky V G, Dubovskaya I Ya and Grubich A O 1980 *Phys. Lett.* A **77** 61

[12] Mkrtchyan A R, Gasparyan R H and Gabrielyan R G 1986 *Phys. Lett.* A **115** 410

[13] Mkrtchyan A R, Gasparyan R H, Gabrielyan R G and Mkrtchyan A G 1988 *Phys. Lett.* A **126** 528

[14] Dedkov G V 1994 *Phys. Status Solidi* B **184** 535

[15] Avakyan R O *et al.* 1988 *Pis'ma Zh. Eksp. Teor. Fiz.* **48** 577
Avakyan R O *et al.* 1990 *Nucl. Instrum. Methods* B **48** 266

[16] Grigoryan L Sh *et al.* 2000 *Radiat. Eff. Defects Solids* **152** 225; 269 **153** 13; 212; 289; 307

[17] Grigoryan L Sh *et al.* 2001 *Nucl. Instrum. Methods* B **173** 132; 184

[18] Grigoryan L Sh *et al.* 2003 *Nucl. Instrum. Methods* B **201** 25

[19] Wagner W, Azadegan B, Grigoryan L Sh and Pawelke J 2007 *EPL* **78** 56004

[20] Azadegan B, Grigoryan L Sh, Pawelke J and Wagner W 2008 *J. Phys.* B **41** 325101

[21] Wagner W, Azadegan B, Büttig H, Pawelke J, Sobiella M and Grigoryan L Sh 2010 *Proc. Int. Conf. "Channeling 2008" (Erice, Italy)* ed Dabagov S B and Palumbo (Singapore: World Scientific) p. 378

[22] Wagner W, Azadegan B, Büttig H, Mkrtchyan A R, Grigoryan L Sh and Pawelke J 2010 *Proc. Int. Conf on Electrons, Positrons, Neutrons and X-Rays Scattering under External Influences (Yerevan-Meghri, Armenia, 2009)* ed Mkrtchyan A R, Bushuev V A and Roshchupkin D V (Yerevan: NAS IAPP Press) p. 35

[23] Wagner W, Azadegan B, Büttig H, Mkrtchyan A R, Grigoryan L Sh and Pawelke J 2011 *Nuovo Cimento* C **34** N. 4, 133

[24] Mkrtchyan A R, Grigoryan L Sh, Mkrtchyan A H, Khachatryan H F, Wagner W and Azadegan B 2010 *Proc. Int. Conf on Electrons, Positrons, Neutrons and X-Rays Scattering under External Influences (Yerevan-Meghri, Armenia, 2009)* ed Mkrtchyan A R, Bushuev V A and Roshchupkin D V (Yerevan: NAS IAPP Press) p. 44

[25] Bogdanov O V, Korotchenko K B and Pivovarov Yu L 2008 *J. Phys.* B **41** 055004

[26] Grigoryan L Sh, Mkrtchyan A H, Khachatryan H F, Wagner W, Saharian A A and Baghdasaryan K S 2003 *Nucl. Instrum. Methods* B **212** 51

[27] Doyle P A and Turner P S 1968 *Acta Crystallogr.* A **24** 390

[28] Backe H, Krambrich D, Lauth W, Buonomo B, Dabagov, S B, Mazzitelli G, Quintieri J, Lundsgaard Hansen J, Uggerhøj U I, Azadegan B, Dizdar A and Wagner W, *Nuovo Cimento* C **34** N. 4, 175

Incoherent bremsstrahlung in flat and bent crystals

N F Shul'ga[1], V V Syshchenko[2] and A I Tarnovsky[2]

[1] Akhiezer Institute for Theoretical Physics of the NSC "KIPT", Akademicheskaya Street, 1, Kharkov 61108, Ukraine

[2] Belgorod State University, Pobedy Street, 85, Belgorod 308015, Russian Federation

E-mail: shulga@kipt.kharkov.ua, syshch@bsu.edu.ru, syshch@yandex.ru

Abstract. Incoherent bremsstrahlung by high-energy particles in crystal is due to the thermal spread of atoms in relation to their equilibrium positions in the lattice. The simulation procedure developed earlier for the incoherent radiation is applied to the case of the electrons and positrons motion in the sinusoidally bent crystal. The results of simulation are in agreement with the data of recent experiments carried out at the Mainz Microtron MAMI. The possibility of use of the sinusoidally bent crystals as undulators is discussed.

1. Introduction

The bremsstrahlung cross section for relativistic electrons in a crystal is split into the sum of the coherent part (due to the spatial periodicity of the atoms' arrangement in the crystal) and the incoherent one (due to the thermal motion of atoms in the crystal) [1, 2, 3]. Although the spectrum of incoherent radiation in crystal is similar to one in amorphous medium, the incoherent radiation intensity can demonstrate substantial dependence on the crystal orientation due to the electrons' flux redistribution in the crystal (channeling etc.). The simulation based on the semiclassical description of the radiation process [4, 5, 6] confirms that viewpoint. The results of simulation are in a good agreement with the corresponding early [7] and recent [8] experimental data.

Here we present the results of simulation using the improved procedure taking into account the crystal deformations. The simulation was carried out under the conditions of the recent experiment performed at the Mainz Microtron MAMI [9] to explore the radiation emission from periodically bent crystal. The possibility of application of such crystals as undulators is discussed during last years [9, 10, 11, 12].

For the reader convenience, in the next section we outline some theoretical ideas of our approach.

2. Computation method

Let us consider the high energy electron incidence on the atomic string in the crystal. The two-dimensional multiple scattering angle $\boldsymbol{\vartheta}$ is equal to the sum of individual scattering angles on the atoms:

$$\boldsymbol{\vartheta} = \sum_n \boldsymbol{\vartheta}(\boldsymbol{\rho}_n), \tag{1}$$

where $\boldsymbol{\rho}_n$ is the impact parameter of the collision with the n-th atom of the crystal. The mean square of the multiple scattering angle (averaged over the thermal vibrations of atoms) can be expressed as the sum of two blocks of terms:

$$\left\langle \left| \sum_n \boldsymbol{\vartheta}(\boldsymbol{\rho}_n) \right|^2 \right\rangle = \sum_{n,m} \langle \boldsymbol{\vartheta}(\boldsymbol{\rho}_n) \rangle \langle \boldsymbol{\vartheta}(\boldsymbol{\rho}_m) \rangle + \sum_n \left\{ \langle \boldsymbol{\vartheta}(\boldsymbol{\rho}_n)^2 \rangle - \langle \boldsymbol{\vartheta}(\boldsymbol{\rho}_n) \rangle^2 \right\}. \qquad (2)$$

The first block describes the coherent scattering which can be interpreted as a motion in the uniform string potential [2, 3, 13]. The second one describes the incoherent scattering of the electron on the thermal vibrations of the lattice atoms.

The bremsstrahlung spectrum of the electrons passing through the crystal also can be expressed as the sum of the coherent and incoherent contributions [1, 2, 3]. For the electrons of the energy $\varepsilon \sim 1$ GeV the main contribution of the coherent effect is made to the soft range of the spectrum (the photon energy $\hbar\omega$ is less or of the order of dozens of MeVs). In the medium and hard ranges of the spectrum the incoherent radiation is predominant.

The spectral density of the incoherent radiation from the individual electron moving on the given trajectory is described by the formula [5, 6]

$$\left(\frac{d\mathcal{E}}{d\omega} \right)_{incoh} = \frac{2e^2\varepsilon(\varepsilon - \hbar\omega)}{3\pi m^2 c^5} \left\{ 1 + \frac{3}{4} \frac{(\hbar\omega)^2}{\varepsilon(\varepsilon - \hbar\omega)} \right\} \sum_n \left\{ \langle \boldsymbol{\vartheta}(\boldsymbol{\rho}_n)^2 \rangle - \langle \boldsymbol{\vartheta}(\boldsymbol{\rho}_n) \rangle^2 \right\}, \qquad (3)$$

where m and e are the electron's mass and charge, c is the velocity of light. It is convenient to compare the incoherent radiation intensity of the uniform electron beam in the crystal to the corresponding intensity in the amorphous medium (described by Bethe-Heitler formula). The ratio of these two values is equal to [5, 6]

$$N_\gamma = \frac{1}{2\pi N R^2 \ln(mRc/\hbar)} \int d^2\rho_0 \sum_n \left\{ \langle \boldsymbol{\vartheta}(\boldsymbol{\rho}_n)^2 \rangle - \langle \boldsymbol{\vartheta}(\boldsymbol{\rho}_n) \rangle^2 \right\}, \qquad (4)$$

where N is the total number of the electron's collisions with atoms under its motion through the crystal, R is Thomas-Fermi radius of the atom, integration over $d^2\rho_0$ means the integration over all possible points of incidence of the electron on the crystal in the limits of one elementary cell.

The values of the function $F(\rho) = \langle \boldsymbol{\vartheta}(\boldsymbol{\rho})^2 \rangle - \langle \boldsymbol{\vartheta}(\boldsymbol{\rho}) \rangle^2$ are determined by linear interpolation of the values pre-calculated on the regular grid of impact parameters. The impact parameters $\boldsymbol{\rho}_n$ are found using the electron's trajectory obtained by numerical integration of the equation of motion in the field of the set of parallel uniform strings.

The incoherent scattering of the electron on the thermal vibrations of atoms (described by the second term in (2)) is taken into into account in our algorithm as follows. We add to each component of the two-dimensional vector $\boldsymbol{\vartheta}$ (that describe the electron's motion direction) the random value with Gaussian distribution with the dispersion $F(\rho_n)/2$ after each collision with the lattice atom. In this case the dispersion of the angle of multiple incoherent scattering will be right equal to the second term of (2).

For the further computational details see [5, 6].

3. Results of simulation

The origin of the orientation dependence of the incoherent radiation intensity is illustrated on the figure 1. When the electron is incident to the atomic plane of the crystal under angle θ less than some critical angle θ_c, it can be captured by the attractive potential of the plane. The finite motion in such potential is called as the planar channeling (see, e.g., [2, 3, 13]).

Under planar channeling the electrons collide with atoms under small impact parameters more frequently than in amorphous medium, that leads to the increase of the incoherent

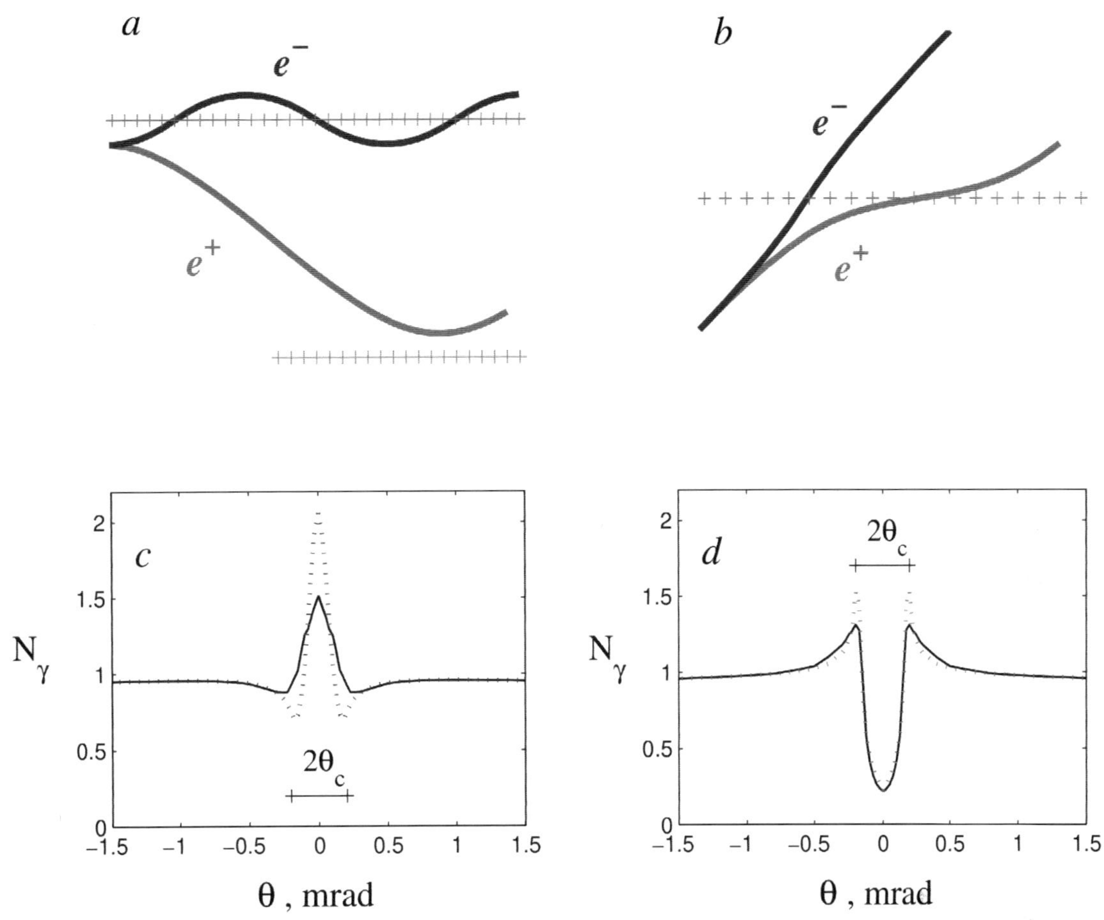

Figure 1. (a) Typical trajectories of the electrons and positrons under planar channeling. Pluses mark the positions of atomic strings (perpendicular to the figure plane) forming the atomic planes of the crystal. The horizontal scale of the figure is highly compressed. **(b)** The same for above-barrier motion. **(c)** Simulated incoherent bremsstrahlung intensity (in ratio to the Bethe-Heitler intensity in amorphous medium, see (4)) from 1 GeV electrons vs incidence angle θ to (110) plane of 30 μm thick Si crystal [4]. Dotted line corresponds to the trajectories simulated neglecting thermal vibrations of atoms. **(d)** The same for positrons.

bremsstrahlung intensity; for above-barrier motion the situation is opposite. The account of the incoherent scattering of the particles on the thermal vibrations of the atoms leads to dechanneling and, hence, to the smoothing of the described orientation dependence (compare solid and dotted lines on the figure 1 (c) and (d)).

In the present article the simulation was carried out under the conditions of the experiment [9], where the radiation from $\varepsilon = 855$ MeV electrons under their incidence onto the silicon crystal with sinusoidally bent (110) planes had been studied. The yield of photons with the energy $\hbar\omega = \varepsilon/2$ (for which the incoherent mechanism of bremsstrahlung is predominant) had been registered.

The results of simulation (figure 2) demonstrate the qualitative agreement with the experimental data [9]. We can see characteristic structures similar to one on the figure 1 (c), generated by different crystallographic planes with the common $[1\bar{1}0]$ axis.

Journal of Physics: Conference Series **357** (2012) 012026 doi:10.1088/1742-6596/357/1/012026

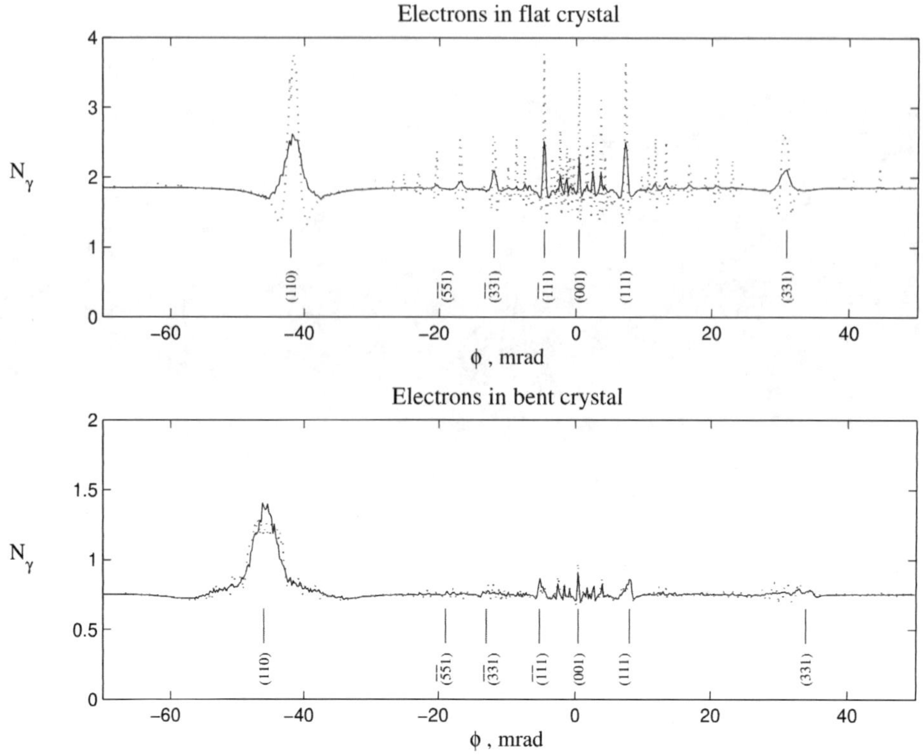

Figure 2. The incoherent bremsstrahlung intensity (in ratio to Bethe-Heitler intensity in amorphous medium) from 855 MeV electrons in flat (upper plot) and sinusoidally bent (lower plot) silicon crystals under scanning of the goniometric angle ϕ like in the experiment [9].

The decrease of the radiation intensity in comparison to the reference flat crystal permits to estimate the reduction of the dechanneling length due to the crystal bending. We can see that the bending of the crystallographic planes increases the dechanneling rate so highly that the scattering on the thermal vibrations of atoms already have no substantial influence on the incoherent radiation intensity (compare solid and dashed curves on the lower panel of the figure 2).

The main cause of the rapid dechanneling in the bent crystal lies in the arising of the centrifugal addition to the planar potential [13]:

$$U_{eff} = U(x) - \varepsilon \frac{x}{R_b} \quad \text{under} \quad |x| \gg R_b \tag{5}$$

(where R_b is the bending radius) and, as a consequence, to the reduction of the potential barriers between which the channeling could take the place. For the crystal [9] the bending radius $R_b \gtrsim 6.2 \cdot 10^{-3}$ m. Assuming the planar potential has the shape of quadratic parabola with the depth $U_0 \approx -23.5$ eV for (110) plane, the effective potential (5) in the domain of maximal curvature of the crystal would have the shape presented by the curve 2 on the figure 3. We see substantial decrease of the potential wells depth comparing to the reference flat crystal (curve 1). So, only a small part of incident particles could be captured into the channeling regime.

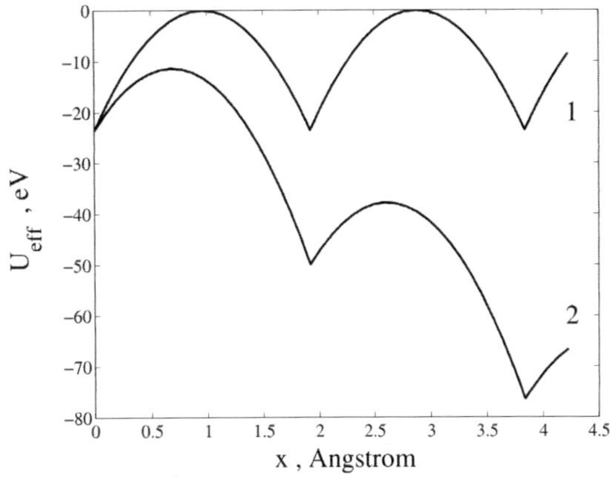

Figure 3. Potential energy of the electron in the planar potentials of the flat (curve 1) and bent (curve 2) crystals.

4. Conclusion

The incoherent bremsstrahlung yield exceedance (deficiency) from the electron (positron) beam in the oriented crystal in comparison to the value predicted by Bethe-Heitler formula for the amorphous medium can be used as an indicator of the relative amount of the particles moving in the channeling regime.

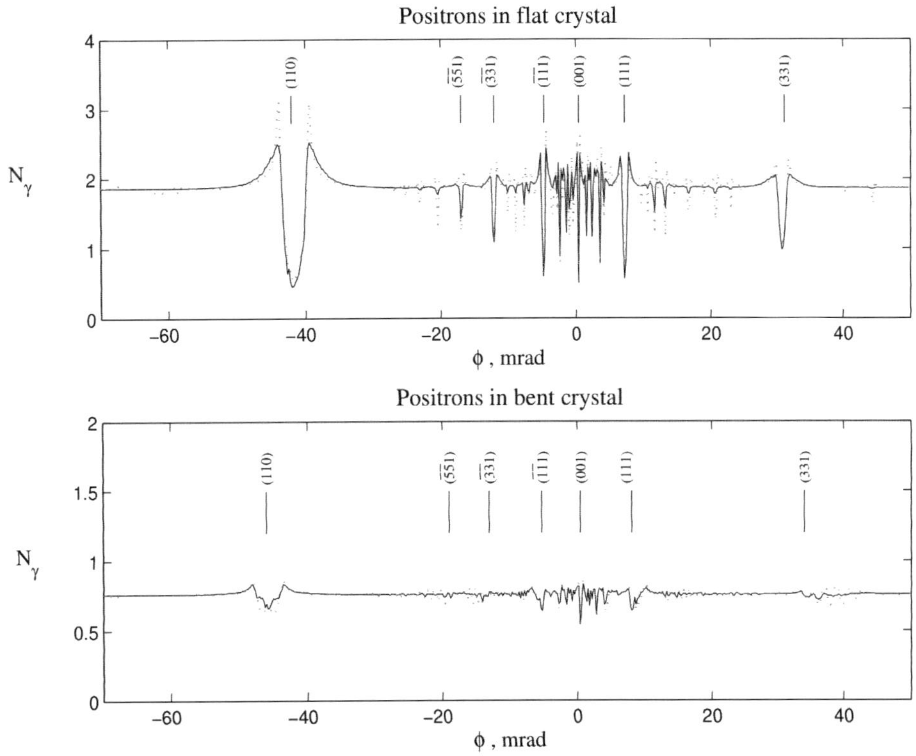

Figure 4. The same as on the figure 2 for positrons.

The results of simulation demonstrate that the planar channeling effect leads to the orientation dependence of the incoherent bremsstrahlung intensity not only in flat, but also in bent crystals. However, in the bent crystal the electrons rapidly leave the planar channels. The figure 4 demonstrates that for the positron beam instead of the electron one the dechanneling rate would be almost the same. The effective capture of particles into planar channels (and so the use of sinusoidally bent crystals as undulators) is possible under changing the bending parameters to ensure smaller curvature than in the crystal [9].

Acknowledgments

We are grateful to Professor Hartmut Backe for the attention to our work and very helpful discussions.

This work is supported in part by internal grant of Belgorod State University and the federal program "Academic and Teaching Staff of innovation Russia", the government contract 16.740.11.0147 dated 02.09.2010.

References

[1] Ter-Mikaelyan M L 1972 *High-Energy Electromagnetic Processes in Condensed Media* (New York: Wiley-Interscience)
[2] Akhiezer A I and Shul'ga N F 1996 *High-Energy Electrodynamics in Matter* (Amsterdam: Gordon and Breach)
[3] Uggerhøj U I 2005 *Rev. Mod. Phys.* **77** 1131
[4] Shul'ga N F and Syshchenko V V 2005 *Nucl. Instr. and Meth.* B **227** 125
[5] Shul'ga N F, Syshchenko V V and Tarnovsky A I 2008 *Nucl. Instr. and Meth.* B **266** 3863
[6] Shul'ga N F, Syshchenko V V and Tarnovsky A I 2010 *J. Phys.: Conf. Series* **236** 012027
[7] Sanin V M, Khvastunov V M, Boldyshev V F and Shul'ga N F 1992 *Nucl. Instr. and Meth.* B **67** 251
[8] Backe H, Kunz P, Lauth W and Rueda A 2008 *Nucl. Instr. and Meth.* B **266** 3835
[9] Backe H, Krambrich D, Lauth W, Lundsgaard Hansen J and Uggerhøj U I 2011 *Nuovo Cimento* C **134** 04 157
[10] Bellucci S et al. 2004 *Phys. Rev.* ST-AB **7** 023501
[11] Tabrizi M, Korol A V, Solov'yov A V and Greiner W 2007 *Phys. Rev. Lett.* **98** 164801
[12] Backe H, Lauth W, Kunz P, Rueda A, Esberg J, Kirsebom K, Lundsgaard Hansen J and Uggerhøj U I 2010 *Charged and Neutral Particles Channeling Phenomena* ed S B Dabagov and L Palumbo (Singapore: World Scientific) p 281
[13] Akhiezer A I, Shul'ga N F, Truten' V I, Grinenko A A and Syshchenko V V 1995 *Physics-Uspekhi* **38** 1119

Journal of Physics: Conference Series **357** (2012) 012027 doi:10.1088/1742-6596/357/1/012027

Planar channeling radiation by relativistic electrons in different structures of silicon carbide

B Azadegan[1], **S B Dabagov**[2,3] **and W Wagner**[4]
[1] Sabzevar Tarbiat Moallem University - P.O. 397, Sabzevar, Iran
[2] INFN, Laboratori Nazionali di Frascati - Via E. Fermi 40, 00044 Frascati (RM), Italy
[3] RAS - P.N. Lebedev Physical Institute - Leninsky Pr. 53, 119991 Moscow, Russia
[4] Helmholz-Zentrum Dresden-Rossendorf - P.O.Box 510119, 01314 Dresden, Germany

E-mail: azadeganb@gmail.com

Abstract. Spectral distributions of channeling radiation by relativistic electrons channeled in different crystallographic planes of different types of polytype silicon carbide crystals such as hexagonal, zincblende and rhombohedral are presented. For every structure we have found the planes where the emission of channeling radiation is possible. Using the Doyle-Turner approximation to the atomic scattering factor and taking in to account thermal vibrations of crystal atoms, the continuum potentials for different planes in different structures of a SiC single crystal have been calculated. In the framework of quantum mechanics, the theory of channeling radiation was applied to calculate transverse electron states, transition energies, radiation linewidths, the depth dependence of the occupation of quantum states and spectral-angular distributions of channeling radiation. At electron energies larger than 100 MeV, the trajectories, velocities and accelerations of channeled electrons have been obtained by means of classical mechanics, and the spectral-angular distributions were calculated by means of classical electrodynamics using realistic trajectories, velocities and accelerations of channeled electrons. Specific properties of planar channeling radiation generated in different structures of SiC are discussed.

1. Introduction

Channeling radiation (CR) is emitted by relativistic charged particles during passing through a single crystal nearly parallel to some crystallographic axis or plane. Electron CR has mainly been studied on monatomic crystals such as diamond [1, 2, 3], Si [4, 5, 6], Ge [7, 8] and on metals (*e.g.* Be, Ni, Sb, W [9, 10, 11]). Binary or polyatomic crystals (*e.g.* LiH [12], LiF [13, 14], GaAs, ruby [15]) have scarcely been utilized. Most of the above mentioned crystals are characterized by a cubic lattice. In such crystals, intense CR is observed from channeling along planes with small indices such as (100), (110) and (111). In our previous papers [16, 17] we investigated CR generated in different planes of a quartz crystal (SiO_2) which has a hexagonal structure. The more complicate structure of the binary quartz crystal (compared with cubic crystals such as, *e.g.*, diamond, Si or Ge) allows for planes with relatively large indices, which have rather deep continuum potentials favouring CR emission. One may observe some specific properties of CR also on other types of crystals. The most remarkable feature of the silicon carbide (SiC) crystal structure is its polytypism. There is a great variety of SiC polytypes. Therefore, it has become accepted practice to refer to the cubic polytype as b-SiC and to all non-cubic structures (hexagonal and rhombohedral) as a-SiC. Despite the large number of well documented SiC polytypes, only very few of them are considered to be thermodynamically stable. However, there appear to be five small-period polytypes (2H, 3C, 4H, 15R and 6H) which are considered as the basic SiC structures because they are found to occur most frequently. In the present paper, planar CR on 3C-SiC (zincblende), 4H-SiC (hexagonal) and 9R-SiC (rhombohedral) is investigated.

2. Theory

2.1. Planar channeling radiation

For planar channeling, the relativistic particle must enter the crystal under a small incidence angle with respect to the crystal plane considered. The description of channeling is based on the division of the particle motion into a longitudinal and a transverse component. Since in first approximation the longitudinal component along the channeling plane is not affected by forces, the velocity in this z-direction is actually nearly constant, $v_z \approx c$. However, under channeling condition, the transverse momentum of the channeled particle, p_x, is small compared to the longitudinal one, p_z. Therefore, its transverse energy may be defined by the expression

$$E_x = \frac{p_x^2}{2m\gamma} + V(x) \tag{1}$$

where $m\gamma$ is the relativistic mass and γ is the Lorentz factor [18]. In order to describe the radiation emitted by the charged particle during channeling, the applicability of both the classical and the quantum theory has to be considered.

2.2. Continuum potential

Since the longitudinal component of the electron velocity approaches the speed of light, c, the crystal plane is assumed to be charged continuously, and $V(x)$ is approximated by the continuum potential [19]. This potential has the periodicity of the lattice. Therefore, its expansion into a Fourier series represents its most general form which reads

$$V(x) = \sum_{n=1,2,...} V_n \exp(ingx) \qquad n = ... -1, 0, 1, 2, ... \tag{2}$$

where V_n denote the Fourier coefficients of the periodic potential. Using the Doyle-Turner approach [21] for the electron-atom interaction, they can be written as

$$V_n = -\frac{2\pi a_0 e^2}{V_c} \sum_j \exp[-M_j(\vec{g})] \cdot \exp(-i\vec{g} \cdot \vec{r_j}) \sum_{i=1}^{4} a_i \exp[-\frac{1}{4}(\frac{b_i}{4\pi^2})(ng)^2] \tag{3}$$

where V_c is the volume of unit cell, a_0 is the Bohr radius, e is the electron charge, $\vec{r_j}$ represents the coordinates of the j atoms in the unit cell, a_i and b_i are tabulated coefficients [21], and $M_j(\vec{g}) = g^2\langle u_j^2\rangle/2$ denotes the Debye-Waller factor which describes the thermal vibration of the jth atom by a mean-squared amplitude $\langle u_j^2\rangle$. The term $\sum \exp(-i\vec{g} \cdot \vec{r_j})$ in (3) is the structure factor of the crystal. The structure factor determines those sets of Miller indices, (hkl), for which the reflections from the corresponding planes are strong.

2.3. Classical description of planar channeling radiation

In the framework of the classical model, the scattering from the ordered crystal atoms becomes coherent now, causing an oscillatory motion of the electrons along the corresponding plane of atoms. Considering this oscillatory motion as an accelerated one within the rest frame of the electron, it emits electromagnetic radiation called channeling radiation. Although the oscillation frequency, ω_0, is rather low and corresponds to a radiation energy, $\hbar\omega_0$, in the optical region, relativistic effects such as the Lorentz contraction of the longitudinal coordinate and the Doppler effect transform the energy of emitted CR photons observed in beam direction into the domain of X-rays.

The equation of motion for the relativistic particle in the one-dimensional continuum potential, $V(x)$, has the form

$$\gamma m\ddot{x} = -\frac{\partial V(x)}{\partial x}. \tag{4}$$

The initial conditions are: the point of incidence into the crystal, $x(0) = x_0$, and the transverse momentum $p_x(0) = p\theta_0$ (θ_0 is the angle of incidence with respect to the plane) which define the initial transverse energy (1) as

$$E_x = \frac{p^2\theta_0^2}{2\gamma m} + V(x_0). \tag{5}$$

The energy density radiated into a solid angle $d\Omega$ and a frequency interval $(\omega, \omega + d\omega)$ has the following form [18]

$$\frac{d^2E}{d\omega d\Omega} = \frac{e^2}{4\pi^2 c}\left|\int_0^\tau \exp[i(\omega t - \vec{k}\cdot\vec{r})]\frac{\vec{n}\times((\vec{n}-\vec{\beta})\times\dot{\vec{\beta}})}{(1-\vec{\beta}\cdot\vec{n})^2}dt\right|^2 \tag{6}$$

where $\vec{\beta} = \dot{\vec{r}}(t)/c$ is the particle velocity, $\vec{r}(t) = \vec{v_z}t + x(t)$ is its trajectory, $\vec{k} = \omega\vec{n}/c$ is the wave vector, $\vec{n}$ is the unit vector defining the direction of photon emission, and τ is the time-of-flight of the particle through the crystal. Classical calculations of CR spectra are based on (6).

2.4. Quantum description of planar channeling radiation

In the framework of quantum mechanics, the transverse motion of planar channeled electrons of mass γm may be described by a one-dimensional Schrödinger equation which contains the averaged potential of the crystal. This motion is restricted to discrete (bound) channeling states of the planar continuum potential. Spontaneous transitions between these eigenstates lead to the emission of CR, the energy spectrum of which is consequently characterized by a line structure of possible photon energies. For electrons of relatively low energy, the quantum theory of CR is based on a Bloch-wave solution

$$\psi(x) = \exp(ikx)\sum_n c_n \exp(ingx) \qquad n = ... -1, 0, 1, 2, ... \tag{7}$$

of the Schrödinger equation

$$-\frac{\hbar^2}{2\gamma m}\frac{d^2\psi(x)}{dx^2} + V(x)\psi(x) = E\psi(x) \tag{8}$$

for incident particles passing through the periodic planar potential $V(x)$. Here $\vec{k}$ is the electron wave vector, and $\vec{g}$ is the reciprocal lattice vector of the plane. Substitution of (2) and (7) into (8) reduces the problem to the calculation of the eigenvalues of a matrix **A**, which in the planar case consist of the components

$$A_{nm} = V_{n-m} \;\; for\,(n \neq m)\,; \qquad A_{nn} = \frac{\hbar^2}{2\gamma m}(k+ng)^2 + V_0. \tag{9}$$

Since the potential as well as the eigenvalues converge very fast, one can restrict the solution to 21 Bloch functions what leads to a 21×21 matrix problem.

For planar CR directed along the electron motion, the spontaneous transition rate per unit of solid angle, per unit of photon energy, per unit of crystal thickness and per electron is given by

$$\frac{d^3N_{CR}(i\to f)}{d\Omega_\gamma dE_\gamma dz}=\frac{\alpha\lambda_c^2}{\pi c\hbar}\int dkE_k^0\mid\langle\psi_{kf}\mid\frac{d}{dx}\mid\psi_{ki}\rangle\mid^2\frac{(\Gamma_k/2)}{(E_\gamma-E_k^0)^2+0.25\Gamma_k^2}P_{ki}(z) \qquad (10)$$

where $\langle\psi_{kf}\mid\frac{d}{dx}\mid\psi_{ki}\rangle$ is the transition matrix element, $P_{ki}(z)$ gives the occupation probability of the channeling state i as function of initial population and crystal thickness, $E_k^0=2\gamma^2(\epsilon_{ki}-\epsilon_{kf})$ means the photon energy of the transition $i\to f$, and Γ_k is the intrinsic CR linewidth. It should be underlined that the integral in (10) has to be taken over the Bloch momenta, k. A second integral may be taken over the occupation P_i of the state i along a given crystal thickness $z=L$ to calculate the spectral-angular distribuition of CR. Both the intrinsic CR linewidth and the occupation probability of the state i are calculated as given in [3] and [17], respectively.

2.5. Classical versus quantum model for planar CR on SiC

The applicability of classical or quantum approach for CR calculations may be checked by estimation of the number of states within the continuum potential [18]. At low energies (units up to tens of MeV) the particle motion has quantum character. The line structure of the CR spectrum is clearly pronounced, while band-structure effects appear for high lying levels. Calculations and measurements for main crystallographic directions of Si and diamond crystals show that the number of bound states is relatively small for electron energies less than 100 MeV. Therefore, a quantum mechanical description of CR is necessary. With increasing electron energy, the density of bound states increases what eventually leads to an overlap of states while band-structure effects become less important. This situation can be handled by a classical calculation [20].

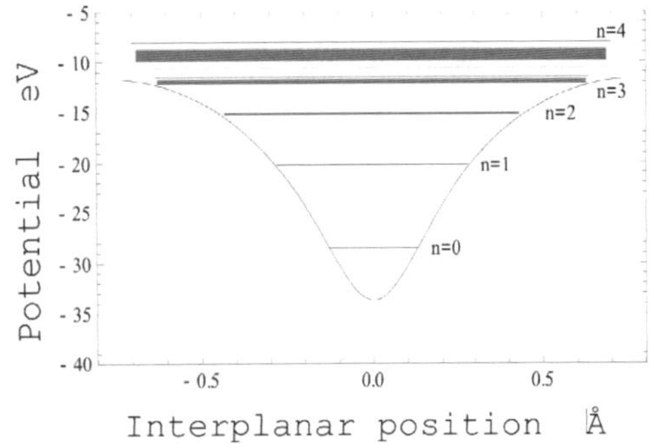

Figure 1. Continuum potential of the (11$\bar{2}$0) plane of a 4H-SiC crystal and bound states for electrons of energy 20 MeV.

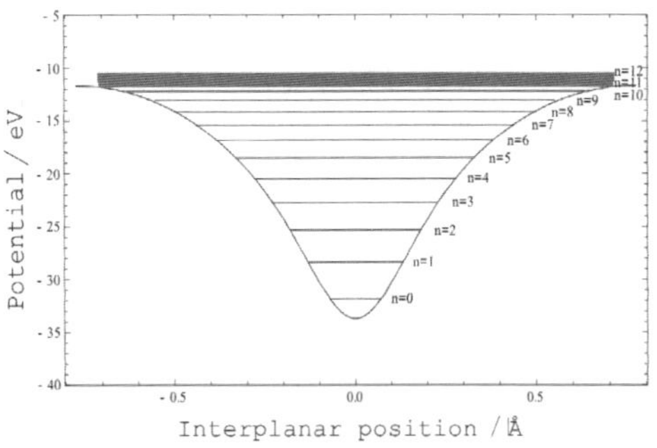

Figure 2. Continuum potential of the (11$\bar{2}$0) plane of a 4H-SiC crystal and bound states of channeled electrons of energy 200 MeV.

3. Crystals

Polytypes of SiC differ in the ordered sequence of closely packed Si-C biplanes along the stacking axis (c-axis). Each polytype has a symbol consisting of a number followed by a letter, where the number denotes the number of layers in one repeat unit along the c-axis and the letter denotes the symmetry. There is one cubic (C) polytype: 3C-SiC. Other polytypes have either hexagonal (H) or rhombohedral (R) symmetry. The primitive unit cell of 3C-SiC contains one Si-C unit, whereas the primitive unit cells of the nH or 3nR polytypes contain n Si-C units.

Journal of Physics: Conference Series **357** (2012) 012027 doi:10.1088/1742-6596/357/1/012027

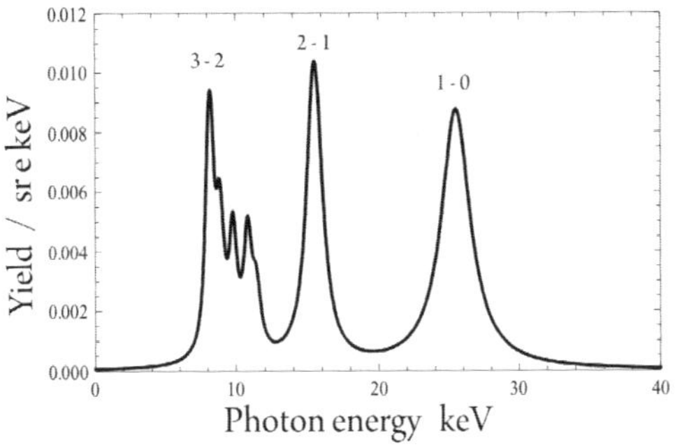

Figure 3. Simulation of the CR spectrum for electrons of energy 20 MeV channeled in the $(11\bar{2}0)$ plane of a 5μm thick 4H-SiC crystal.

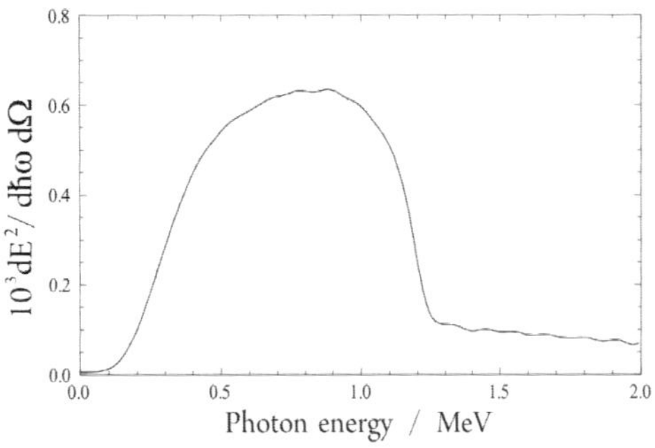

Figure 4. Averaged over points of incidence spectral-angular distribution of CR for electrons of energy 200 MeV channeled in the $(11\bar{2}0)$ plane of a 5μm thick 4H-SiC crystal.

3.1. SiC(4H)

A 4H-SiC crystal is characterized by a hexagonal structure containing 4 Si and 4 C atoms in the unit cell. The space group is C^4_{6v}-P6$_3$mc. The primitive vectors in the Cartesian coordinates system are

$\vec{a_1} = a\frac{1}{2}\hat{x} - \frac{\sqrt{3}}{2}\hat{y}$, $\vec{a_2} = a\frac{1}{2}\hat{x} + \frac{\sqrt{3}}{2}\hat{y}$, $\vec{a_3} = c\hat{z}$ with $a = 3.0805\,\AA$ and c=10.0848 Å [22].

The coordinates of the Si and C atoms in the hexagonal coordinates system are

$Si(I)\ : \{0\,,0\,,Z_1c\},\ \ \{0\,,0\,,(Z_1+1/2)c\}$

$C(I)\ : \{0\,,0\,,Z_2c\},\ \ \{0\,,0\,,(Z_2+1/2)c\}$

$Si(II)\ : \{\frac{1}{2}a\,,\frac{1}{\sqrt{12}}a\,,Z_3c\},\ \ \{\frac{1}{2}a\,,\frac{-1}{\sqrt{12}}a\,,(Z_3+1/2)c\}$

$C(II)\ : \{\frac{1}{2}a\,,\frac{1}{\sqrt{12}}a\,,Z_4c\},\ \ \{\frac{1}{2}a\,,\frac{-1}{\sqrt{12}}a\,,(Z_4+1/2)c\}$

with Z_1=0.0, Z_2=0.1875, Z_3=0.249825 and Z_4=0.437325. In this form, a crystal plane is given by a set of four indices $(hkil)$ with $i = -(h+k)$ and h, k, l being Miller indices.

3.2. SiC(3C)

The second unit-cell SiC polytype is cubic (space group T^2_d-F43m, unit cell length $a = 4.36\,\AA$). The primitive vectors in the Cartesian coordinates system are

$\vec{a_1} = \frac{1}{2}a(\hat{y}+\hat{z})$, $\vec{a_2} = \frac{1}{2}a(\hat{x}+\hat{z})$, $\vec{a_3} = \frac{1}{2}a(\hat{x}+\hat{y})$

The coordinates of Si and C atoms in the hexagonal coordinates system are

$Si\ : \{0\,,0\,,0\},\ \ C\ : \{\frac{1}{4}a\,,\frac{1}{4}a\,,\frac{1}{4}a\}$

For cubic crystals, the three Miller indices h, k, l describe crystallographic axes and planes.

3.3. SiC(9R)

The third unit-cell SiC polytype is rhombohedral (space group C^4_{3v}-R3m, $a = 3.018\,\AA$, $c = 22.177\,\AA$). The primitive vectors in the Cartesian coordinates system are

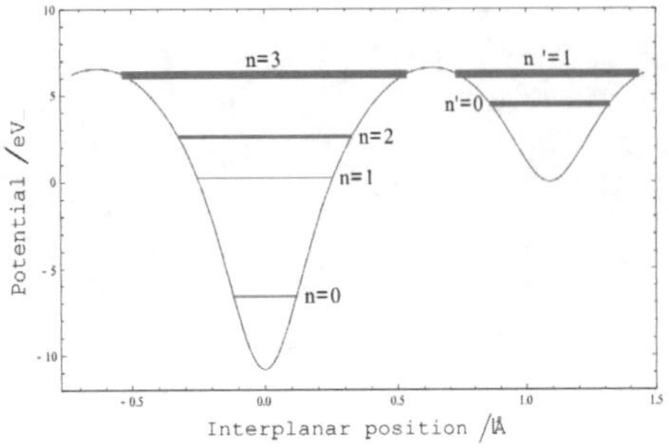

Figure 5. Continuum potential of the (110) plane of a 3C-SiC crystal and bound states of channeled electrons of energy 30 MeV.

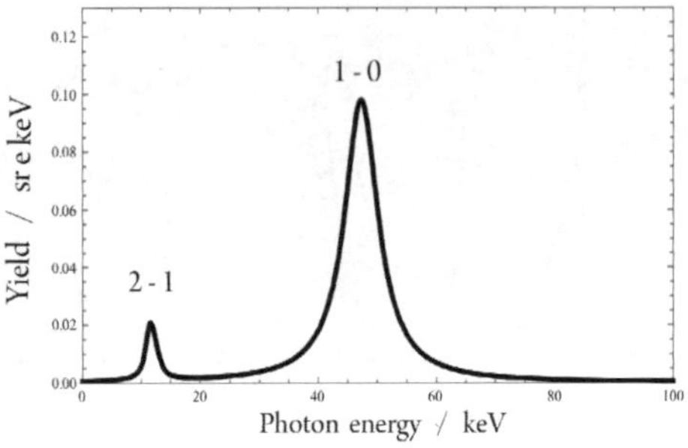

Figure 6. Simulation of the CR spectrum for electrons of energy 30 MeV channeled in the (110) plane of a 5μm thick 3C-SiC crystal.

$\vec{a_1} = \frac{1}{2}a\hat{x} + \frac{1}{2\sqrt{3}}a\hat{y} + \frac{1}{3}c\hat{z}, \;\; \vec{a_2} = -\frac{1}{2}a\hat{x} + \frac{1}{2\sqrt{3}}a\hat{y} + \frac{1}{3}c\hat{z}, \;\; \vec{a_3} = -\frac{1}{\sqrt{3}}a\hat{y} + \frac{1}{3}c\hat{z}$

The coordinates of Si and C atoms are

$C(I) \;\; : \{0,0,x_1c\}, \;\; \{0,0,x_2c\}, \;\; \{0,0,x_3c\}$

$Si(I) \;\; : \{0,0,x_4c\}, \;\; \{0,0,x_5c\}, \;\; \{0,0,x_6c\}$

with $x_1 = 0.222$, $x_2 = 0.0$, $x_3 =$ -0.222, $x_4 = 0.3055$, $x_5 = 0.0833$, $x_6 =$ -0.1388.

4. Calculations

4.1. SiC(4H)

The calculated planar potential for electrons channeled along the $(11\bar{2}0)$ plane of a 4H-SiC crystal, eigenvalues and Bloch bands for the two electron energies of 20 MeV and 200 MeV are shown in figures 1 and 2, respectively. When at 20 MeV the particle motion has clearly quantum character, because there are only four bound states and, hence, three seperate CR lines in the spectrum (figure 3), at 200 MeV one observes 10 bound states. It seems that channeling at this energy has already classical character.

Our calculations of the depth dependence $P_{kn}(z)$ of the occupation of bound and quasi-free states, performed for the electron energy of 20 MeV, showed that the statistical equilibrium of occupation of states is reached at a depth of only about 2μm. The total CR photon yield one obtains by integration of the occupation function over the entire crystal thickness. The result for a 5μm thick 4H-SiC crystal is presented in figure 3.

In order to classically calculate the spectral-angular distribution of CR for electrons of energy 200 MeV, one has to substitute into (6) the trajectories, velocities and accelerations of channeled electrons obtained from numerical solutions of the equation of motion (4) with the potential of the $(11\bar{2}0)$ plane of the 4H-SiC crystal. Since different initial conditions lead to different trajectories and partial spectral-angular distributions of CR as well, one has to average over all possible trajectories. The result for a 5μm thick 4H-SiC crystal is given in figure 4.

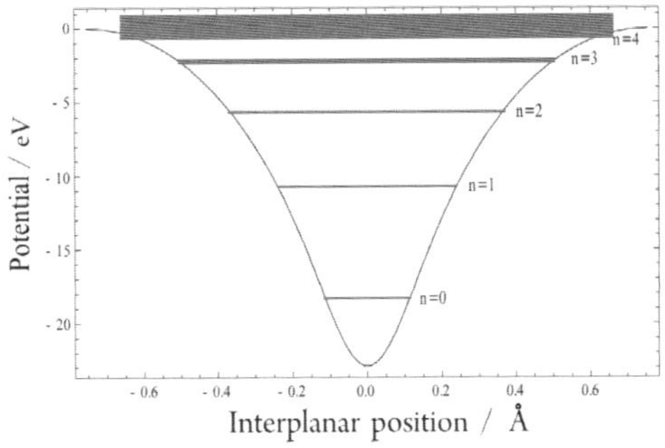

Figure 7. Continuum potential of the $(1\bar{1}0)$ plane of a 9R-SiC crystal and bound states of channeled electrons of energy 30 MeV.

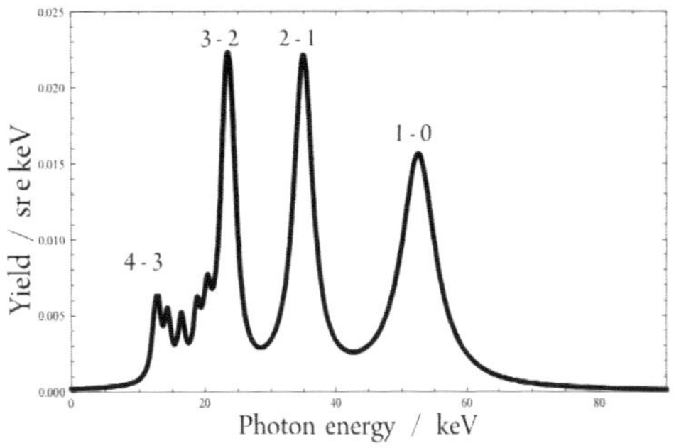

Figure 8. Simulation of the CR spectrum for electrons of energy 30 MeV channeled in the $(1\bar{1}0)$ plane of a 5μm thick 9R-SiC crystal.

4.2. SiC(3C)

The continuum potential calculated for the (110) plane of the cubic 3C-SiC crystal is shown in figure 5. The eigenvalues and Bloch bands correspond to an electron energy of 30 MeV. The potential is formed by a combination of Si and C atomic potentials. It shows alternating segregate potential wells, where the deep ones are formed by Si atoms and smaller modulations are caused by the C atoms. The CR spectrum simulated by means of (10) for electrons of energy 30 MeV channeled in the (110) plane of a 5μm thick 3C-SiC crystal is shown in figure 6.

4.3. SiC(9R)

The continuum potential of the $(1\bar{1}0)$ plane of a 9R-SiC crystal and the corresponding CR spectrum calculated for electrons of energy 30 MeV and a 5μm thick 9R-SiC crystal are presented in figure 7 and 8, respectively.

5. Summary

In the present paper, the quantum mechanical and the classical approach for the description of channeling radiation are explained and applied for simulations of planar electron channeling in different structures of the polytype SiC crystal. At electron energies less than 100 MeV, transverse transition energies, linewidths and spectra of channeling radiation have been calculated numerically applying the well-known many-beam formalism with thermally averaged Doyle-Turner potentials. For higher electron energies, the spectral-angular distributions of channeling radiation were calculated by means of the classical method. Realistic trajectories, velocities and accelerations of channeled electrons have been obtained, and they were used to calculate the photon spectra of channeling radiation according to classical electrodynamics.

References

[1] Gouanere M, Sillou D, Spighel M, Cue N, Gaillard M J, Kirsch R G, Poizat J-C, Remillieux J, Berman B L, Catillon P, Roussel L and Temmer G M 1982 *Nucl. Instrum. Meth.* **194** 225

[2] Reiz I 1999 *Diploma thesis* (Darmstadt: Technical University); Genz H 2001 *Proc. NATO Advanced Research Workshop on Electron-Photon Interaction in Dense Media* (Nor-Hamberd, NAS Armenia)

[3] Azadegan B, Wagner W and Pawelke J 2006 *Phys. Rev.* B **74** 045209

[4] Gouanere M, Sillou D, Spighel M, Cue N, Gaillard M J, Kirsch R G, Poizat J-C, Remillieux J, Berman B L, Catillon P, Roussel L and Temmer G M 1988 *Phys. Rev.* B **38** 4352

[5] Andersen J U and Lægsgaard E 1980 *Phys. Rev. Lett.* **44** 1079

[6] Kephart J O, Berman B L, Pantell R H, Datz S, Klein R R and Park H 1991 *Phys. Rev.* B **44** 1992

[7] Park H, Swent R L, Kephart J O, Pantell R H, Berman B L, Datz S and Fearick R W 1983 *Phys. Lett.* **96A** 45

[8] Chouffani K, *Ü*berall H, Genz H, Hoffmann-Staschek P, Nething U and Richter A 1999 *Nucl. Instrum. Meth.* B **152** 479

[9] Andersen J U, Bonderup E, Lægsgaard E and Sørensen A H 1983 *Phys. Scr.* **28** 308

[10] Gary C K, Fisher A S, Pantell R H, Harris J and Piestrup M A 1990 *Phys. Rev.* B **42** 7

[11] Buschhorn G, Dietrich E, Kufner W, Rzepka M, Genz H, Grf H-D, Hoffmann-Staschek P, Nething U and Richter A 1997 *Phys. Rev.* B **55** 6196

[12] Gary C K, Pantell R H, Özcan M, Piestrup M A and Boyers D G 1991 *J. Appl. Phys.* **70** 2995

[13] Berman B L, Datz S, Fearick R W, Kephart J O, Pantell R H, Park H and Swent R L 1982 *Phys. Rev. Lett.* **49** 474

[14] Swent R L, Pantell R H, Park H, Kephart J O, Klein R K, Datz S, Fearick R W and Berman B L 1984 *Phys. Rev.* B **29** 52

[15] Freudenberger J, Genz H, Groening L, Hoffmann-Staschek P, Knpfer W, Morokhovskii V L, Morokhovskii V V, Nething U, Richter A and Sellschop J P F 1996 *Nucl. Instrum. Meth.* B **119** 123

[16] Wagner W, Azadegan B, Grigoryan L Sh and Pawelke J 2007 *EPL* **78** 56004

[17] Azadegan B, Grigoryan L Sh, Pawelke J and Wagner W 2008 *J. Phys.* B **41** 325101

[18] Kumakhov M A and Weddel R 1991 *Radiation of Relativistic Light Particles during Interaction with Single Crystals* (Heidelberg: Spektrum)

[19] Lindhard J 1965 *Kg. Danske Vid. Selsk. Mat. Fys. Medd.* **14** 34

[20] Bogdanov O V, Korotchenko K B and Pivovarov Yu L 2008 *J. Phys.* B **41** 055004

[21] Doyle P A and Turner P S 1968 *Acta Crystallogr.* A **24** 390

[22] Bernstein N, Gotsis H J, Papaconstantopoulos D A and Mehl M J 2005 *Phys. Rev.* B **71** 075203

Channeling of 20-35 MeV electrons in Si substrates of multilayer structures

V.V. Kaplin, S.R. Uglov, A.A. Voronin

Tomsk Polytechnic University P.O. Box 25, Lenin Ave. 2-a, 634050 Tomsk, Russia

E-mail address: uglov@tpu.ru

Abstract. Studying on X-rays generation by 20 – 33 MeV betatron electrons at its grazing incidence on a 56 ?m Si plate with 1.5 mm length along the electron beam are presented. The plate was prepared from the Si samples used as substrates to create the multilayers on their surfaces. The photos of the angular distributions of generated X-rays show its concentration along atomic axis [110] and atomic planes (110) and (111). That is due to electron channeling when electron moves along atomic axis or plane and radiates in a narrow cones along its directions. As a result, that forms the narrow spot and bands of intensity within the broad cone of ordinary bremsstrahlung. A capture of electrons in the channeling mode at large angles between crystallographic planes and electron beam takes place due to multiple scattering of electrons in the bulk of the crystalline plate. The orientation dependence of angular pattern and spectra of radiation in the intensity spot and bands were measured. Such a complicated "background" is necessary to take into account at studying the X-rays generated in the layered structures created on Si plates, which manifest itself also in the bands of intensity oriented along the layers.

1. Introduction

Studying on X-ray generation by fast electrons in single crystals and periodic stacks of micro-thick crystals or amorphous foils performed at the Tomsk synchrotron [1, 2] led to the investigations of artificial nanostructured radiators such as multilayer X-ray mirrors and X-ray waveguides. Multilayer X-ray mirrors are widely used in X-ray optics for producing the X-ray beams with required parameters under Bragg reflection of real photons. The experiments at the synchrotron showed that they can also be as effective X-rays generators excited by 300–900 MeV electrons. The experiments were carried out with the use of X-ray mirrors consisting of several hundreds of layers of tungsten W and boron carbide B_4C with the thickness of 1.2–4.0 nm. When a fast electron crosses such structure at a certain angle to a set of layers, X-rays are emitted in a narrow cone in the Bragg direction due to the diffraction of pseudophotons of the fast electron field at a periodic layered structure. Later, the studies were carried out with regard to the characteristics of radiation generated by the 15–33 MeV electrons in the X-ray multilayer mirrors and waveguides which were installed inside the chamber of the 35 MeV betatron [3–5]. X-ray waveguides consisting of a pair of thin layers and in-between channels made of lighter material, are actively investigated in X-ray optics [6] as one possible way to generate microbeams for practical purposes. By compressing the broad beams of the primary X-rays that is directed towards the waveguide surface and then passes it through a channel in waveguide modes, a secondary source of narrow focus radiation can be generated at the output waveguide. In our experiments such structures were used as the waveguide radiators with new X-ray emission which occurs when electrons cross the radiator at sliding angles to the layer structure consisting of two 100 nm layers of tungsten and in-between 50 nm carbon gap. X radiation generated by electron in the material of the radiator or at the interfaces is captured into the waveguide channel and propagates it in the waveguide modes, forming the output sharp focus source with a narrow angular distribution of radiation, determined by the angle of total reflection of photons from the waveguide walls. The size of the source focus equals to the waveguide width which can be reduced to the value of about 10 nm, according to Heisenberg's uncertainty relation for the transverse photon momentum and coordinate. We presently investigate the generation of X-rays by fast electrons in more complex structures then a homogeneous multilayer mirror or simple waveguide. These structures consist of layered periodic substructures, i.e. X-ray mirrors, separated by waveguide channels. Currently such structures are referred to as non-uniform X-ray photon crystals in the X-ray optics. The use of these structures as new X-ray source provides the combination of the effect of parametric X-ray radiation in the channel walls layered structures with the waveguide effect for X-rays generated in the substructures and penetrating the radiator in the waveguide modes similar to the Borrmann effect with low absorption. Such structures are created on single-crystal silicon substrates. To determine the influence of possible "disturbing" effects by X-rays generated by fast electrons in a crystalline substrate on the pattern of generation of radiation in the layered structure, the study [7] was carried out. The angular distributions and spectra of radiation generated in the "bare" silicon plate with a thickness of 200 microns and a length of 4 mm along the electron beam were investigated. The sample was identical to those we used to produce layered structures on their surfaces. The influence the plate crystal structure on the formation of the angular distribution of generated radiation was shown. Concentration of radiation intensity in the direction of crystallographic planes (111) was observed. The intensity bands follow the directions of the atomic planes at tilting the crystal relative to the direction of electron beam. This effect was assigned to the fast electron channeling along crystal atomic planes (111). At channeling, electrons move for a time along the atomic plane in the bound states of transverse motion [8], emitting X-rays in a narrow cone along the planes due to spontaneous transitions between the levels of bound states [9] and bremsstrahlung, born on the nuclei of the plane atoms. Therefore, the effect manifests itself within a broad cone of usual bremsstrahlung from electrons scattered in the crystal in chaotic manner. The existence of the bands at electron beam disorientation relative to the atomic plane at the angle exceeding critical channeling angle θ_c shows that electrons are captured into the channeling mode within the crystal volume due to their multiple scattering [10].

This paper presents the results of observing the influence of volume capture of 20 – 33 MeV electrons into the channeling along the atomic planes (111), (110) and along atomic axis [110] of Si plate with the thickness of 56 microns and a length L = 1.5 mm in the direction of the electron beam. That is ultra-thick crystal for the channeling effect because its length exceeds the channeling length L_{ch} in a few hundreds times. This plate was cut from the 0.28 mm Si substrate of X-ray mirror and polished to the required thickness.

2. Experimental setup

Figure 1 shows the scheme of experimental setup established on the basis of the B-35 betatron. The betatron was equipped with a special experimental chamber (1). Under the action of additional magnetic field created by a special coil for 30 ms, the orbit radius for accelerated electrons decreased and the electron beam hit the radiator (2) placed in the goniometer (3). The goniometer was located inside the chamber but had an external control system of the radiator attitude relative to the electron beam. The recurrence rate of acceleration cycles is 50 Hz, the divergence of the electron beam incident on the radiator is 0.3 mrad, and the electron energy spread is about 0.5%. The radiation emitted from the radiator was transmitted through 50 μm Mylar window and hit the X-ray film (4) or the Si detector (5) located at a distance of 105 cm from the radiator. The detector aperture is 13 mm^2, and the energy resolution for the line Am (13.95 keV) is 250 eV. The detector can be displaced relative to the axis of the generated radiation cone.

Silicon wafers are traditionally used as substrates designed for building on their polished surfaces of layered periodic structures by deposition of alternating layers of heavy and light materials. Such plates were used by the authors [5] to fabricate the W-C-W waveguide structures. Figure 2 shows a scheme of the Si plate location relative to the direction of the electron beam (solid arrow) hitting the plate surface. The dotted arrow shows the direction of crystallographic axis [011] of the crystal plate,

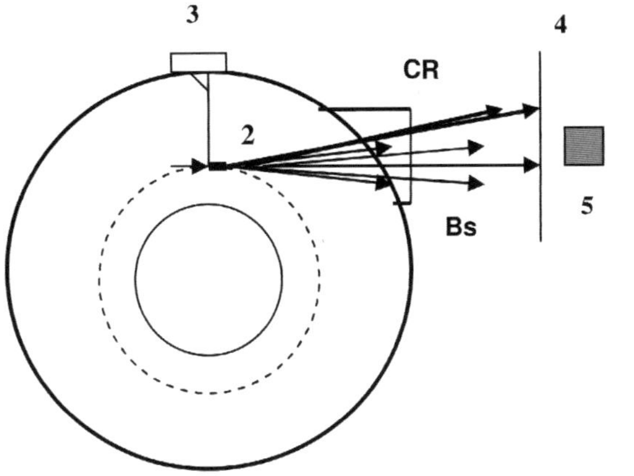

Figure 1. Experimental setup: 1 – betatron chamber, 2 – crystal, 3 – goniometer, 4 – photofilm, 5 – detector, Bs – bremsstrahlung, CR – X-rays from channeling electrons.

and its lateral surface coincided with the (100) crystallographic plane. The crystal plate with a thickness of 56 ?m, vertical and horizontal (along the electron beam) sizes of 5 and 1.5 mm, respectively, was placed on a vertical thin holder of the goniometer that enabled the crystal rotation around the vertical axis. At changing the angle of electron incidence on the crystal, the electron beam

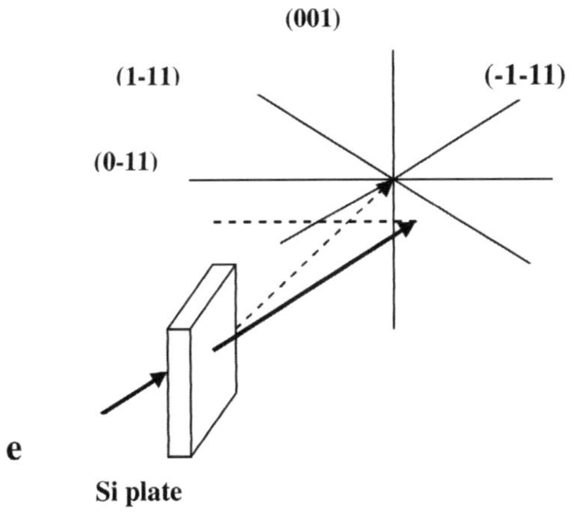

Figure 2. Scheme of [011] Si crystallographic direction and disposition of Si plate with respect to electron beam.

was directed along the horizontal dotted line in such a way that consistently coincided with the direction of the crystallographic planes (-1-11) and (001). In experiment, the angular length of the line was 6.4°.

3. Experimental results

The angular distributions of radiation were photographed with X-ray Retina films at a distance of 44 cm from the crystal plate. Photographs were processed with a scanner to obtain intensity profiles in the angular distribution of radiation. Photos of the radiation angular distributions for three orientations of the crystal plate relative to the direction of electron beam, shown in figure 3, demonstrate the influence of crystal structure on the formation of the angular distribution, with its form varying versus the crystal orientation. The radiation concentration is observed in the directions of crystallographic planes

(011), (111) and axis [011] in such a manner that elements in the form of bands and spots are formed in addition to the usual bremsstrahlung cone. Decreasing the electron energy from 33 to 20 MeV, the width of the bremsstrahlung cone, as well as that of bands and spots are increased in about 1.5 times.

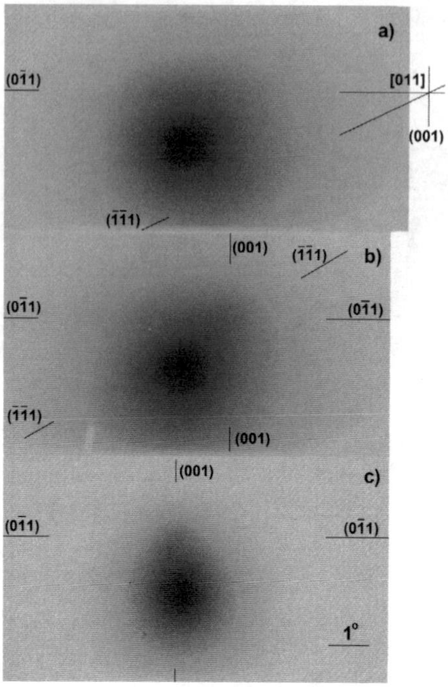

Figure 3. Photographs of X-ray distributions.

This means that both bands and spot are not formed by the diffracted bremsstrahlung. From the experience in studying the orientation effects for a few MeV electrons [9, 11], it was assumed that photos of the angular distributions of X-ray radiation of the accelerated electrons penetrating the crystal show the channeling effect of fast electrons along the atomic planes and axis. The channeling electrons move along the atomic plane or axis and generate radiation in a cone with a solution of about γ^{-1} (γ is an electron relativistic factor) along the axis or plane. Therefore, the effect manifests itself in the radiation angular distribution as an additional intensity band or a spot within the broad cone of bremsstrahlung from electrons that were repeatedly and randomly scattered during the penetration into crystal. As a result, the angular distribution of radiation consists of a cone of bremsstrahlung and the additional contribution of radiation from the channeling electrons, with its elements, given a thick crystal, being much narrower as compared to the bremsstrahlung cone with the width depending, apart from the relativistic factor, on the multiple scattering of electrons in the crystal. In the case of 33 MeV electrons, the angle γ^{-1} = 15 mrad, and the cone width of bremsstrahlung is much greater because the electron multiple scattering is quite large with the given crystal.

According to the scheme shown in figure 2, at crystal plate turning in the goniometer in a horizontal plane, electron beam is directed at different angular distances off the crystallographic planes. Figure 3 shows photographs of the angular distributions of radiation generated by 33 MeV electrons in Si plate at the angle θ = 115, 18 and 4 mrad between the electron beam and the vertical atomic plane (001) – photos a, b and c, respectively. The directions of the main crystallographic planes are shown in each photograph, and the direction of electron beam is in the center of a wide dark spot caused by bremsstrahlung. The bands coincided with the directions of the crystal atomic planes and displaced when tilted. At the same time, the direction of the cone axis of bremsstrahlung generated in the crystal as an amorphous medium remains constant. The top image of X-ray angular distribution generated at $\theta = 115$ mrad demonstrates mainly the effects defined by atomic planes (011) and (111), and the middle photo obtained at $\theta = 18$ mrad mainly contains bremsstrahlung and additional darkening in the direction of the crystallographic axis [011]. On their background, the effects from the crystallographic planes are less pronounced. The lower image shows only an intense spot of ordinary bremsstrahlung and less intense spot in the crystallographic axis [110]. Here, the effects defined by the atomic planes (011) and (111) are not evident.

Figure 4 shows the profiles of the density of blackening in the image of the angular distribution of radiation obtained by processing the image in figure 3c on the scanner, for horizontal scan lines - through a spot of bremsstrahlung, curve 1, and through the top spot, curve 2, formed by radiation from electrons channeling along atomic axis [011]. The width of a curve at half height of about $\Delta_1 = 2.1^o$ is

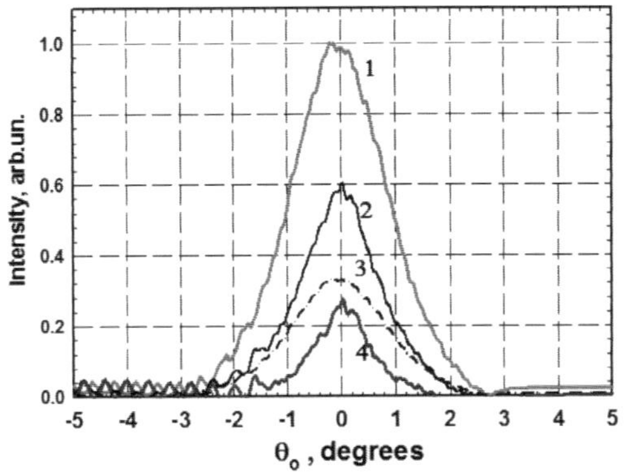

Figure 4. Profiles of the density of blackening on the image of the angular distribution of X-rays shown in figure3c.

greater than the theoretical estimation $\Delta_{Br} = \gamma^{-1} = 0.87^o$ for this parameter of the angular distribution of bremsstrahlung in the neglect of electron multiple scattering. This broadening of the distribution is determined by multiple scattering of electrons in the crystal plate. The width of curve 2 is about $\Delta_2 = 1.6^o$, which is less than Δ_1 because this distribution consists of bremsstrahlung distribution with a width $\Delta = \Delta_1$ from the multiply scattered electrons and a narrower distribution of radiation from the electrons channeled along atomic axis [011] with a width $\Delta_{Ch} < \Delta_1$ due to the bound movement of channeling electrons along the atomic rows without random scattering. They radiate due to spontaneous transitions between the levels of bound states of its transverse motion in a potential of a pair of atomic rows [011], as well as emit bremsstrahlung at interactions with the atomic nuclei of the rows. In order to separate the contribution of radiation generated by the channeling electrons, the subtraction of curve 3 from curve 2 was made. Curve 3 represents the curve 1, but scaled so as to "seam" it with curve 2 on its “tails”. It was assumed that curve 3 could describe the contribution of bremsstrahlung from randomly scattered electrons in this case. The result of subtraction is represented by curve 4. The angular distribution of radiation from channeling electrons has a half-height width of about $\Delta K = 1.2^o$, which is somewhat higher than the estimate of $\gamma^{-1} = 0.87^o$. This is probably due to the fact that the distribution is "smeared" because of the transverse oscillations of electrons channeled along a pair of atomic rows [011].

The existence of bands at the electron beam disorientation relative to the atomic plane at an angle θ significantly exceeding the critical angle θ_{Ch} for channeling shows that electrons are captured into the channeling mode within the crystal volume with their multiple scattering. In addition, the crystal thickness exceeds the length of the electron channeling in about 300 times, which also excludes a significant role of direct electron capture into the channeling mode on the crystal surface.

The experiment also included the measurement of the spectra of radiation generated by 33 MeV electrons channeling along the crystallographic axis [011] and also by electrons multiple scattered in the bulk of the Si plate. Figure 5 shows the spectra of radiation emitted by 33 MeV electrons in the direction of the center of the top spot in figure 3c, curve 1, and in leftward direction of 9 mrad off, curve 2. The spectra are normalized to the yield of radiation at photon energies > 55 keV in order to allow determining changes in the spectrum shape at the alteration of the electrons penetration mode in the plate thickness. The emission spectrum measured along the axis [011] shows no obvious peaks due

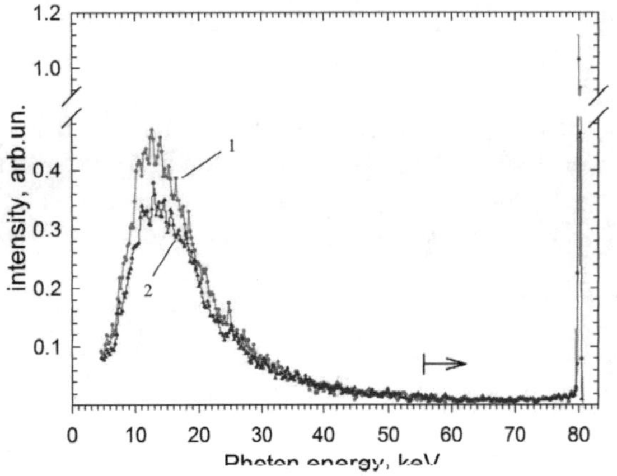

Figure 5. Spectra of X-rays generated by 33 MeV electrons and emitted in upper spot in figure 3c, curve 1, and outside the spot, curve 2.

to spontaneous transitions between the bound states of channeling, which occur in the case of thin crystals [9] with the thickness being of the order of channeling length. But at the same time, the spectra comparison shows the difference in the X-ray photon energy range of about 10–25 keV. Estimates show that it can be defined by the spontaneous transitions of electrons between the topmost levels of their motion states bound with atomic rows [011] during the crystal penetration.

4. Conclusion

The experiments have shown that the effect of electron channeling significantly changes the general form of the distributions of radiation generated in the crystal plate even if angular distances between electron beam and crystallographic planes or axis are greater then θ_c , and $L >> L_{ch}$. But, in this case the electrons can be captured in the channeling along the crystal atomic planes or axis by the effect of volume capture [10] at a multiple small-angle scattering, when the electron trajectory gradually turns to the direction of the atomic plane or axis, as well as at a strong single scattering when an electron abruptly changes its trajectory to the direction along the atomic plane or axis. In both cases, having been capturing into the channeling mode the electrons move for some time along the atomic plane or axis in the bound states of transverse motion in a potential of atomic plane or axis [8]. The channeling electrons generate bremsstrahlung on the nuclei of atoms of the atomic plane or axis, as well as spontaneous radiation due to transitions between the levels of the bound states of transverse motion. The cones of these emissions from the channeling electrons are much narrower than that of ordinary bremsstrahlung of electrons randomly scattered in the bulk of the crystal. The cone of ordinary bremsstrahlung is strongly broadened by multiple scattering of the electrons, and the widths of both cones of radiations from the channeling electrons are defined by the electron relativistic factor. As a result, the channeling effect provides additional elements having the forms of narrow stripes and spots in addition to a wide spot of ordinary bremsstrahlung. These studies were undertaken to determine the possible role of any effects in the crystals, which can prevent the observation of radiation generated by electrons in layered structures created on the surface of these crystals as on substrates. Non-uniform photonic crystals that are currently studied by the authors as a new X-ray radiator, are periodic structures consisting of a lot of alternating layers of tungsten and carbon, which have the non-uniformity inside in the form of the channel through which the radiation generated in periodic structures passes in the waveguide modes and forms the output angular distribution in the form of intensity bands oriented along the structure layers. Therefore, it is important to consider the interfering factors determined by the crystalline substrate when studying the effects in the radiation generated in the layered structure created on the substrate.

Acknowledgments
This work was supported by the Russian Foundation for Basic Research, project 10-02-00942, "The non-uniform photonic crystal as a new source of coherent submicrofocus X-rays from relativistic electrons".

References

[1] Andreyashkin M Yu, Zabaev V N, Yoshida K Kaplin V V, Rozum E I, Uglov S R and Endo I 1995 *Russian Pisma v JETP* 62 770
[2] Kaplin V V, Uglov S R, Zabaev V N, Piestrup M, Gary C and Fuller M 2000 *Appl. Phys. Lett.* 76 3647
[3] Kaplin VV, Uglov S R, Bulaev O F, Piestrup M, Gary C and Fuller M 2002 *Appl. Phys. Lett.* **80** 3427
[4] Kaplin V V, Sohoreva V V, Uglov S R, Bulaev O F, Voronin A A, Piestrup M, Gary C and Fuller M 2009 *Nucl. Instr. Meth. Phys. Res.* B 267 777
[5] Kaplin V V, Sohoreva V V, Uglov S R, Bulaev O F, Voronin A A, Piestrup M and Gary C 2011 *Nucl. Instr. Meth. Phys. Res.* B 269 1518
[6] Jark W, Cedola A, Di Fonzo S, et al. 2001 *Appl. Phys. Letters.* **78(9)** 1192
[7] Kaplin V V, Uglov S R and Voronin A A 2010 *Russian Izvestia VUS. Fizika* **10/2** 241
[8] Gridnev V I, Kaplin V V, Khlabutin V G, Rozum E I and Vorobiev S A 1987 *Phys. Stat. Sol.* (b) **122** 49
[9] Kephart J O, Pantell R H, Berman B L, Datz S, Park H and Klein R K 1989 *Phys. Rev.* B **40** 4249
[10] Gridnev V I, Kaplin VV, Khlabutin V G, Rozum E I and Vorobiev S A 1989 *Phys. Stat. Sol.* (b) **152** 61

Radiation Spectra of Channeled Electrons in Thick Si (111) Crystals

O V Bogdanov[1,2], and S B Dabagov[1,3]

[1]INFN Laboratori Nazionali di Frascati, Via E. Fermi 40, I-00044 Frascati (RM), Italy
[2]Tomsk Polytechnic University, Lenin Ave. 30, Tomsk, 634050 Russia
[3]PN Lebedev Physical Institute & NRNU MEPhI, Moscow, Russia

Email: bov@tpu.ru

Abstract. Dechanneling processes for electrons in a (111) Si crystal based on the solution of Fokker-Planck equation have been studied. The dynamics of particle distribution density has been investigated in dependence on both energy and initial scattering distribution of electron beams. The influence of dechanneling process on spectral intensity of channeling radiation for electrons in a crystal is investigated.

1. Introduction

Studies on relativistic electron/positron interaction in solids under the channeling conditions is of continuous interests due to its ability to produce electromagnetic radiation well-known as channeling radiation. This type of radiation is characterized by high both monochromaticity and intensity; moreover, such sources are tunable within broad energy spectrum. Due to the principal difference in interaction potentials at channeling for electrons and positrons, the radiation sources based on positron channeling might be much powerful for positrons because of the large dechanneling length in comparison with the beams of electrons. However, till now there is an uncertainty in calculations of dechanneling processes for moderate energies [1].

General features of electron/positron channeling radiation at moderate energies have been recently analyzed without the details of beam dechanneling [2]. As known at the energies higher then 100–150 MeV the description of electron/positron channeling in most of the crystals (or crystallographic directions) can be done within the classical approximation (Some details of the analysis for classical and quantum approximations at electron channeling for a specific crystal are presented in [3]), the motion is characterized by the particle trajectories, which the projectiles form within the continuous potentials of crystal planes or axes at either planar or axial channeling, respectively. However, the motion under the channeling regime is rather unstable due to the strong scattering processes caused by various interactions in a crystal (the perturbations by thermal atomic vibrations, electron subsystem, etc.). That is why in order to calculate the radiation intensity we have to know the beam both transverse-spatial and energy distributions in the field of continuous potential, in other words, we have to solve the diffusion equation for the beam. For light relativistic projectiles the typical diffusion equation describing the kinetics of the beam evolution down the crystal can be reduced in well-known Fokker-Planck equation [3, 4].

Another solution approach to this problem based on the binary collisions method of simulations for the trajectories of channeled electrons in a crystal [5].

In our work we have studied the dechanneling processes for electrons in a (111) Si crystal, based on the solution of Fokker-Planck equation [6, 7]. The dynamics of particle distribution density has been investigated in dependence on energy as well as initial scattering distributions of electron beams. This method, combined with the theory developed in [8, 9], was used to calculate the radiation spectra at planar channeling of relativistic electrons and positrons in thick crystals.

2. Basics of the channeling motion

The electron motion in the channeling regime in a crystal is determined by the average potential of planes / axes that can be modeled within various well-known approximations [10, 11]. The potential energy for interaction of electrons with (100), (110) and (111) crystal planes in the transverse direction are given in figure 1. Herewith, the longitudinal motion of charged particles (down the crystal) is considered to be free.

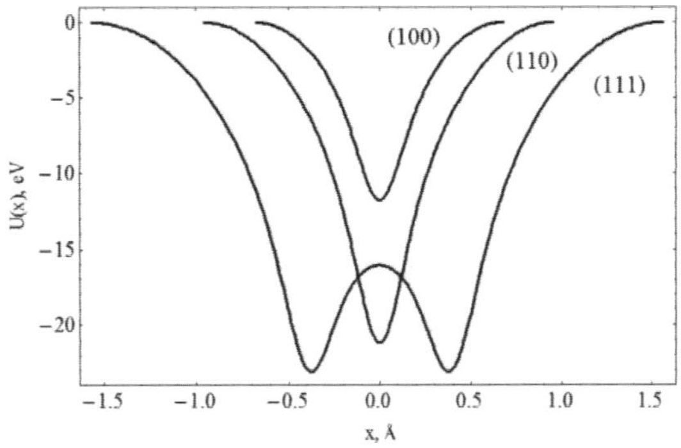

Figure 1. Potential energy of electron in the planes (100), (110), (111) Si crystals.

At Si (100) and (110) electron channeling the motion of projectile can be distinguished as either under-barrier or above-barrier (for details, see in [11, 12]). However, for the case of Si (111) channeling, the motion becomes more complex, and we can define three various types of the electron motion [8]. The analysis below is done for this specific case, which is presently less studied.

The equation of transverse motion for a relativistic electron in the potential U(x) (averaged planar potentials of the main Si crystallographic directions are shown in figure 1 is described by classical equations

$$\gamma m\ddot{x} = F = -\frac{\partial U(x)}{\partial x}, \gamma = \frac{1}{\sqrt{1-\upsilon_{\parallel}^2/c^2}}. \tag{1}$$

in the assumption of nonrelativistic transverse motion $\upsilon_\perp \equiv \dot{x} << \upsilon_\parallel$; $\upsilon_\perp / c << 1/\gamma$. For the initial conditions: the inlet point into the crystal $x(0) \equiv x_0$, the transverse momentum of electron $p_\perp(0) \equiv p_\parallel \theta_0$, where θ_0 is the incidence angle of an electron momentum with respect to the channeling planes, the transverse projectile energy in a crystal can be defined as follows

$$\varepsilon_\perp = U(x) + \frac{\mathbf{p}_\perp^2(0)}{2\gamma m} = U(x_0) + \frac{\mathbf{p}_\parallel^{2}\theta_0^2}{2\gamma m}, \tag{2}$$

$$\mathbf{p}(0) = \mathbf{p}_\perp(0) + \mathbf{p}_\parallel(0)$$

Depending on the ratio between θ_0 and critical channeling angle $\theta_c = \sqrt{2U_0/\varepsilon}$, as well as on the height of internal potential barrier $\Delta U = U_0 - U_1$ (figure 1), three solution types of Equation (1) could be resolved. As seen from figure 1 two solutions describing electron channeling in neighbor potential wells are of under-barrier origin, and the third solution describes electron quasichanneling for over-barrier motion.

3. Beam kinetics at channeling

Let assume that the initial angular distribution of relativistic electrons in the beam satisfies the normal law (known also as the Gaussian distribution):

$$\Phi(\theta)=\frac{1}{\sqrt{2\pi}\sigma_y}\exp\left[-\frac{(\theta-\theta_0)^2}{2\sigma_y^{\,2}}\right]. \tag{3}$$

Here, σ_y is the standard deviation.

The relation for transverse energy of a particle with energy E_0: $E_\perp(x)=\frac{E_0\,\theta^2}{2}+U(x)$ defines, as known, the angular dependence of the projectiles at channeling

$$\theta(E_\perp)=\sqrt{\frac{2(E_\perp-U(x))}{E_0}}, \tag{4}$$

Hence, the projectile density distribution function will be given by the following integral

$$F_0(E_\perp)=\frac{1}{\sqrt{2\pi\sigma_y^{\,2}}}\int_{x_u}^{x_d}\frac{1}{\theta(E_\perp)}\left(\exp\left[-\frac{(\theta(E_\perp)-\theta_0)^2}{2\sigma_y^{\,2}}\right]+\exp\left[-\frac{(-\theta(E_\perp)-\theta_0)^2}{2\sigma_y^{\,2}}\right]\right)dx. \tag{5}$$

The points $x_{d,}\ x_u$ correspond to the solutions of equation $E_\perp=U(x)$ for $E_\perp/U_0<1$, and $x_u=d_p/2,\ x_d=-d_p/2$ for the $E_\perp/U_0\geq 1$ case.

Figure 2 shows the initial distribution at the parallel to channeling planes incidence $\theta_0=0$ for various incident beam divergences $\sigma_y=1/2\theta_C$, $\sigma_y=\theta_C$, $\sigma_y=1.5\theta_C$. Obviously, due to the non-zero divergence, a part of particles undergoes the above-barrier motion even at $\theta_0=0$ (unlike a model with equal probability of the particles distribution over transverse energy where all particles get to the channel [8, 14]).

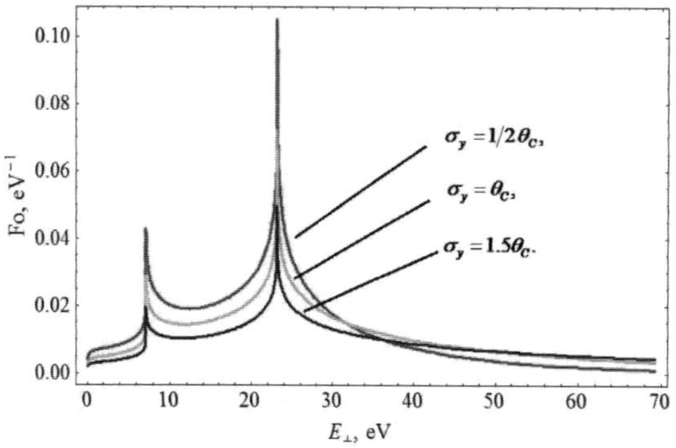

Figure 2. The initial angular distributions of relativistic electrons channeled in Si (111) at zero incident angle $\theta_0=0$. Various curves correspond to different beam divergences: $\sigma_y=1/2\theta_C$, $\sigma_y=\theta_C$, $\sigma_y=1.5\theta_C$.

The kinetic description of the channeling phenomenon was first suggested by Lindhard using the equation of motion of the diffusion type (see [3]). Successfully, Kitagava and Ohtsuki have first reduced the diffusion at channeling to the solution of the Fokker-Planck equation in phase space of transverse coordinates and velocities [4]. General expression of the Fokker-Planck equation can be written in the following from

$$\frac{\partial F(z,E_\perp)}{\partial z}=\frac{\partial^2}{\partial E_\perp^{\,2}}\left[D_e^{(2)}(E_\perp)F(z,E_\perp)\right]-\frac{\partial}{\partial E_\perp}\left[D_e^{(1)}(E_\perp)F(z,E_\perp)\right], \tag{6}$$

$$\frac{\partial F(z,0)}{\partial E_\perp}=0,\quad \frac{\partial F(z,E_{\perp,C})}{\partial E_\perp}=0,$$

in which the drift coefficient $D_e^{(1)}(E_\perp) = \overline{\left\langle \frac{\Delta E_\perp}{\Delta z} \right\rangle_T}$ is responsible for the mean transverse energy increase, while the term $D_e^{(2)}(E_\perp) = \frac{1}{2}\overline{\left\langle \frac{(\Delta E_\perp)^2}{\Delta z} \right\rangle_T} = \left\langle 2\frac{\overline{\Delta E_\perp}}{\Delta z}(E_\perp - U(x)) \right\rangle_T$ is the diffusion coefficient. The coefficients are averaged by the period of motion in a channel (The period of motion may vary in dependence of the motion type, above-barrier or under-barrier, with the difference factor of 2.).

Both drift and diffusion coefficients have been calculated within the Kitagava-Ohtsuki approximation [3] evaluating the integrals

$$D_e^{(1)}(E_\perp) = \aleph \int_{x_d}^{x_u} \frac{\exp\left(-x^2/2u_1^2\right)}{\theta(E_\perp, x)} dx \tag{7}$$

and

$$D_e^{(2)}(E_\perp) = \aleph E \int_{x_d}^{x_u} \theta(E_\perp, x) \exp\left(-x^2/2u_1^2\right) dx \tag{8}$$

where $\aleph = \frac{E_s^2}{2EvX_0} \frac{4}{T(E_\perp)} \frac{d_p}{\sqrt{2\pi} u_1}$ is the function defined by the projectile as well as crystal parameters, with is the total electron energy, $u_1 = 0.076$Å is the thermal vibration amplitude, the coordinates x_d, x_u are as above defined, X_0 is the characteristic radiation length of the crystal, for the (111) crystal planes the continuum potential depth and interplanar distance are $U_0 = 21.21 eV$ and $d_p \approx 1.92$ Å, respectively. The time parameter can be reduced from the solution of motion equation:

$$T(E_\perp) = \sqrt{\frac{E}{2c^2}} \int_{-x_d}^{x_u} \frac{dx}{\sqrt{E_\perp - U(x)}}, \tag{9}$$

where x_0 is defined by $E_\perp = U(x_0)$ at $E_\perp / U_0 < 1$; and $x_0 = d_p/2$ - at the case of $E_\perp / U_0 \geq 1$. Figure 3 shows the dependence of time parameter on electron transverse energy.

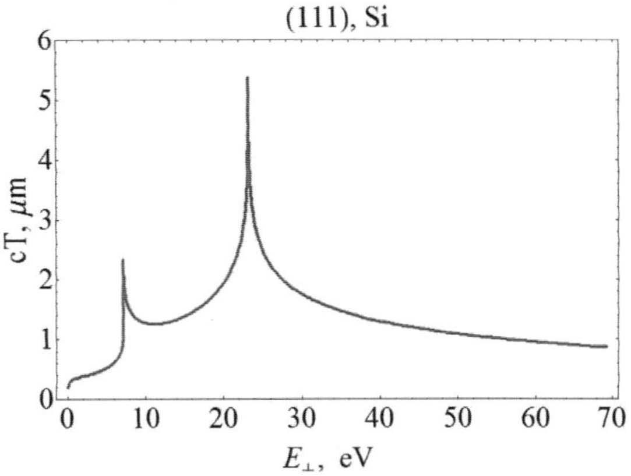

Figure 3. Time parameter of electron for (111) planar channeling in a Si crystal.

Taking into account the above mentioned conditions, the dependences of both drift and diffusion coefficients on transverse energy are shown in figure 4.

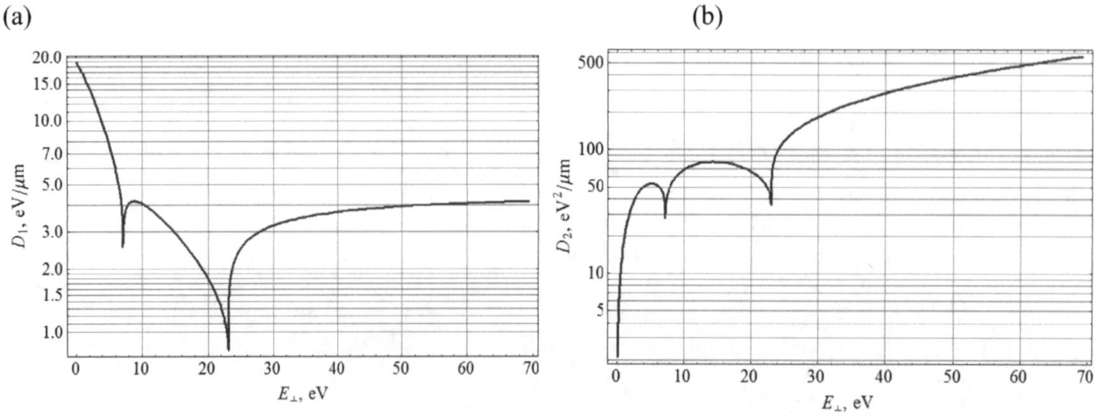

Figure 4. The drift (a) and diffusion (b) coefficients of electron at planar channeling in a Si (111) crystal.

It should be underlined that at transverse energy $E_\perp = U_0$, for which the time parameter has a singularity, both drift and diffusion coefficients are characterized by deep minima. It takes place due to the fact that the electron density (its probability) at the potential maximum is very high. Since the nuclear density in the mid-plane between the crystallographic planes is rather low (mainly negligible), the probability of projectile scattering on nuclei becomes small. That is why, kinetic coefficients reach their minima at that plane that is in contrast to the condition $E_\perp = 0$, for which electron appears at a crystal plane. For $E_\perp = 0$ the drift coefficient approaches the constant value for amorphous matter, since the continuous potential influences only very little the trajectory of an electron, while the diffusion coefficient increases linearly as function of $E_\perp$.

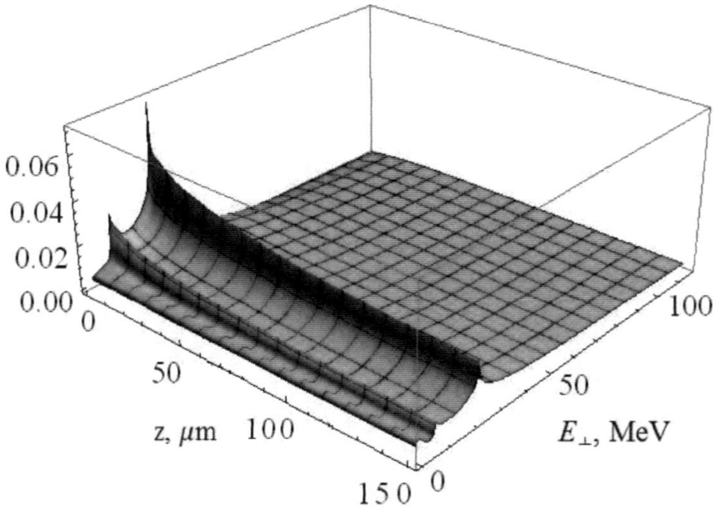

Figure 5. Probability of electron density at planar channeling in a Si (111) crystal.

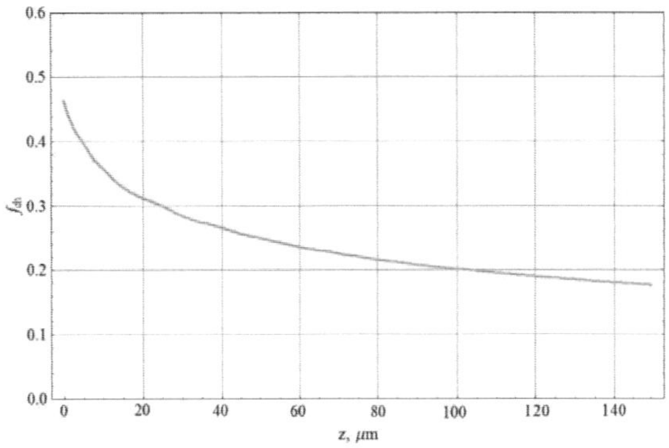

Figure 6. Electron dechanneling function for the case of Si (111) planar channeling.

$$J(z,\ E_\perp) = -\frac{\partial}{\partial E_\perp}\left[D_e^{(2)}(E_\perp)F(z,\ E_\perp)\right] + D_e^{(1)}(E_\perp)F(z,\ E_\perp) = J_{diff}(z,\ E_\perp) + J_{drift}(z,\ E_\perp) \quad (10)$$

$$L_d(z) = \frac{f_{ch}(z)}{J_{drift}(z,\ E_\perp = U_0)} \quad (11)$$

z is chosen from the condition that particle flux determined by the decrease of dechanneling function $f_{ch}(z) = \int_0^{U_0} F(z,\ E_\perp)dE_\perp$ in e times (figure 6): $\frac{f_{ch}(0)}{f_{ch}(z)} = e$.

This definition for dechanneling length does not depend on standard deviations in angular distribution, thus becoming the channel characteristics. For 855 MeV electrons channeling in a Si (111) crystal the dechanneling length calculated by (11) equals to 18.8 μm.

4. Dependence of channeling radiation on beam-crystal parameters

As known the length-normalized radiation intensity of relativistic charged particles at channeling in crystals is defined by the following expression [12, 13]

$$I_{thin}\frac{dW}{d\omega dz} = \frac{e^2\omega}{c^4T^2}\sum_{n=1}^{\infty}\Theta\left(1-\frac{\omega T}{4\pi n\gamma^2}\right)\left(1-\frac{T\omega}{2\pi\gamma^2 n}+\frac{1}{2}\left(\frac{T\omega}{2\pi\gamma^2 n}\right)^2\right)\left|\upsilon_{\tilde{\omega}}^T\right|^2,$$

$$\tilde{\omega} = \frac{2\pi n}{T},\quad \upsilon_{\tilde{\omega}}^T = \int_0^T \upsilon_\perp(t)e^{i\tilde{\omega}t}dt \quad (12)$$

where $\upsilon_{\tilde{\omega}}^T$ is the Fourier-component of transverse particle velocity. Successively, channeling radiation taking into account the particle initial angular distribution can be evaluated as

$$I_{F_0}(\hbar\omega,\ E_\perp) = I_{thin}(\hbar\omega,\ E_\perp)\cdot F_0(z,\ E_\perp) \quad (13)$$

However, this approach is valid only for very thin crystals. Indeed, for thick crystals we have to pay attention to the projectile dechanneling (with the crystal thickness increase the projectiles become unbound by the channel potential due to the increase of their transverse energies), which contributes to variation of the distribution function $F(z,\ E_\perp)$. Hence, the expression for radiation intensity in thick crystals can be written as follows

$$I_{thick}(\hbar\omega,\ E_\perp) = \frac{1}{L}\int_0^L I_{thin}(\hbar\omega,\ E_\perp)\cdot F(z,\ E_\perp)dz \quad (14)$$

Above presented equations are used for a single particle case; moreover, they describe the features of radiation for given transverse energy $E_\perp$. Thus, total radiation spectrum is formed by the projectiles over complete set of the transverse energies at channeling $E_\perp$ that results in the necessity of additional averaging by transverse energy, i.e. $\langle I_{thin}\rangle$, $\langle I_{F_0}\rangle$, $\langle I_{thick,L}\rangle$.

Journal of Physics: Conference Series **357** (2012) 012029 doi:10.1088/1742-6596/357/1/012029

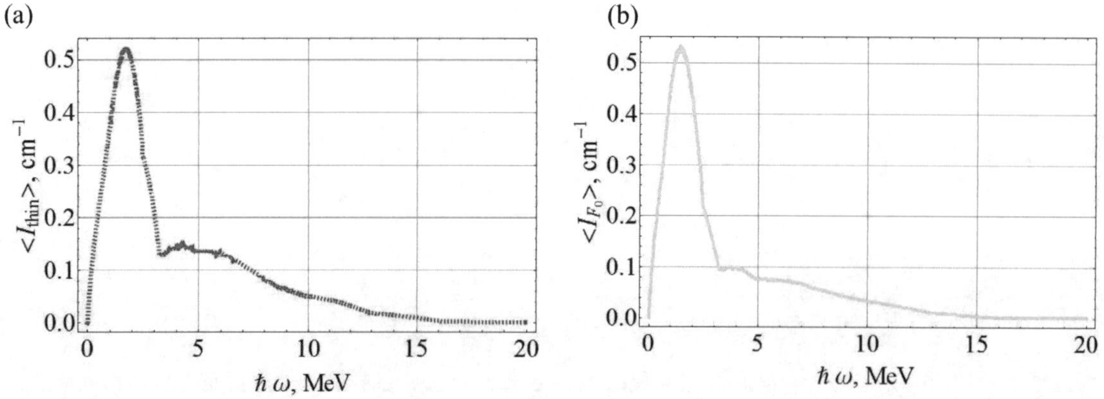

Figure 7. Channeling radiation spectrum for a $E = 855$ MeV electron channeled in Si (111) (a) $I_{thin}(\hbar\omega, E_{\perp})$ at homogeneous transverse energy distribution (a), and $I_{F_0}(\hbar\omega, E_{\perp})$ with initial distribution of electrons F_0 defined as in figure 2 (b).

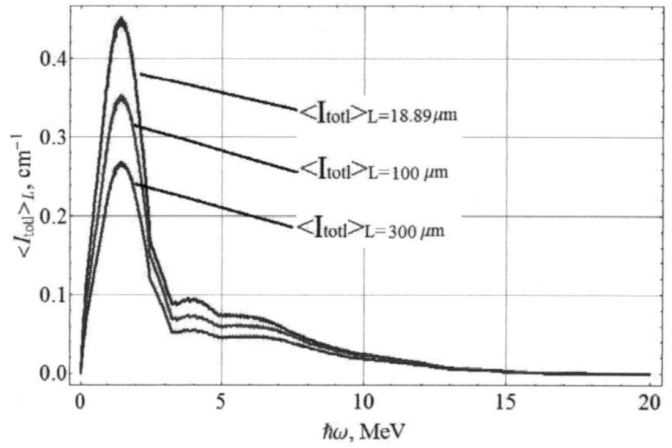

Figure 8. Thickness dependence for radiation spectrum at electrons channeling in a Si (111) crystal.

Figure 8 shows the radiation spectra of channeled electrons per unit length of the crystalline target Si.

5. Conclusion

In this work we have presented a model for the dual-channeling of 855 MeV electrons in a Si (111) crystal, taking into account the effects of dechanneling. The calculations were performed for the initial angular distributions of relativistic electrons in the beam as a function of transverse energy for various angles of incidence with respect to the crystal (111) planes. Within developed model both drift and diffusion coefficients of the beam density distributions for trapped electron motion in the system of double planes (111) have been evaluated.

For 855 MeV electrons channeling in a Si (111) crystal the dechanneling length evaluated from the solution of the Fokker-Planck equation equals to 18.8 μm. Spectral characteristics of channeling radiation by electrons of moderate energies in thick crystals have been first obtained taking into account the dynamics of beam scattering at channeling as well as the initial angular distribution of channeled electrons.

Acknowledgements

We are grateful to Profs. H. Backe and Yu. Pivovarov for stimulating discussions on the subject. The work was partially supported by the Grant of the President of RF № MK-2059.2011.2.

References

[1] Carrigan R. A., Negative Particle Planar and Axial Channeling and Channeling Collimation, in the book "Charged and Neutral Particles Channeling Phenomena - *Channeling 2008*", Dabagov S.B., and Palumbo L., Eds., *World Scientific Publ.* (2010) 129.

[2] Dabagov S.B., and Zhevago N.K., *Nuovo Cimento* **31(9)** (2008) 491.

[3] Kumakhov M.A., and Shirmer G., Atomic Collisions in Crystals, *Moscow: Atomizdat,* 1980 (in Russian).

[4] Beloshitsky VV, Komarov FF, and Kumakhov MA, Dechanneling, flux-peaking and energy losses of fast charged particles penetrating through thick crystals, *Phys. Rep.* **139 (6)** (1986) 293-364.

[5] Kostyuk A., Korol A., Solov'yov A.and Greiner W., *J. Phys. B: At. Mol. Opt. Phys.* **44** (2011) 075208.

[6] Ohtsuki Y.-H., Charged Beam Interaction with Solids, *New York,* 1983.

[7] Backe H.et al. Nucl. *Instr. Meth. in Phys. Res.* **B266** (2008) 3835.

[8] Bogdanov O.V., Korotchenko K.B. and Pivovarov Yu.L., *J. Phys. B: At. Mol. Opt. Phys.* **41** (2008) 055004.

[9] Babaev A.A., et al., On Crystal-Assisted Processes by Means of 20–800 MeV e_/e+ *LNF Beams, Preprint* LNF **22 (IR)** (2008) pp. 1-42.

[10] Gemmell D.S., *Rev. Mod. Phys.* **46** (1974) 129.

[11] Doyle P.A. and Turner P. S., *Acta Crystallogr.* **A 24, 390** (1968).

[12] Baier V. N., V. M. Katkov and V. M. Strakhovenko 1998, Electromagnetic Processes at High Energies in Oriented Single Crystals. *World Scientific, Singapore*

[13] Akhiezer A.I., and Shu'lga N.F. High Energy Electrodynamics in Matter, *Gordon and Breach, Luxemburg,* 1996

[14] Azadegan B., and Dabagov S.B., *European Physical Journal* - **Plus 126** (2011) 58.

Journal of Physics: Conference Series **357** (2012) 012030 doi:10.1088/1742-6596/357/1/012030

Peculiarities of Angular Distribution of Electrons at Si <100> Channeling

O V Bogdanov[1,2], Yu L Pivovarov[1], Y Takabayashi[3] and TA Tukhfatullin[1]

[1]Tomsk Polytechnic University, Lenin Ave 30, 634050 Tomsk, Russia
[2]INFN Laboratori Nazionali di Frascati, Via E. Fermi 40, I-00044 Frascati (RM), Italy
[3]SAGA Light Source, 8-7 Yayoigaoka, Tosu 841-0005, Japan

E-mail: TTA@tpu.ru

Abstract. The properties of both angular and spatial distribution of 255 MeV electrons at <100> channeling in silicon crystal has been investigated experimentally at the linac injector of SAGA light source and by computer simulations. The simulation of trajectories, angular and spatial distributions of electrons on the screen monitor has been performed taking into account initial spatial as well as angular beam divergence of electron beam. Both experimental data and simulations show the brilliant effect of so-called «doughnut scattering».

1. Introduction

The «doughnut scattering» (DS) (scattering of charged relativistic particles by crystal axes in the plane perpendicular to the axes) was first observed in earlier high-energy channeling experiments (1976 - 1980) using the proton and meson beams at CERN, see, e.g. in [1]. Theoretical consideration and Monte-Carlo calculations of similar effects but for the case of electrons and positrons, was performed in [2-3]. The review of crystal-assisted processes at high energies, including DS, was given in [4]. Rather recently (2006), similar effects have been observed using 150 MeV electrons at REFER (Hiroshima University) [5]. Dependence of electrons deflection on the beam divergence, which indicated the possibility of uLsing such a technique for beam diagnostics as well as beam collimation, was observed unless the beam divergence exceeded several times the critical channeling angle. This dependence was only qualitatively explained by computer simulations. In our work, more detailed investigation of DS for 255 MeV electrons in a thin Si crystal using another technique as well as more correct computer simulations based on BCM model [6] have been performed.

2. The experimental setup of SAGA-LS and results of experiments

The experiments were performed at the linac injector of SAGA light source (SAGA-LS) facility, see in figure 1. The electrons with an energy 255 MeV passed trough a single silicon crystal of 20 μm thickness along <100> axis. The vertical and horizontal sizes of the incident beam were σ_x=0.2 mm and σ_y=0.7 mm, respectively. The spatial the distribution of the electrons in the incident beam is considered as a normal and σ is the dispersion of this distribution. The horizontal and vertical angular divergences of the beam were 0.2 mrad and 0.3 mrad, respectively. The beam profile after passing through the crystal was measured by the screen monitor placed at 182 cm from the crystal.

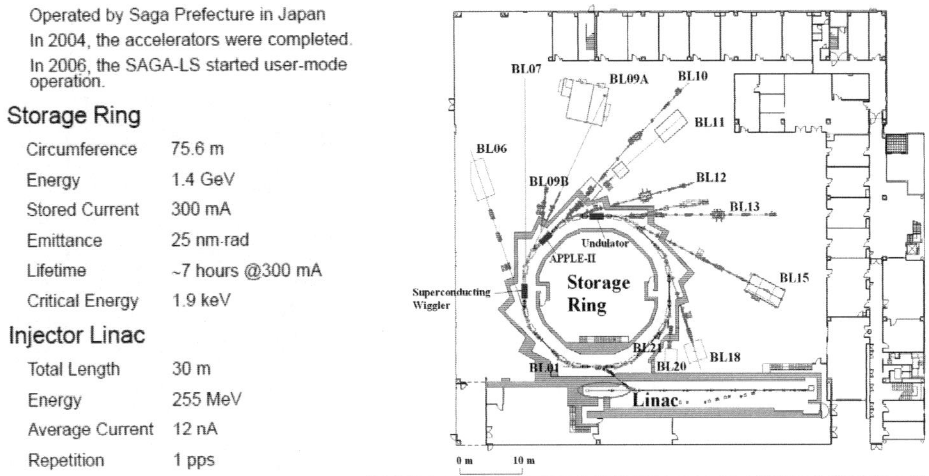

Figure 1. SAGA light source facilities.

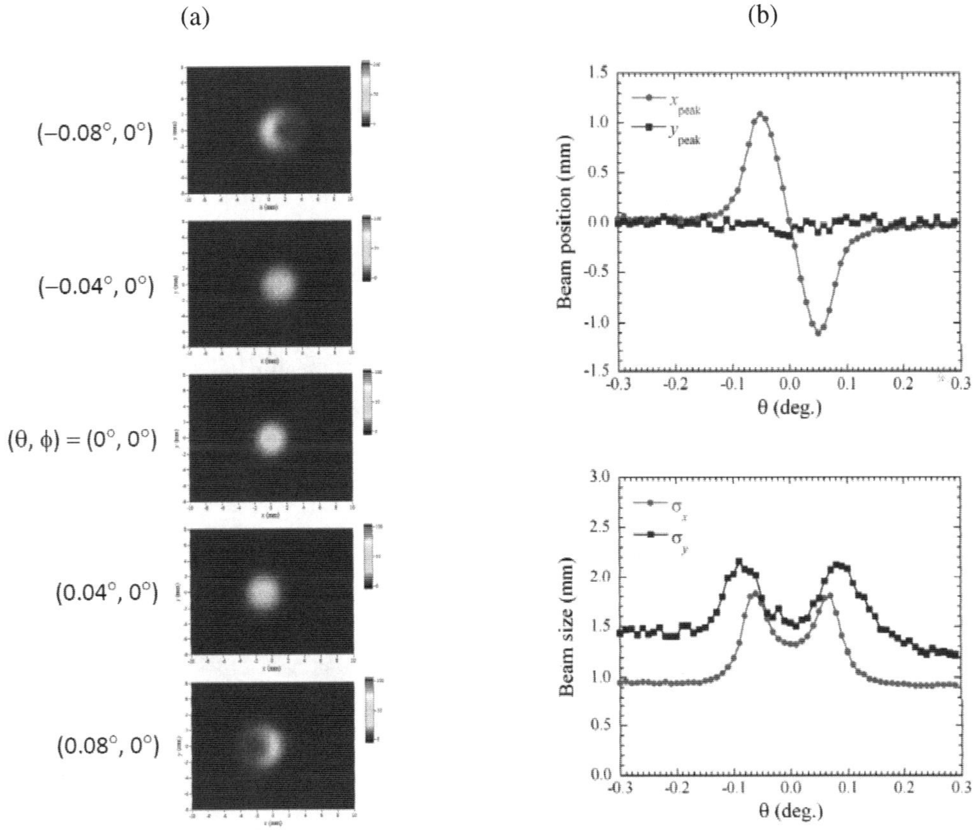

Figure 2. The experimental results for 255 MeV electrons penetration through <100> 20 μm Si crystal. (a) Profiles of electron beam transmitted through Si crystal for different angles of incidence θ into the crystal. (b) Vertical and horizontal beam positions x_{peak}, y_{peak} and beam sizes σ_x, σ_y versus angle of incidence θ.

The compact two axes goniometer was installed in a way allowing the crystal to be rotated around vertical and horizontal axes in the plane perpendicular to the beam. The angular step size for both axes was 0.001 degree. The spatial distribution of the particles was observed on the screen monitor, see in figure 2(a). To perform the data analysis, a 2-dimensional beam profile was projected onto the vertical and horizontal directions. Than these profiles were fitted with a Gaussian function, like in Ref. [5]:

$$I_x = I_{ox} \exp\left[-\frac{\left(x - x_{peak}\right)^2}{2\sigma_x}\right] + a_x x + b_x, \tag{1}$$

and the beam sizes σ_x, σ_y and the beam positions x_{peak}, y_{peak} are defined in figure 2 (b).

3. Electron trajectories simulation

For the calculation of electrons trajectories we used the classical description (a quantum approach is applied for electron energies less than ~100 MeV). Under the channeling condition the longitudinal component of relativistic electrons velocity $\upsilon_\parallel$ is nearly equal to the speed of light. The transversal component of velocity is much less than the longitudinal one: $\upsilon_\perp \equiv \sqrt{\dot{x}^2 + \dot{y}^2} << \upsilon_\parallel$; $\upsilon_\perp / c << 1/\gamma$ (γ denotes the relativistic factor). Under these conditions, equations of motion are reduced to the non-relativistic equations with the replacement of the particle mass m to the relativistic mass γm, see e.g. [7].

$$\gamma m\ddot{x} = -\frac{\partial U(x,y)}{\partial x}, \quad \gamma m\ddot{y} = -\frac{\partial U(x,y)}{\partial y}, \quad \gamma m\ddot{z} = 0, \tag{2}$$

where x and y are the transversal coordinates, z is the longitudinal coordinate, $U(x, y)$ here describes the potential energy of the electrons in the field of periodically arranged crystal axes.

The initial conditions for this system contain the electron entrance point into the crystal $x(0) \equiv x_0$, $y(0) \equiv y_0$ and transversal components of initial velocity:

$$\upsilon_x(0) = c\sqrt{1 - \frac{1}{\gamma^2}}\sin(\theta)\cos(\varphi), \quad \upsilon_y(0) = c\sqrt{1 - \frac{1}{\gamma^2}}\sin(\theta)\sin(\varphi), \tag{3}$$

where θ, φ are the angles between electron momentum and OZ and OX axes correspondingly.

The potential energy $U(x, y)$ of electron interaction with crystallographic axes within the Doyle-Turner approximation [8] for single atom potential (obtained by fitting the electron scattering form factor) is periodic in (x, y) plane:

$$U(x,y) = eV(x,y) = e \cdot \sum_k V_{ax}(x - kd_{xp}, y - kd_{yp}), \tag{4}$$

where $V_{ax}(x, y)$ is the potential of a single axis, d_{xp}, d_{yp} are the inter axes distances. The potential energy $U(x, y)$ for Si <110> direction is shown in figure 3. The dashed line shows the region where the entrance points of electrons were located.

Thus, to calculate every individual electron trajectory $x(t)$, $y(t)$ it is necessary to solve the system of classical equations of motion (2). The trajectories and velocities of electrons are obtained by numerical integration of the equation of motion, at this stage without taking into account of the dechanneling process. For the numerical simulation we used a new computer code "Basic Channeling with Mathematica©" BCM–1.0 developed by the authors [6].

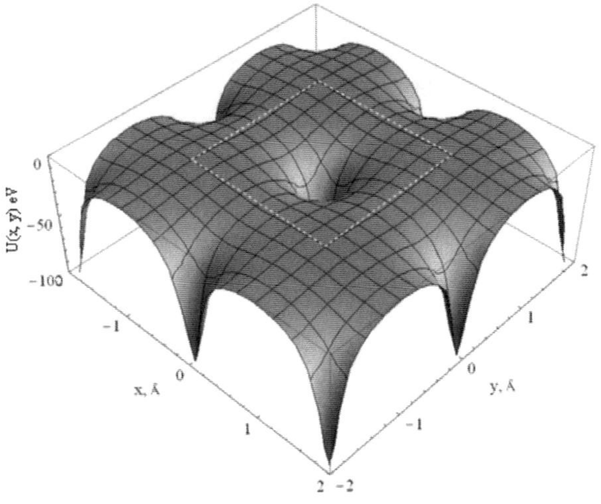

Figure 3. The potential energy $U(x, y)$ for <110> direction in Si crystal. The dashed line shows the region where the entrance points of electrons were located.

(a)

(b)

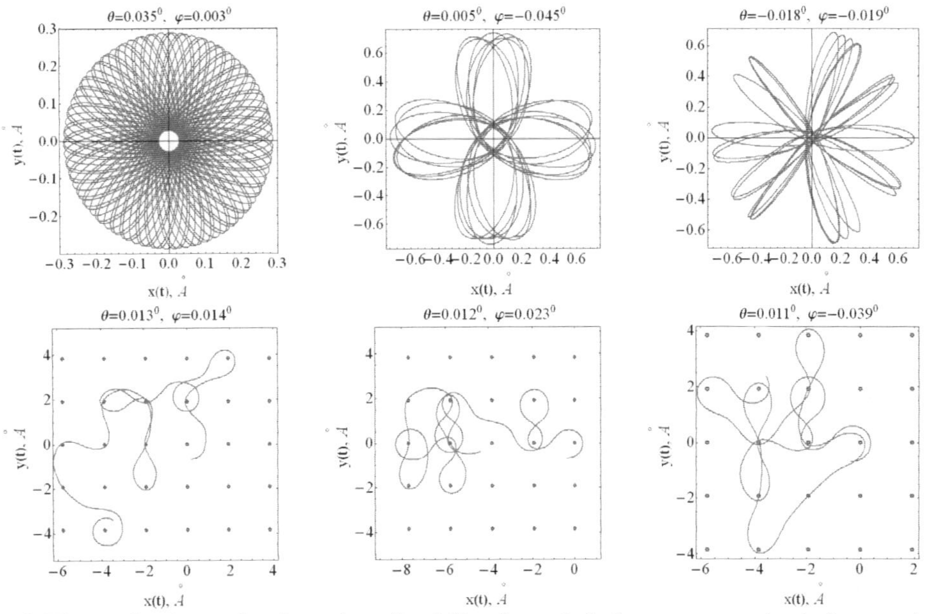

Figure 4. The projections of trajectories of axially channeled electrons onto *(x,y)* plane under penetration through a 20 μm thick Si crystal, for the different incident angles and entrance points: (a) – periodic motion; (b) - non periodic motion. The points depict crystallographic axes.

The projections of trajectories of axially channeled electrons onto *(x,y)* plane strongly depend on the incident points into the crystal and incident angles with respect to the channeling axes. Several examples are shown in figure 4. The trajectory showed in figure 4 (a) represents rosette motion when electron is trapped by the potential well of crystal axis. At larger angles of incidence the electron is scattered by the axes and trajectory of electrons becomes more complicated, see in figure 4 (b).

4. Simulation of spatial and angular distributions

The procedure for calculation of both angular and spatial distributions of electrons passed through a silicon single crystal was as follows.

1. The calculation of trajectories, the entrance points of which are situated within the square shown in figure 3. For each point of entry, 10 values of incident angles θ and φ were generated by a

random number generator with normal distribution. The dispersion of the normal distribution corresponds to the experimental ones (horizontal – 0.2 mrad, and vertical – 0.3 mrad). The total number of calculated trajectories was more than 10 000. Next, these data were used to calculate spatial distribution of electron beam during penetration through the crystal using the periodicity of the crystal lattice potential.

2. The calculation of initial incident points for electrons into a crystal was performed in accordance with the profile of experimental beam by the random number generator. We assume that the initial distribution of a primary electron beam was the normal one with vertical dispersion of 0.2 mm and horizontal dispersion of 0.7 mm. The calculated SAGA-LS incident beam profile is shown in figure 5.

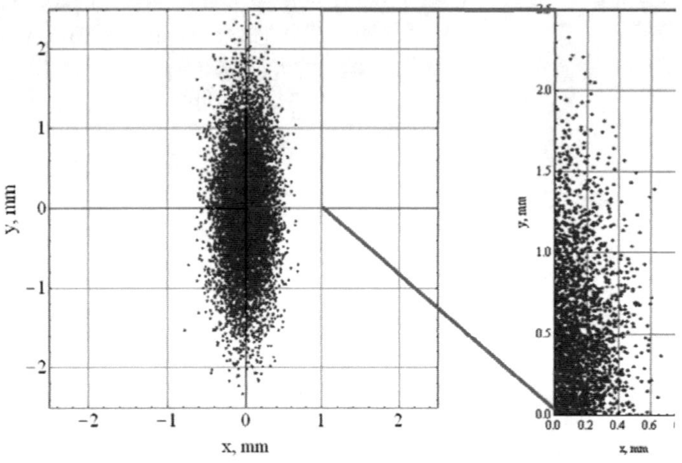

Figure 5. SAGA-LS incident beam profile simulation results (100 000 electrons).

Because of the large number of trajectories, the computer code [6] was parallelized and further calculations were performed at the supercomputer cluster of Tomsk Polytechnic University [9]. We calculated the exit angles of electrons from the crystal and spatial distribution of electrons on the screen monitor, which is located at the distance of 1.82 m from the crystal. The results of angular distributions calculation are shown in figure 6. (a). The exit angles are calculated according the formulae $\theta_x = \upsilon_x / c, \theta_y = \upsilon_y / c$, where υ_x and υ_y - are the transversal components of the particle velocity at the exit from the crystal. The figures illustrate that if an incidence angle is less than the critical channeling angle (0.05 degree for this condition), than the distributions are quite symmetrical. For the large angles of incidence the angular distributions became crescent-shaped. The spatial distributions shown in figure 6 (b) clear illustrate the brilliant effect of (DS).

To perform comparison with experiment, 2-dimensional beam profile was projected onto the vertical and horizontal directions. The one-dimensional beam profiles were fitted by the Gaussian function (1), and the beam size and beam position have been estimated. The dependences of the beam position x_{peak}, y_{peak} and beam size σ_x, σ_y on the incidence angles θ are shown in figure 7.

From figure 7, one can conclude that results of computer simulations are in rather good quantitative agreement with experimental data. It should be mentioned, that when we calculated the beam size in figure 7, we took into account the contribution to total dispersion arising due to multiple scattering of electrons, with dispersion equal to that in amorphous target σ_{ms} [10]. Then, then the total beam size was calculated according the equation $\sigma = \sqrt{\sigma^2_{simulation} + \sigma^2_{ms}}$.

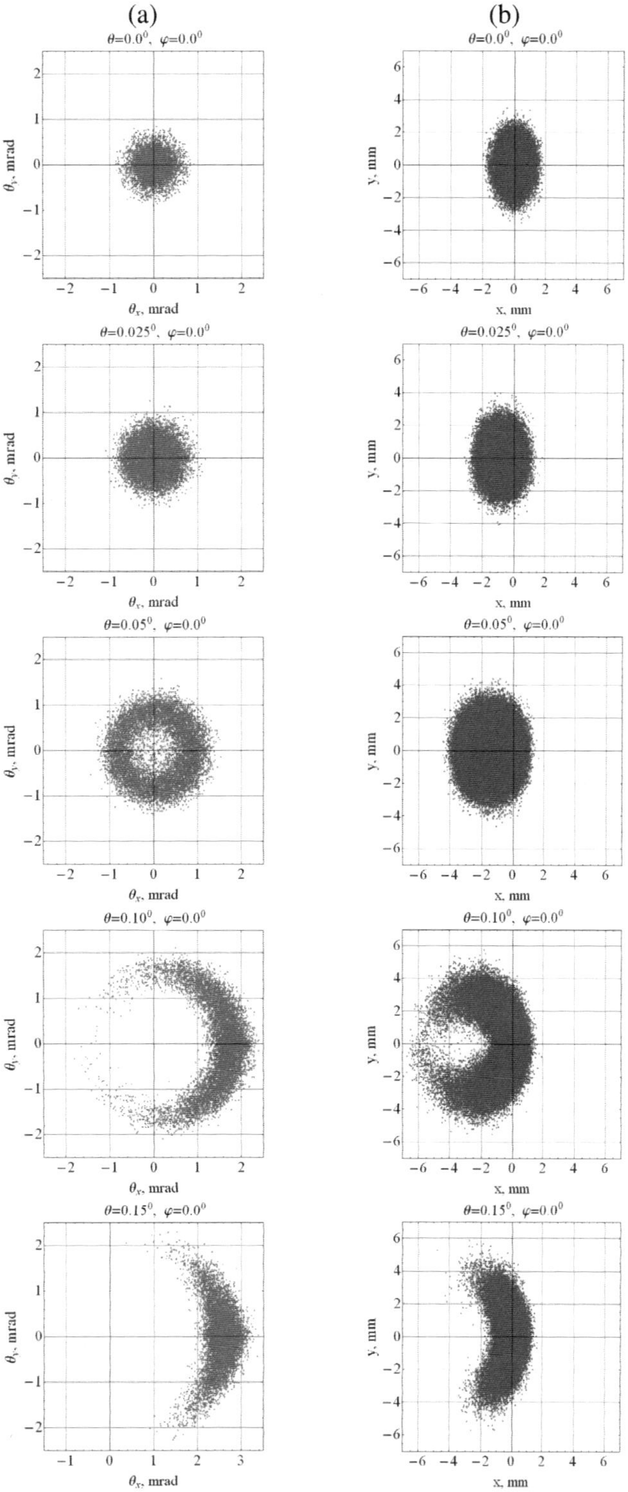

Figure 6. Angular distribution of electrons penetrated through the crystal (a), and spatial distribution of electrons on the screen monitor for different incident angles θ. The exit angles were calculated according the formulae: $\theta_x=\upsilon_x/c$, $\theta_y=\upsilon_y/c$.

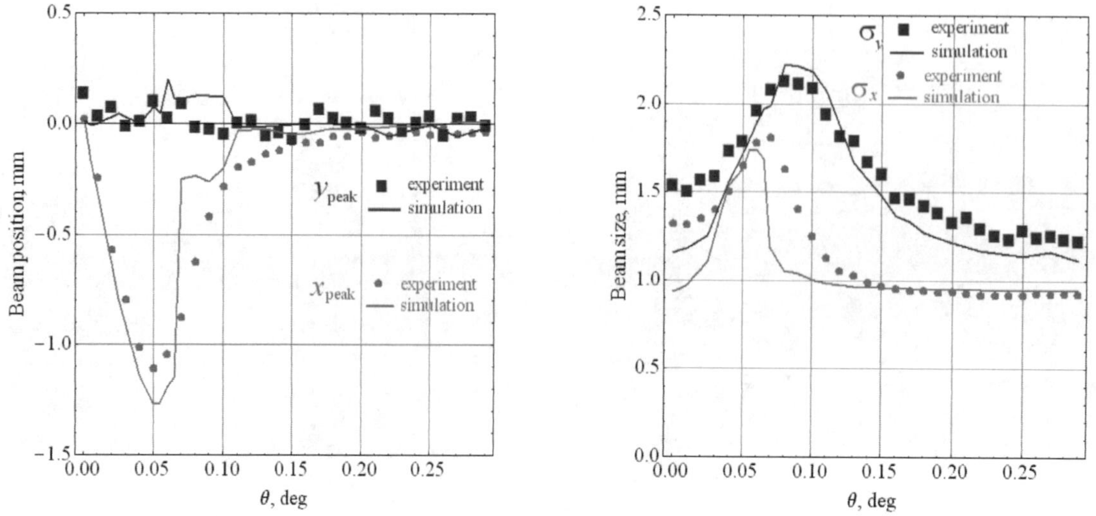

Figure 7. Comparison of the simulation results with experimental data (experimental data – filled squares and circles, simulation results - solid lines).

5. Conclusions

The first experiments on 255 MeV electrons channeling and DS in a thin aligned Si crystal were performed at SAGA-LS injector Linac. The simulation of trajectories at the <100> axial channeling in Si, angular and spatial distributions of electrons on the screen monitor have been performed taking into account initial spatial and angular beam divergence of the electron beam. The comparison of experimental and theoretical results shows a good quantitative agreement. Both experimental data and simulation results show the brilliant effect of doughnut scattering - DS. In addition, we may definitely conclude, that DS of a few hundred MeV electrons in aligned thin crystals can be used for the diagnostics of the incident beam angular divergence.

6. Acknowledgements

The authors are grateful to N. Shul'ga, S. Fomin and K. Korortchenko for helpful discussions. This work was supported by the Russian Fund for Basic Research within the grant No 10–02–08607 and within the Grant of the President of RF № MK-2059.2011.2.

7. References

[1] Andersen S K, et al. 1980 *Nucl. Instr. and Meth. in Phys. Res.* **B 167** 1
[2] Shul'ga N F, Truten' S P 1982 *Sov. Phys. Techn. Phys.* **27** 1399
[3] Artru X, Fomin S P, Shul'ga N F 2005 *Nucl. Instr. and Meth. in Phys. Res.* **B 230** 476
[4] Uggerhoj U 2006 *Nucl. Instr. and Meth. in Phys. Res.* **B 252** 16
[5] Strokov S, Takahashi T, Endo I, Iinuma M, Kuroiwa K H, Ohnishi T, Sawada S 2006 *Nucl.Instr. and Meth. in Phys. Res.* **B 252** 16
[6] Bogdanov O V, Fiks E I, Korotchenko K B, Pivovarov Yu L and Tukhfatullin T A 2010 *J. of Phys.: Conf. Ser.* **236** 1 012029
[7] Bazylev V A, Zhevago N K 1982 *Sov. Phys. Usp.* **25** 565
[8] Doyle P A and Turner P S 1967 *Acta Cryst.* **A24** 390
[9] http://cluster.tpu.ru/
[10] Bichsel H, Groom D E, Klein S R 2010*J. of Phys. G. Review of particle physics* **37** 7a 285

Journal of Physics: Conference Series **357** (2012) 012031 doi:10.1088/1742-6596/357/1/012031

Acoustic "pumping effect" for quartz monochromators

A Gogolev[1], S Stuchebrov, A Wagner, Yu Cherepennikov, A Potylitsyn

Tomsk Polytechnic University, Tomsk, Russia

E-mail: alextpuftf@tpu.ru

Abstract. There are investigations of dependences scattered at quartz crystals X-ray in Bragg direction of intensity on amplitude and frequency external alternating electromagnetic field, which initiates acoustic waves in crystal, in spectral mode.

Nowadays the X-rays are widely used for different applied problems of science, technology and medicine. Today X-ray tubes are used as a prevalent source of X-ray. However in some applications, such as medical roentgenology, the standard methods of diagnostics using X-ray tubes run into certain difficulties, caused by radiation of white spectrum by X-ray tubes. Some difficulties can be obviated by using monochromatic X-ray radiation.

The monochromatic X-ray source on the basis of compact electron accelerator is of great interest nowadays. One of the common methods to generate monochromatic X-ray is monochromatization of bremsstrahlung. In the other method the Compton effect is used. The both methods have specific disadvantages. The first case is characterized by low efficiency, because a considerable part of initial X-ray beam is lost due to partial reflection and absorption of radiation by a monochromator. In the second case the installation has low frequency.

In this report it is offered to use a "pumping effect" [1] to improve the parameters of monochromatic X-ray beam on the basis monochromatization of bremsstrahlung. This effect appears in the crystals with ordered deformation. Deformation can be obtained with using constant electrical fields, elastic stresses, temperature gradient, acoustic wave fields etc. For example, in case of using acoustic field the effect appears when a frequency of the excited wave coincides with the resonant frequency determined by:

$$f_n = nv/2l, \tag{1}$$

where l is crystal thickness, v is sound velocity, n is odd integer number.

Using the optical elements with "pumping effect" allows adjusting the intensity of monochromatic X-rays source based on diffraction and implement dynamic control and transportation of X-ray beams with minimal loss in the optical elements. Besides, it is possible to control the intensity and form of diffracted beam in time and space with a controlled external influence.

In the papers [2–4] the authors demonstrated the possibility to control the intensity of the diffracted beam in case of application of the acoustic wave field and non-uniform temperature field at the piezo-electric crystal monochromator. Theoretical explanation of "pumping effect" and independent experimental results of investigations of acoustic wave field influence on the process of X-rays coherent scatter at the integral mode are presented in the paper [5].

The purpose of this report is the investigation of "pumping effect" in a spectral mode.

The experimental setup is shown in figure 1. X-ray tube RAP 160-5 was used as a source of X-rays. All measurements were performed with the same parameters of the tube, the voltage and average current were 48 kV and 1 mA, respectively. X-ray tube was placed in lead with 5 cm wall.

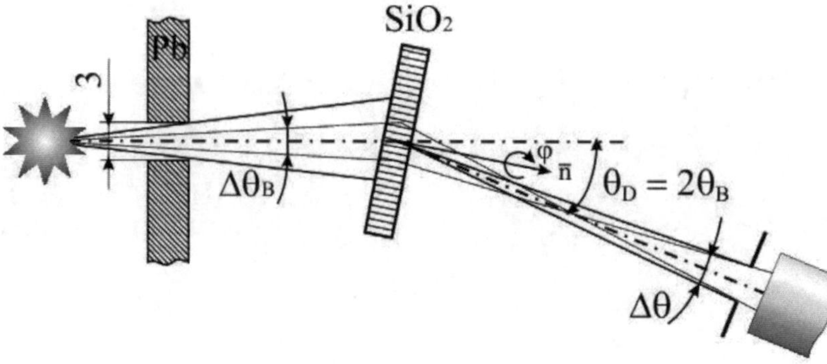

Figure1. Scheme of the experiment. θ_B – Bragg angle, θ_D – observation's angle, $\bar{n}$ – direction normal to the surface of the crystal.

The *X*-cut quartz crystals with different thickness (0.3 and 0.9 mm) were used in the experiment. Aluminium coating was applied on crystal surface for more effective stimulation of acoustic fields in crystal.

Radiation from X-ray tube was formed by 3 mm collimator, which was placed at the distance 90 mm from exit window and after that scattered by quartz crystal, then fixed in the goniometer at the distance of 215 mm from collimator.

Goniometer has three translational and three rotational freedom degrees with accuracy of orientation not worse than 0.5 mm and 10 μrad, respectively. It is possible to install the crystal at the Bragg angle for any families of reflecting atomic planes, so that a detector located in the horizontal plane could register only beams propagating in this plane. The reflection from atomic planes oriented perpendicular to the large surface of the sample $(10\bar{1}1)$ with interplanar distance d = 3.3429 Å is studied.

Radiation was registered at the symmetric Laue geometry at distance from crystal monochromator equal to 300 mm by semiconductor detector BDER-KI-11K with resolution about 280 eV at the 5.6 keV line. A round lead lengthy collimator with 2 mm diameter and 50 mm length was placed in front of the detector.

Crystal was oriented for the position to satisfy the maximum of rocking curve according to techniques from paper [5].

The theoretical values of resonant frequency were calculated by formula (1) and it is equal to 3.16 and 9.50 MHz for thickness 0.9 and 0.3 mm accordingly. Also frequency was determined by scanning in around theoretical value (figure 2) and after fitting were determined real values of f_1 (table 1).

Table 1. Theoretical and experimental frequency values for different crystal thickness.

Crystal thickness (mm)	Resonant frequency (MHz)	
	Theoretical	Experimental
0.90±0.02	3.16	3.18+0.01
0.30±0.02	9.50	9.43±0.03

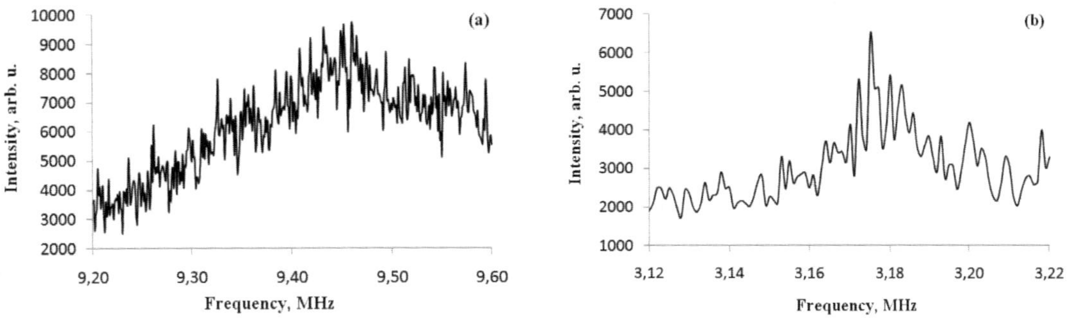

Figure 2. Dependencies of diffracted beam intensity on the frequency of electromagnetic field affect on the monochromator for crystal thickness equal to 0.3 (a) and 0.9 (b) mm.

Dependencies of diffracted beam intensity on voltage value on crystal are presented in figure 3. Data are shown relative to the mean level of diffracted beam intensity without excitation. The initial parts of these dependencies are linear. The length of a linear part depends on crystal's thickness and equals to 40 V for 0.9 mm and 12.5 V for 0.3 mm.

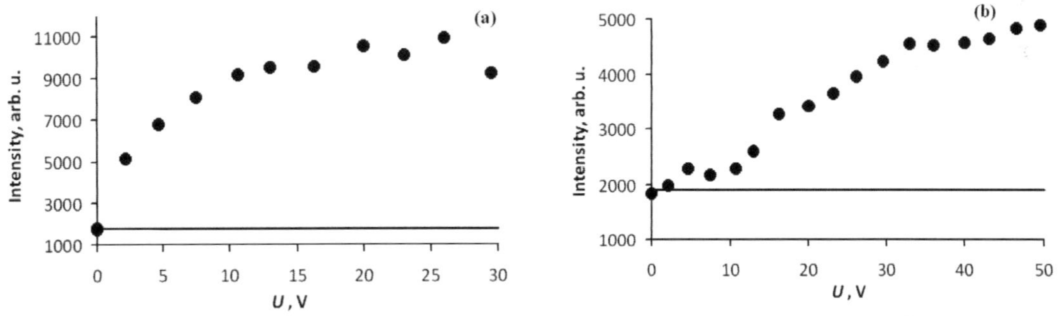

Figure 3. Dependencies of diffracted beam intensity on amplitude of the voltage for crystals thickness equal to 0.3 (a) and 0.9 (b) mm. Line is mean level of diffracted beam intensity without excitation.

Radiation spectra for cases of oriented and disoriented crystal are shown in figure 4. At the oriented case is changing only the diffracted beam intensity when deformation field turn on, and spectral line is lost when crystal is disoriented, as it was expected.

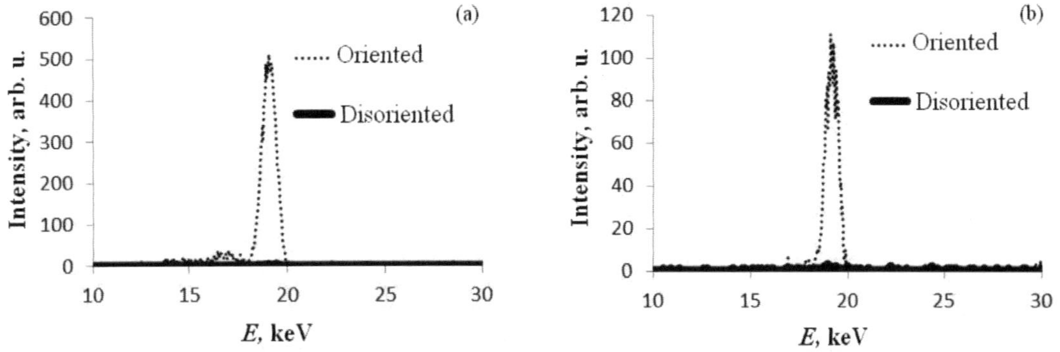

Figure 4. Radiation spectra for cases of oriented and disoriented crystal with acoustic wave field (a) and without (b). Thickness of crystal is 0.3 mm

Spectra in the maximum of rocking curve for different orientation angle for cases of excited and unexcited crystal are presented in figure 5. The Bragg angle was equal to 4.4° and 5.5° that corresponds to energy 24.3 and 19.3 keV. Value of amplitude was equal to 12 and 40 V for crystals with thickness of 0.3 and 0.9 mm accordingly.

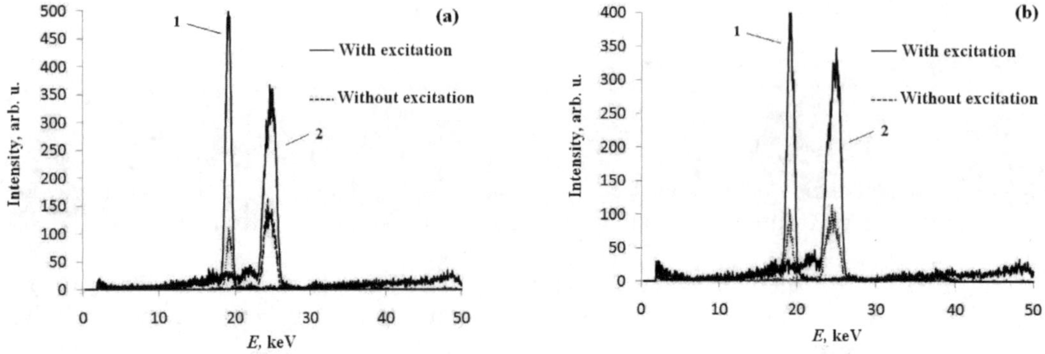

Figure 5. Comparison of diffracted radiation spectra for quartz crystals with and without excitation. Thickness of crystals is 0.3 (a) and 0.9 (b) mm. 1 – Bragg angle is 5.5°, 2 – Bragg angle is 4.4°.

According to the experimental results the intensity of X-ray beam into diffraction direction was increased 5.3 and 4.7 times for thickness equal to 0.3 and 0.9 mm, correspondingly. The Width of energy line is determined by experiment conditions and width of detector instrument line.

As figure 5 shows, with the change of orientation angle the energy line changes, and when turning on the external deformation field the intensity of a diffracted beam increases, as it was expected, that allows to speak about prospectivity of using "pumping effect" when creating brighter sources of monochromatic X-ray radiation with controlled parameters of radiation on the basis of bremsstrahlung monochromatisation.

The presented experiments demonstrated an efficiency of the pumping effect. With the help of external influences on the crystal monochromator increasing and tunability of the diffracted X-ray beam intensity are shown. The enhancement of intensity of 5 times under acoustic pumping has been observed. The obtained results suggest the possibility of creating a fast tunable adaptive X-ray optics.

Acknowledgment
The authors express their gratitude to the staff of the Institute of Applied Problems of Physics NAN RA (Yerevan, Armenia), Mkrtchyan A R, Mkrtchyan A G, Kocharian V R for numerous stimulating discussions and for the opportunity to conduct research on quartz monochromator. This work was partly supported by AVTSP RNPVSH № 2.1.2/13162 and Government of the Russian Federation "On the Federal Target Program" GK № 11.519.11.2030.

References

[1] Gabrielyan R G *et al* 1985 *Phys. Stat. Sol*.V (a) **92** 361-8
[2] Mkrtchyan A R, Navasardian M A and Mirzoyan V K 1982 *JETP Letters* **8** 677-80 (in Russian)
[3] Mkrtchyan A R, Navasardian M A, Gabrielyan R G *et al* 1983 *JETP Letters* **9** 1181-4 (in Russian)
[4] Mkrtchyan A R, Navasardian M A, Gabrielyan R G *et al* 1986 *Solid State Communications* **59** 147-9
[5] Gogolev A S, Popov Yu A, Wagner A R, Potylitsyn A P, Kocharyan V R, Movsiyan A E and Mkrtchyan A G 2010 X-ray scattering in deformed crystals *Universities news. Physics* **53** 33–8 (in Russian)

Inelastic nuclear interactions at protons multiple passage in bent crystals

A Babaev[1] and S B Dabagov[2,3]

[1] Tomsk Laboratory of Computational Physics, Tomsk, Russia
[2] INFN Laboratori Nazionali di Frascati, Frascati (RM), Italy
[3] RAS P.N. Lebedev Phys. Inst. & National Research Nuclear University, Moscow, Russia

E-mail: krass58ad@mail.ru

Abstract. Inelastic nuclear interactions for relativistic channeled and quasi-channeled protons in a bent crystal are theoretically studied. Multiple passage of projectiles through experimental setup was in details analyzed. Paying attention to the features observed, simulation results have been compared with known experimental data.

1. Introduction

Planar channeled and quasi-channeled projectiles travel in a crystal under small angle to the crystallographic planes interacting with the averaged field of plane's atoms instead of separate atomic fields (see, for example, [1] and references therein). The channeling phenomenon takes place when incident angle θ_0 between projectile velocity and crystallographic planes does not exceed a certain critical value θ_L known as the Lindhard angle [2]. Channeled projectiles, penetrating into the crystal at angles $\theta_0 < \theta_L$, travel between two parallel planes, which form so-called planar channel. Till the projectile transverse energy remains less than the potential barrier of a planar channel, a particle is channeled, but once the transverse energy exceeds the barrier, we deal with quasi-channeling mode of the motion (obviously, for transverse energies close to the barrier) [1]. On the contrary to the channeled particles, quasi-channeled ones are not constrained within any single channel.

For a bent crystal the averaged crystal field depends on a distance to the center of a bent plane curvature (see [3, 4] for details). The beam directed along the bent planes under the channeling or quasi-channeling conditions is split onto two beams in a crystal. The first beam corresponds to channeled particles, which deflect down to bent channel at large angles from the initial direction of motion, while the second one - to quasi-channeled particles. For quasi-channeled particles, moving initially to the center of crystal curvature, the volume reflection phenomenon might be observed. Namely, on the particle trajectory we can define a "turn" point where the travel direction changes with respect to the center of curvature. Finally, the reflected particles are deflected by crystal field in the direction opposite to one of channeled particles. When quasi-channeled particles initially travel from the center of curvature, they are slightly deflected mainly due to multiple scattering on the crystal nuclei. Multiple scattering leads to the transition of projectiles from a channeled state to quasi-channeled one and vice versa.

[1] At transverse energies over the barrier the particles are known as dechanneled

Published under licence by IOP Publishing Ltd

The first one when the projectile transits from the channeling to quasi-channeling regimes of motion is named "dechanneling'. The second when the quasi-channeled projectile is captured in a channeling state - "volume capture".

The main goal of this work is to investigate the probability of inelastic nuclear interactions (INI) between channeled (as well as quasi-channeled) protons and nuclei in a bent crystal. Namely, we have analyzed the case for 120 GeV protons moving through Si crystal bent along (220) crystallographic planes at the angle $\alpha_R = 150$ μrad with the radius $R = 1333$ cm (the conditions of experiment [5]). The crystal thickness is $z_{cr} = 2$ mm, and the critical channeling angle for this case - $\theta_L = 19$ μrad.

Below the averaged probability of INI will be introduced, and, successfully, its dependence on crystal orientation will be investigated. Moreover, we suggest a model of experimental scheme that allows the features of multiple passage of projectiles through the crystal to be simulated. Finally, the simulation results together with the analysis of experimental data [5] will be presented.

2. Averaged probability of INI

Let designate σ the INI cross-section. The INI probability P when proton passes the way z in the amorphous solid is $P = \sigma N z$ (which is valid while $P << 1$) where N is the nuclear density[2]. In aligned crystal, for both channeling and quasi-channeling conditions, the nuclear density N strongly depends on the distance from crystallographic planes due to the thermal vibrations of nuclei near the equilibrium position located on a plane. In this manuscript we suppose the Gaussian distribution for the displacement of nuclei from the plane [1]. Then for aligned crystal the INI probability can be rewritten as follows:

$$P = \sigma \langle N \rangle_{tr} z,$$

where $\langle N \rangle_{tr}$ is the averaged nuclear density over the projectile trace.

Now let consider nondivergent proton beam hitting the bent crystal at the incident angle θ_0 (see, in [4, 6] and references therein for details). To define the mean (averaged) nuclear density $\langle N \rangle_0$ at given crystal orientation (which is defined by the angle θ_0) we have to average $\langle N \rangle_{tr}$ by all proton trajectories in a crystal [7].

Assuming that the number of protons penetrating into the crystal is large, $n_p >> 1$, we should not analyze the trajectory for each proton, but we can suggest that the averaged INI probability in a crystal volume is defined by

$$P = \sigma \langle N \rangle_0 z \tag{1}$$

for one arbitrary proton and, successfully, the full number of INI - by

$$Z = P n_p. \tag{2}$$

Hence, if the counter is mounted anywhere near the crystal for the INI registration, it records the quantity proportional to the value Z.

The results of simulations for the probability P revealed from Eq.(1) are presented in figure 1. The trajectories were evaluated using a model presented in [4, 6]. The multiple scattering of protons on both crystal electron and nuclei was introduced based on [8, 9]. Let define $\theta_0 > 0$ for the projectiles moving from the center of curvature initially, and $\theta_0 < 0$ - in opposite case. The value $\theta_0 = 0$ corresponds to the projectile entering a crystal strongly along the tangent line to bent planes at the crystal entrance end. The INI cross-section $\sigma = 0.506 \cdot 10^{-24}$ cm^2 was

[2] The nuclear density in this manuscript coincides with the concentration

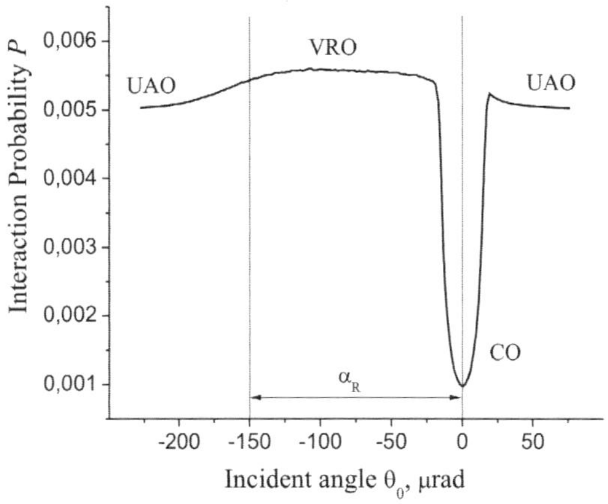

Figure 1. Dependence of the INI probability P on the incident angle θ_0 (120 GeV protons, (220) Si bent crystal, $\alpha_R = 150\ \mu$rad, $R = 1333$ cm).

taken from [10]. The designations in the figure 1 are: UAO - unaligned crystal orientation, when the deflection of protons is determined mainly by multiple scattering; VRO - volume reflection orientation, when the deflection of protons is determined mainly by the volume reflection; CO - channeling orientation, when the deflection of protons is determined mainly by the channeling.

When the parallel proton beam is directed initially along the crystallographic planes (CO) the probability of INI is reduced five times in comparison to one at unaligned crystal orientation (UAO). This result is in agreement with the experimental data [5]. At the conditions of VRO, the INI probability is about 10% higher than at UAO. The authors of [10] came to the similar conclusion. On other hand, the experimental curve in [5] for the condition of VRO appears below the curve at UAO. As further shown this effect is related to the particles multiple passage through the experimental setup. It is notable that in a whole considered interval of the angles θ_0 the INI probability is less than 1%.

The shape of the curve $P(\theta_0)$ is related to the nuclear density in the area where proton moves. Channeled protons move in a central area of the channel where nuclear density is lowest. The protons traveling at the UAO conditions feel the same averaged nuclear density as in amorphous medium. On the contrary, near the turn point the reflected protons move in the area of extremly high nuclear density during long time (see, for details, in [7]). The latter proves the fact that the INI probability at VRO exceeds the ones at UAO.

3. Multiple passage of projectiles through a bent crystal

3.1. The idealized experimental setup

In experiments [5, 10] the holder with bent crystals was mounted into the pipe of storage ring resulting in the multiple passage of projectile through the setup. Obviously, the INI value should vary in comparison with a single passage case above examined. On one side, multiple passage of the projectile through the crystal should be characterized by similar features that we can observe with the crystal thickness increase, i.e. one can expect the growth of the INI number. On other side, projectiles might be large deflected from its initial direction of motion due to both channeling and volume reflection effects, and, finally, projectiles can leave the portion of a beam hitting a crystal. As a result, the deflection of particles must restrict the growth of the INI number in the multiple passage regime.

To model the situations described above, let consider a simple idealized experimental scheme (see, in figure 2). A parallel beam 1 maintains initially the number $I >> 1$ of protons. The absorber systems 5 are placed to cut the tails of a beam. The focusing system 6 is mounted behind the absorber. Keeping the beam profile, it delivers the beam to the setup entrance. Let

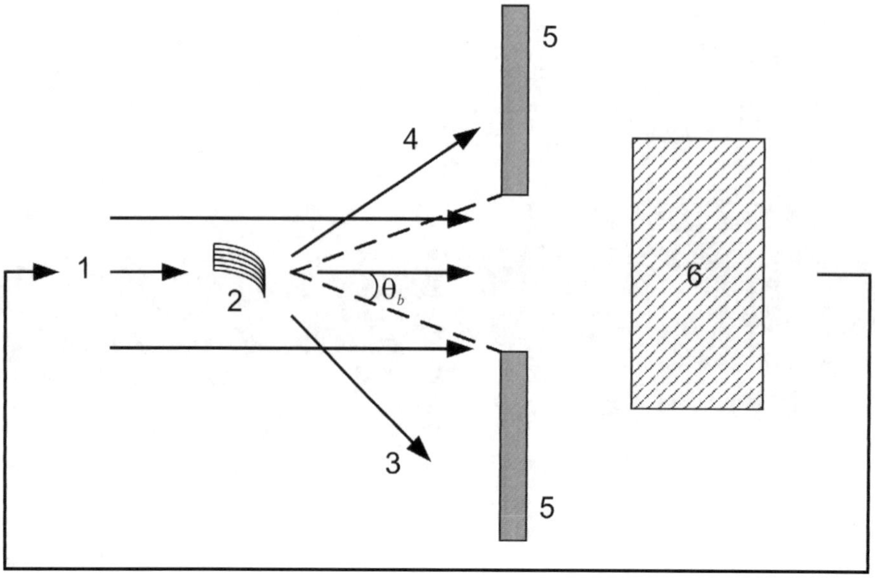

Figure 2. Experimental setup scheme: 1 - the initial parallel proton beam, 2 - the bent crystal, 3 - the channeled protons, 4 - the reflected protons, 5 - the absorber, 6 - the focusing system.

mount the bent crystal 2 into the beam before the absorber 5; the number αI of protons hits the crystal. Hence, particles can be deflected from the initial direction due to channeling 3 or volume reflection 4 at large angles exceeding the cutting angle θ_b. These particles cut by the absorber 5 will leave the beam. Other particles, namely, which either did not hit the crystal or being deflected at angles less than the cutting angle θ_b, pass to the focusing system 6. The focusing system makes the beam parallel again and restores initial beam profile. After that a steered beam is turned to the crystal collimation setup.

Let introduce coefficient β defining the part of protons hitting the crystal and being deflected by bent crystal on the angles $|\theta| < \theta_b$. Namely, a coefficient β is the part of protons crossing the crystal and successfully passing to the focusing system. The coefficient β can be evaluated from the angular distribution. To obtain angular distributions we used the model [4, 6] where the multiple scattering of projectiles is introduced [8, 9]. While the focusing system restores the beam profile, both coefficients α and β remain unchanged from the turn to the turn; they depend on orientation as well as on position of a crystal.

3.2. The INI estimation

First of all, let define the dependence for particles number in a beam on the number of turns under the INI neglecting condition. The initial number of particles in a beam is $I_{0,1} = I_0$ at the experimental setup entrance. The number of particles hitting the crystal is $\alpha I_{0,1}$. The part of these particles that passes to the focusing system is defined by $\alpha\beta I_{0,1}$. Therefore, the full number of particles at the focusing system entrance is $I_{t,1} = (1-\alpha)I_{0,1} + \alpha\beta I_{0,1}$. These particles are delivered to the setup entrance again. Successfully the described procedure repeats. Obviously, for the T-th turn the number of particles entering the setup will be $I_{0,T} = (1-\alpha+\alpha\beta)^{T-1}I_0$.

Let now consider the problem of INI in the crystal volume. As shown in Section 2, the INI probability is quite small for one proton during single passage through the crystal. Therefore, the proton leaving the beam due to INI does not affect the full number of particles of a beam, and Eq.(2) is valid for the INI evaluation over T turns. We can define $Z_i = P\alpha I_{0,i}$ as the INI number for the i-th turn, and the full INI number over T turns as follows:

$$Z = \sum_{i=1}^{T} Z_i = P\alpha I_0 \sum_{i=1}^{T} (1 - \alpha + \alpha\beta)^{i-1}.$$

To make the further simplifications, one can assume that the crystal does not accept the beam completely, therefore, $\alpha < 1$. Additionally, the coefficient $\beta < 1$, if the absorber cuts protons, whereas the value $\beta = 1$ corresponds to the case when all protons pass to the focusing system. Therefore, at $\beta < 1$, we have

$$Z = \frac{1 - (1 - \alpha + \alpha\beta)^T}{1 - \beta} P I_0. \tag{3}$$

For the case $\beta = 1$ one can find that

$$Z = T\alpha P I_0. \tag{4}$$

Indeed, when all particles pass to the focusing system the full number of particles does not change while condition $Z_i << I_{0,i}$ is fulfilled. Hence, Eq.(4) concludes that in each turn the same number of INI, $P\alpha I_0$, takes place.

On the contrary, as follows from Eq.(3), when number of turns increases the full number of nuclear reactions Z reaches the saturation characterized by the following limit

$$Z_{\text{inf}} = \frac{P}{1 - \beta} I_0, \tag{5}$$

which is α independent. As the matter of fact, α defines the time to reach the saturation.

4. INI simulation results

To eliminate the dependence on crystal position (i.e., on coefficient α) we will consider further the INI number in the saturation regime. From considerations above one can conclude, the INI number depends on both incident θ_0 and cutting θ_b angles. It should be underlined that the coefficient β is strongly correlated with the angular distribution shape of protons traveled through the crystal. The detailed explanation of that fact can be found in [7]. The dependencies $\beta(\theta_0)$ at different cutting angles θ_b are presented in figure 3 where to higher curve by the position in a figure corresponds larger angle. All curves in figure 3 are revealed the deep in the CO area. Indeed, channeled projectiles are deflected by bent crystal at large angles $\theta \approx -\alpha_R < -\theta_b$. Hence, all channeled particles hit the absorber and leave the beam.

At small angle $\theta_b = 1$ μrad the small part of projectiles scattered by crystal passes to the focusing system and it reveals a slight variation at angular distribution (i.e. θ_0) change. Therefore, the corresponding curve in figure 3 is the lowest and the coefficient β changes slightly. On the contrary, at the largest angle $\theta_b = 80$ μrad almost all projectiles pass to the focusing system at any crystal orientation except the CO one. Therefore, one can expect $\beta \approx 1$ except the sharp decrease at CO.

For intermediate angles $\theta_b = 10$ μrad and 50 μrad the coefficient β in figure 3 demonstrates quite different behavior. Namely, for the case $\theta_b = 10$ μrad one can see the deep in the area $\theta_0 \approx -\alpha_R$ whereas for the case $\theta_b = 50$ μrad this deep is absent. The matter is connected with the shape of angular distribution in the transition area from VRO to UAO. Actually, when crystal is mounted at VRO, the majority of protons form the single sharp peak (see, for example, experimental data [11]). This peak is scanned at both $\theta_b = 10$ μrad and $\theta_b = 50$ μrad. On the

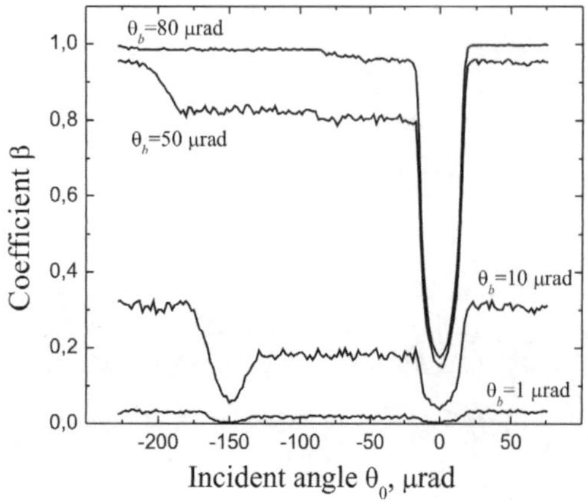

Figure 3. The coefficient β as the function of the incident angle θ_0 at different cutting angles θ_b (120 GeV protons, (220) Si bent crystal, $\alpha_R = 150$ μrad, $R = 1333$ cm).

contrary, when crystal is mounted at angle $\theta_0 \approx -\alpha_R$ outgoing beam is split onto two beams. One part of projectiles moves under volume reflection condition whereas other part is deflected mainly due to multiple scattering. The observer watching the projectiles behind the absorber (at the entrance of a focusing system) scans both peaks at $\theta_b = 50$ μrad simultaneously, i.e. he observes the same intensity of projectiles as in the VRO area. But under condition $\theta_b = 10$ μrad the observer registers one peak only or even the deep between peaks. As the matter of fact, the number of particles passing to the focusing system decreases with respect to VRO. When the crystal orientation is changed to UAO (angle $\theta_0 < -\alpha_R$), the single peak forms once again (see, in [7] for details).

Both dependencies $\beta(\theta_0)$ (figure 3) and $P(\theta_0)$ (figure 1) define the INI number in the crystal volume (5). The number Z_{inf} is presented in figure 4 for the same conditions as for coefficient β in figure 3. The cutting angle θ_b increases from figure 4(a) to figure 4(d). For all parts of figure 4 the deep at CO appears. It is caused by deflection of channeled protons at large angles (see, in figure 3) as well as by the reduce of the INI probability (see, in figure 1). The INI number at VRO becomes larger with the increase of angle θ_b. It proves the expectation: the more this angle the more protons pass many times through the crystal.

In figure 4(a) the cutting angle is very small ($\theta_b = 1$ μrad). Therefore, every proton passes through the crystal one time only and after that is deflected by the crystal to the absorber. In this case the ratio of INI number Z_{inf} to the full number of projectiles I_0 is, actually, the INI probability P for one proton (figure 1). The same conclusion follows directly from the formula (5) at small β.

In figure 4(b) ($\theta_b = 10$ μrad) one can see the deep in the area $\theta_0 \approx -\alpha_R$ whereas the similar deep is absent in figure 4(c) ($\theta_b = 50$ μrad) that agrees with the $\beta(\theta_0)$ analysis above presented.

At the cutting angle $\theta_b = 80$ μrad (figure 4(d)) the number of protons, which suffer INI is about 40% from its initial number I_0 at VRO and this number approaches 100% at UAO. It should be underlined that here the saturation state has been considered. Obviously, to distinguish these quantities the huge number of turns is needed.

Let now explain the sharp oscillations of the curves Z_{inf}/I_0 in Figs. 4(c,d). One can notice from figure 3, the coefficient β in the corresponding areas is $0.98 \div 0.99$. Further, from Eq.(5), the small variation in the coefficient β leads to significant changes of the INI number Z_{inf}. Due to the fact that the coefficient β is determined by the shape of corresponding angular distribution, small change of the distribution shape results in large variations of the INI number at large cutting angles. On one hand, actually, the distribution shape changes at different crystal orientation. On other hand, the Monte-Carlo method used in our simulations to obtain the angular distribution

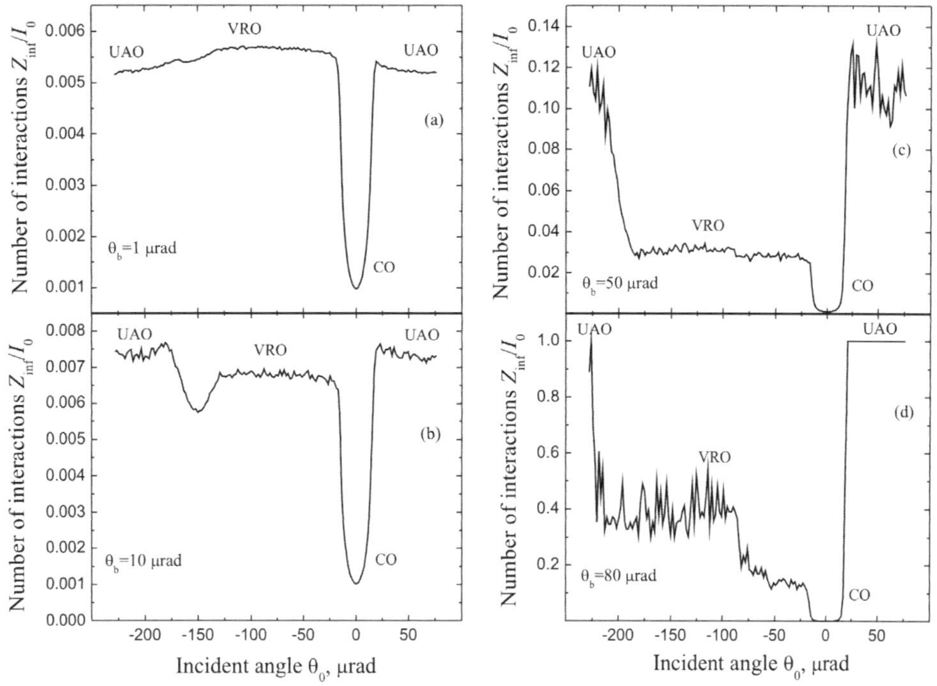

Figure 4. The INI number for the saturation state Z_{inf} with respect to the initial number of protons I_0 in dependence on the incident angle θ_0 at different cutting angles θ_b: (a) $\theta_b = 1$ μrad, (b) $\theta_b = 10$ μrad, (c) $\theta_b = 50$ μrad, (d) $\theta_b = 80$ μrad. The case of 120 GeV protons passing through Si crystal bend along (220) planes over the angle $\alpha_R = 150$ μrad with the radius $R = 1333$ cm is considered.

can also contribute to small variation in the distribution shape. Both contributions form the final oscillations at UAO shown in figure 4(c) as well as at VRO shown in figure 4(d).

For the range $\theta_0 = (-75)\div(-25)$ μrad there is an interval in the VRO area in figure 4(d) where the sharp oscillations are absent. These crystal orientations are characterized by significant deflection (over angles $\theta > \theta_b$) of protons due to the volume capture effect [7, 11]. As a result, a few percents of protons leave the beam and small changes of the distribution shape does not significantly influence the INI number.

Finally, let examine next feature of the results obtained. When projectiles pass through the crystal one time, the INI probability is more at VRO than at UAO (see, figs.1 and 4(a)). When the turn's number increases, on the contrary, this probability is more at UAO than at VRO (figure 4(b-d)). Indeed, at UAO protons are deflected on the fewer angles with respect to the case of VRO (see [7, 11]). Therefore, a proton has a chance to make more turns before it will be deflected over the angle $|\theta| > \theta_b$ at UAO than at VRO. Due to the fact that the probability of interaction growths with the increase of number of turns, at multiple passage we will have the reduced INI probability at VRO with respect to UAO. This result is in agreement with [5, 10].

5. Conclusion

In this work we have evaluated the probability of inelastic nuclear interactions when relativistic protons travel through the bent crystal. The main attention has been paid to the multiple passage of protons through the experimental setup. The idealized experiment was simulated in order to investigate the INI at the given conditions. The scheme of experiment is "designed" to involve both a bent crystal for projectiles deflection and an absorber for deflected particles cut-off. These elements might be used to collimate a charged particle beam, and, hence, the suggested setup, idealized for simplicity, in general, works as a real collimator facility. Moreover, this theoretical experiment is free of the technical details, which make the nature of experimental data not so clear. Hence, our idealized setup can be used to better understand the real experiment features.

Presented analysis enables explaining the difference between results of simulations and experimental data published in [5]. The published data are not adapted for direct analysis. Nevertheless, the shape of our simulated curves in figure 4 is in qualitative agreement with experimental data [5]. Additionally, we would like to underline that the developed model allowed us to explain observed deeps in the INI counts when the crystal is mounted in the transition area from volume reflection to unaligned crystal orientation. Actually, it is not related to inelastic interactions themselves. It is purely the effect of the beam collimation. The mentioned deep can be either enhanced less or more or can be even absent in dependence on the cutting angle at the setup exit. The sharp oscillations at both VRO and UAO discussed above might be smoothed in experiment and unobservable due to the reasons described in [5], namely, the crystal torsion and other imperfections. Also, it should be mentioned here, in our considerations we deal with the saturation state, whereas this state could occur unreachable in experiments [5, 10] where the condition $\alpha << 1$ was realized.

The model suggested in this work based on simple mathematical considerations. It is not very exacting to the computational resources, but it gives the results which are in agreement in the well-known experimental data [5, 10]. Therefore, this model can be useful to plan the future experiments and to simulate the results.

Acknowledgments

The work was supported by the FAI (LNF & Roma I INFN) grant and the RF President grant MK-2059.2011.2. We are particularly grateful to Dr. G. Cavoto for stimulating discussions.

References

[1] Gemmel D S 1974 *Rev. Mod. Phys.* **46** 129
[2] Lindhard J 1965 *Mat. Fys. Medd. Dan. Vid. Selsk.* **34**(14)
[3] Taratin A M and Vorobiev S A 1987 *Nucl. Instr. Meth.* B **26** 512
[4] Babaev A and Dabagov S B 2011 *Nuovo Cimento* C **34** 417
[5] Scandale W et al 2010 *Phys. Lett.* B **692** 78
[6] Babaev A and Dabagov S B 2011 Analysis of relativistic proton deflection by bent crystals *Preprint* LNF-11/04(P). www.lnf.infn.it/sis/preprint/pdf/getfile.php?filename=LNF-11-4(P).pdf
[7] Babaev A and Dabagov S B 2011 On proton multiple inelastic nuclear interactions in bent crystals *Preprint* LNF-11/18(R). www.lnf.infn.it/sis/preprint/pdf/getfile.php?filename=LNF-11-18(R).pdf
[8] Kitagawa M and Ohtsuki Y H 1973 *Phys. Rev.* B **8** 3117
[9] Taratin A M 1998 *Physics of Elementary Perticles and Atomic Nuclei* **29** 437
[10] Scandale W et al 2010 *Nucl. Instr. Meth.* B **268** 2655
[11] Scandale W et al 2008 *Phys. Rev. Lett.* **101** 234801

Structure of below-barrier and above-barrier energy bands and wave functions of a particle in the periodic potential of crystalline lattice

Erine A. Babakhanyan
National Research Laboratory after A.I. Alikhanian (Yerevan Physics Institute), Armenia, 0036, Alikhanian Br. Str. 2

E-mail: eriphys@mail.ru

Abstract. The knowledge of energy spectrum structure allows us to find the Bloch state wave functions. During the interaction of charged particles moving in a crystal, the wave functions represent the superposition of Bloch functions. The population of every Bloch state is characterized by Fourier-component squared. The analytical results received in the given work for periodic potential of a crystal lattice in the model of modified Pöschl-Teller's potential make a basis for studying the phenomena in a channeling mode of particles in crystals (the behaviour of particles at different depths of penetration into crystal, angular distributions of particles, both inside the crystal, and at its exited, etc.). The account of the zone structure of energetic spectrum of channeled electrons results in qualitative change of angular distributions of X-ray radiation under the Bragg angles. Therefore it is interesting within the framework of our approach to study the radiation, as well as an output of the nuclear reactions appearing in case of interaction of charged particles moving in a crystal.

1. Introduction

The channeling effect of various particles (ions, protons, positrons, electrons) was investigated in many theoretical and experimental studies. For protons, ions and other heavy particles it is possible to apply a classical consideration, but for electrons and positrons, which mass is less than the mass of the abovementioned particles, a quantum approach is required because of the existence of diffraction effects. Moreover, for small angles θ between the direction of passing through the crystal fast particles and the crystallographic plane or axis, in so-called sliding motion, the transverse component of the particle energy becomes comparable to the barrier height $E_{\perp} \sim E\sin^2\theta \sim E\theta^2 \sim V$. Hence, an abnormally large or minimal passage of the incident particles, in comparison with amorphous sample, may take place.

By means of channeling effect, which is the most known direction in physics of orientation phenomena, it is possible to study the thermal fluctuations and displacements of atoms in a lattice, the distribution of electron density in the interatomic space in crystals, as well as the possibility the use of channeling for creation the efficient control systems for beams of high energy particles.

2. Channeling of light particles

The channeling of light particles (electrons and positrons) is attractive, especially, in connection with an opportunity of using of monochromatic hard radiation emitted at channeling, which presents practical interest.

A range of the practical applications of orientation effects has been broadening continuously in process of knowledge accumulation. So, the channeling effect has served as a base for revealing the new experimental methods to study the solid state composition and structure. Due to the unique possibilities to determine to a big accuracy the location of impurity atom and self-interstitials in crystalline lattice, to find the profiles of radiation defects and to classify them, to study the structural imperfection in the surface and near-surface layers of crystals and thin monocrystalline films, these methods find more and more wide application in various areas of science and engineering: nuclear physics, solid-state physics, semiconductor technology, microelectronics, etc.

In the earliest works concerning the computer modeling of passage of ions in monocrystals, the anomalous high penetration of ions in comparison with an amorphous sample was received. The main point of this effect is observation of an exit dependence of particles on the incidence angle in case of their incidence by the small angle to a crystallographic plane or axis.

3. Quantum theory of channeling

Many theoretical and experimental works, where it was investigated the channeling of various particles (*p*, e^+, e^-, π^+, π^- etc.), have followed thereupon (see review of Gemmel D.S. [1]).

The consequent quantum theory of channeling effect is developed in a series of works by Yu.M. Kagan and Yu.V. Kononets [2].

M.A.Kumahov's work [3] has stimulated the appearance of a number of studies on the radiation arising in a mode of channeling (see review of Wedell R. [4]).

In the research of channeling effect of charged particles in crystals the big importance is to find the energy spectrum structure (zones and gaps) in the periodic potential formed by chains or planes of atoms of the crystalline solid-state lattice.

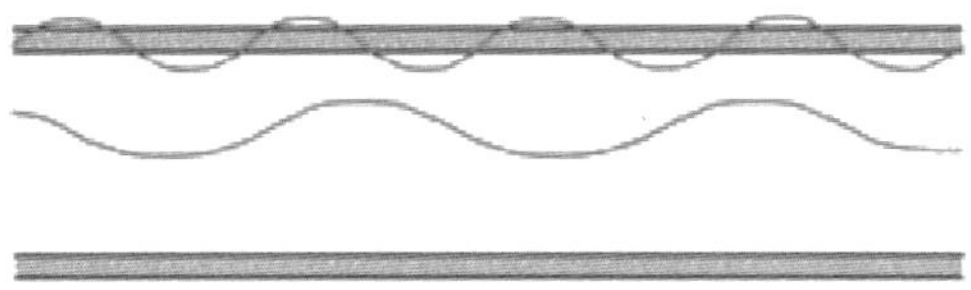

Figure 1. The scheme of trajectories of both positrons and electrons in a crystal planar channel

The channeling of electrons differs from the channeling of heavy particles. Features of electron channeling are caused by influence of their wave properties and negative charge.

For light particles, in particular, for electrons, the quantum parameter characterizing the number of bound states in the potential of a crystalline lattice, is smaller than for positrons of the same energy, and much less, than for heavy particles.

The wave function of the projectile represents the superposition of corresponding Bloch functions, the properties of which define, practically, the whole spectrum of physical phenomena observed under channeling conditions.

4. Schrödinger equation and Bloch functions

At planar channeling the transverse motion is described by Schrödinger equation with some effective one-dimensional periodic potential. In simple cases it is represented by a series of potential barriers of equal height. States of a particle in such potential split into two groups: states with energy less than the potential barrier (so-called below-barrier, allowing a simple consideration, for example, in the strong-coupling approximation), and above-barrier ones with energies in excess of a potential barrier height. The latter play a special role, for instance, in an abnormally deep penetration of electrons into a crystal under channeling conditions.

For revealing the basic conformities to natural laws of behaviour of particles in periodic potential, and also from the point of view of a possibility of full analysis realization, the most acceptable is the model of the periodic potential formed by rectangular barriers of definite width and height. Such potential named as Krönig-Penney potential has allowed, in particular, to analyze the distinctions in channeling pictures of positively and negatively charged particles that turned out to be possible in case of studying a problem in the uniform scheme, considering the narrow holes for negatively and wide ones – for the positively charged particles that is equivalent to inversion of potential relative to the top of hole.

Within the framework of this model of potential it was possible to carry out the full analysis and to receive the main qualitative results [5].

In the channeling effect of electrons we have established a special role of above-barrier states in the immediate vicinity of a potential barrier [6], what, by the way, allows explaining the anomalous transmission of the above-barrier electrons in crystals, which experimentally was observed at the Nuclear Physics SRI of Tomsk Polytechnic Institute [7]. The work by S.B. Dabagov and L.I. Ognev [8] also shows that the angular distribution of particles passing through the crystal depends on the structure of energy bands, as well as the properties of the below-barrier and above-barrier states.

Realizing the conditions under which the coherent phenomena have not damped (small thickness of a crystal), and, having measured the yield for inelastic processes on the crystal lattice nuclei or interstitial impurity atoms [9], it is possible to find experimentally the periods of appropriate oscillations and depth evolution of structure of density distribution for planar channeled particles in quantum approach that will give the information about the energy band structure for particles moving in a crystal [10].

Since the coefficients of above-barrier reflection for a fraction of electrons showing a deep passage through the crystal, have an oscillatory character, depending on the parameters of the particle and the crystal [11], to obtain a more accuracy by comparison with experimental results it is necessary to consider more smooth potential being closer in shape to the real one, for example, composed of single modified potentials of Pöschl–Teller's type for an isolated link.

D.V. Mescheryakov and V.B. Tverskoy, for example, studied the wave functions of a discrete spectrum for separate isolated Pöschl–Teller's potential [12].

In presented work, for the reason to achieve the real potential form of the crystal, the smooth periodic potential is built from the separate modified potentials of the Pöschl-Teller's type for isolated chain, which has the following form:

$$V(x) = -\frac{V_0}{ch^2 \alpha x} \tag{1}$$

In general shape (i.e. simultaneously for negatively and positively charged particles) it can be presented as follows:

$$V(x) = -\xi \frac{V_0}{ch^2 \alpha x} - V_1, \tag{2}$$

where ξ is two-valued, $\xi = \pm 1$: $\xi = 1$ for the potential well, and $\xi = -1$ for the barrier; V_0 and $d=1/\alpha$ are the height (depth) and width of a potential well (barrier); V_1 is a normalization factor.

The Schrödinger's equation

$$\left[-\frac{\hbar^2}{2m} \frac{d^2}{dx^2} + V(x) \right] \psi(x,t) = E\psi(x,t) \tag{3}$$

after the separation of variables $\psi(x,t) = g(t)u(x)$ and introducing new parameters

$$\chi^2 = \frac{2m}{\hbar^2} V_1; \qquad k^2 = \frac{2m}{\hbar^2} E; \qquad \lambda^2 = \frac{2m}{\hbar^2} V_0 \tag{4}$$

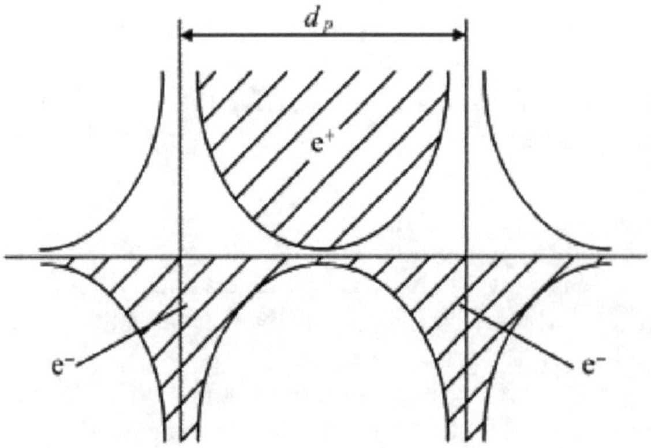

Figure 2. Continuous potential of atomic planes for positrons and electrons

has received the form:

$$u'' + \left(k^2 + \chi^2 - \frac{\lambda^2}{ch^2 \alpha x}\right) u = 0 \tag{5}$$

Solving the Schrödinger's equation with the specified potential, we come to the hypergeometric differential equation. For the smooth continuous potential the Bloch's wave functions and its derivatives have been joined in point of contact of barriers of the separate potential chains. The wave functions and the dispersion equation for calculation of energy bands are found.

The computer calculation for the nearest above-barrier band is carried out for the concrete values of potential parameters: $V_0 = 50$ eV, $d = 2$ Å (figure 3).

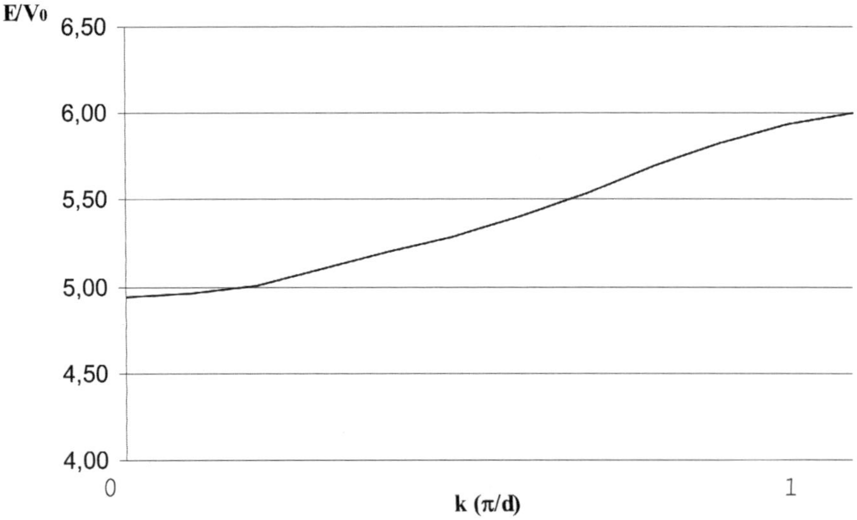

Figure 3. Energy dependence of quasi-momentum describing the Bloch states, for the first zone in the planar channeling of electrons in crystal with the Pöschl-Teller's model of periodic potential.

The development of the newest technologies for growth of quantum points (QP) has resulted in the opportunity of growing of the high-quality QP of various forms and sizes. At present there are intensively investigated the semi-conductor nanostructures similar to the structures of the $GaAs{=}Ga_{1\text{-}x}Al_xAs$ type. During the QP-growth owing to diffusion the limiting potential, which in most cases is approximated with the big accuracy by parabolic potential, is formed. In work [13] as such one is just offered the use of modified Pöschl-Teller's potential.

A number of new experiments on investigation of radiation spectra in case of planar channeling of electrons in crystals are carried out recently. Besides, new experiments with beams of relativistic electrons and positrons on the LNF Frascati (Italy) with energies 20÷855 MeV are planned within the framework of TPU-LNF collaboration and for interpretation of the experimental data received in 2007 on the race-track microtron MAMI (Mainz, Germany) [14, 15].

The works of scientists from Tomsk O.V. Bogdanov et al. [16, 17] show that the account of the band structure of energetic spectrum of channeled electrons leads to a qualitative change of angular distributions of X-ray radiation under the Bragg angles. Therefore it is interesting within the framework of our approach to study the radiation, as well as an output of the nuclear reactions occurring in case of interaction of charged particles the incident moving in a crystal.

Obtained analytical results for periodic potential of a crystal lattice in the model of modified Pöschl-Teller's potential make a basis for the further study of phenomena in a channeling mode of particles in crystals (the behavior of particles on the different depths of penetration into a crystal, the angular distribution of particles, both inside a crystal, and on exit of it, etc.).

References

[1] Gemmel D S 1974 *Rev. Mod. Phys.* Channeling and related effects in the motion of charged particles through crystals **46**(1) pp 129-227

[2] Kagan Yu and Kononets Yu V 1970; 1973; 1974 *JETP* Theory of channeling **58**(1) 226-46; **64**(3) 1042-64, **66** (5) pp 1693-1711 (*in Russian*)

[3] Kumakhov M A 1976 *DAN USSR* On the possibility of existence of the effect of spontaneous emission γ-quants by relativistic channeled particles **230**(5) pp 1077-80 (*in Russian*)

[4] Wedell R 1980 *Physica Status Solidi* Electromagnetic radiation of relativistic positrons and electrons during axial and planar channeling in monocrystals **99b**(1) pp 11-49

[5] Babakhanyan E A, Kononets Yu V 1980 *Physica Status Solidi* Energy band structure and Bloch functions in the planar channeling model with the Krönig-Penney potential **98b**(1) pp 59-77

[6] Kagan Yu, Babakhanyan E A and Kononets Yu V 1980 *JETP Letters* The special role of the above-barrier states in the pattern of channeling of electrons in crystals **31**(12) pp 776-80 (*in Russian*)

[7] Babakhanyan E A, Vorob'ev S A, Kononets Yu V and Popov D E 1982 *JETP Letters* Energy dependence of the structure of the angular distributions of electrons in planar channeling **35**(5) pp 184-7 (*in Russian*)

[8] Dabagov S B and Ognev L I 1988 *Nucl. Instr. Meth.B* Passage of MeV-energy electrons through monocrystals **30** pp 185-90.

[9] Babakhanyan E A 2011 *Nuovo Cimento* The depth evolution of density distribution structure for planar channeled particles in quantum approach **34**(4) pp 119-24

[10] Babakhanyan E A 2010 *Proc. Int. Conf. "Electrons, Positrons, Neutrons and X-rays Scattering under the External Influences" (Yerevan–Meghri, Armenia)* The evolution of a planar channeling picture of electrons and positrons with thickness of crystal pp 56-61.

[11] Babakhanyan E A and Hayrapetian G A 1986 *Radiation Effects* On the above-barrier reflection by the planar channeling of electrons and positrons in crystals **91** pp 287-91

[12] Meshcheryakov D V and Tverskoy V B 2000 *Vestnik MGU Ser 3. Theor. & Mathem. Physics* The wave functions of excited states of discrete spectrum of integrable quantum systems of N particles in the Pöschl-Teller potential **6** pp 8-11 (*in Russian*)

[13] Hayrapetyan D B, Dvoyan K G, Ghazaryan E M and Tchanchapanyan A A 2008 *Reports of NAS RA, Physics* Electronic states of a cylindrical quantum point with a thin crescent-shaped cross-section in case of modified Pöschl-Teller potential **108**(4) pp 320-30 (*in Russian*)

[14] Backe H, Kunz P, Lauth W and Rueda A planar channeling experiments with electrons at the 855 MeV Mainz microtron MAMI 2008 *Nucl. Instr. Meth. Phys. Res. B* **266** (17) pp 3835-51

[15] Backe H, Rueda A, Lauth W, Clawiter N, El-Ghazaly M, Kunz P and Weber T 2005 *Nucl. Instr. Meth. (Mainz U., Inst. Phys.)*. Forward diffracted parametric X-radiation from a silicon single crystal **B234** pp 138-47

[16] Bogdanov O V, Korotchenko K B and Pivovarov Yu L 2007 *JETP* Angular distribution of diffracted X-ray radiation in case of (111)-channeling of electrons in Si: the Influence of structure of energetic bands **85**(11) pp 684-8 (*in Russian*)

[17] Bogdanov O V 2008 *Autoreferat PhD (Tomsk)* Features of radiation of relativistic electrons and positrons in case of (111)-planar channeling in thin crystals 16 p (*in Russian*)

Journal of Physics: Conference Series **357** (2012) 012034 doi:10.1088/1742-6596/357/1/012034

The THz radiation source at SPARC

E. Chiadroni[1], A. Bacci[1], M. Bellaveglia[1], P. Calvani[2], M. Castellano[1], A. Cianchi[3], G. Di Pirro[1], M. Ferrario[1], G. Gatti[1], O. Limaj[2], S. Lupi[2], B. Marchetti[3], A. Mostacci[4], D. Nicoletti[2], A. Nucara[2], E. Pace[1], C. Ronsivalle[5], A. R. Rossi[1], C. Vaccarezza[1]

[1] INFN-LNF, Via E. Fermi, 40 - 00044 Frascati, Rome, Italy
[2] Phys. Dept. and INFN University of Rome "La Sapienza", Rome, Italy
[3] INFN-Roma Tor Vergata & University of Rome "Tor Vergata", Rome, Italy
[4] SBAI Dept. University of Rome "La Sapienza", Rome, Italy
[5] ENEA C.R. Frascati, Italy

E-mail: enrica.chiadroni@lnf.infn.it

Abstract. The linac driven coherent THz radiation source at the SPARC facility is able to deliver broadband THz pulses with femtosecond shaping. In addition, high peak power, narrow-band THz radiation can be also generated, taking advantage from advanced electron beam manipulation techniques, able to generate a train of short electron bunches with THz repetition rate.

1. Introduction

The motivation for developing a linac-based THz source at SPARC stays in the ever growing interest of filling the so-called THz gap with high peak power radiation. From simulations, the peak power expected at SPARC is in the order of 10^8 W. This result has been confirmed by measurements presented in [1]. The corresponding energy per pulse is of the order of tens of μJ, that is well above standard table top THz sources. Applications of this kind of source concern mainly time domain THz spectroscopy and frequency domain measurements on novel materials [2]. Beyond these applications, coherent THz radiation is also used as longitudinal electron beam diagnostics to reconstruct the beam charge distribution [3].

SPARC is an electron photo-injector deliverying high current, low emittance beams, so called high brightness electron beams (HBEB). SPARC was originally conceived for both electron beam dynamics studies [4, 5] and Free Electron Laser (FEL) experiments at 500 nm in the Self Amplified Spontaneous Emission (SASE) regime [6, 7] and down to shorter wavelengths with novel emission schemes, e.g. cascaded FEL configuration seeded by harmonics generated in gas [8].

Recently, SPARC is extending its prospects to a wider range of high brightness electron beams applications, e.g. Plasma and Laser Wakefields Acceleration (PWFA-LWFA), Thomson scattering and, not least, high peak power THz radiation. The generation of THz radiation and its characterization as radiation source is the main topic of the paper.

2. Theory

Radiation extending up to the THz range can be generated as coherent radiation, e.g. transition radiation (TR), from ultra-short ($\approx$ sub-ps) electron bunches. The total intensity emitted by a bunch of electrons, is the sum of two terms [9, 10]: the incoherent one, proportional to the number of particles in the bunch through the single particle radiation intensity, and the coherent one, proportional to the number of particles squared, which dominates the emission at wavelengths equal and longer than the bunch length. The coherent intensity depends on the bunch longitudinal dimension through the longitudinal form factor, defined as the Fourier transform of the bunch longitudinal profile.

2.1. Broad-band THz radiation

High peak power broad band THz radiation is produced by means of ultra-short high brightness electron beams. Few hundreds of femtosecond duration pulses are obtained by longitudinally compress the beam. As alternative to standard magnetic compression, RF compression, named as velocity bunching (VB) [11], is used at SPARC. This technique relies on the fact that the beam is injected in the first accelerating section at the phase of zero crossing field. Since the beam is slightly slower than the phase velocity of the RF wave, it slips back to phases where the field is accelerating and at the same time it is chirped and compressed. The difference between the maximum energy phase and the injection phase is called compression phase. The bunch length at the end of the linac changes by varying the compression phase (Fig.1, black data). Therefore, moving along the RF wave is equivalent to move along the compression curve shown in Fig.1 (red dashed curve). Due to non linearities in the longitudinal phase space (LPS), the

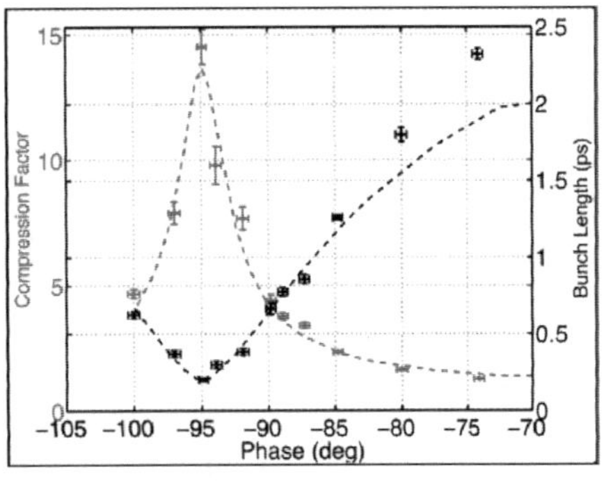

Figure 1. Measurements of bunch length as function of the RF phase (black dots), compared to TSTEP simulation (black dashed line). The red dots are the measured compression factors, compared to TSTEP simulation (red dashed curve).

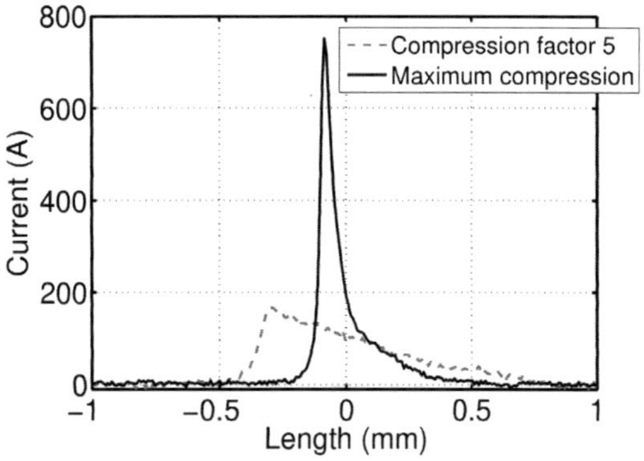

Figure 2. Beam current profile for compression factor 5 (red dashed curve) and maximum compression (black line).

rising edge of the current profile gets steeper and steeper as the compression factor increases. Figure 2 shows the current profile for two compression factors, that is the ratio between the length of maximum energy beam and the length of the compressed one, which depends on the compression phase. From the point of view of THz emission, the sharper the rising edge of the longitudinal profile the higher the THz frequency emission.

Journal of Physics: Conference Series **357** (2012) 012034 doi:10.1088/1742-6596/357/1/012034

2.2. Narrow-band THz radiation

Narrow band THz radiation is produced by means of longitudinally modulated electron beams, i.e. comb beams [12, 13]. If a train of laser pulses with picosecond (ps) inter-distance illuminates the photocathode, assuming instantaneous emission at the cathode, each pulse will produce a disk of electrons at the same ps distance. Due to longitudinal space charge, the bunches lengthen and the original pulse structure is lost in the current profile. Nevertheless, some memory of the original structure is kept in the longitudinal phase space as sawtooth energy profile. In the VB regime, a rotation in the longitudinal phase space is forced, restoring the original current modulation at the end of the linac. The proper pulses length and inter-distance can be adjusted by changing the compression phase.

If a comb train with THz repetition rate is transported down to the transition radiation target, narrow band emission occurs at the frequency corresponding to the pulse distance. For example, in case of four pulses in the train with time separation of 1 ps and time duration of the order of few hundreds of femtoseconds rms, the longitudinal form factor has emission peaks at the fundamental, i.e. 1 THz, and higher harmonics. The number of harmonics and their intensity depend on the modulation of the temporal profile and on the size of each pulse.

2.3. Tunable THz radiation

The emission frequency in the THz range can be tuned by changing the distance between pulses in the train. The pulse distance can be varied by changing the compression phase. Indeed, as shown

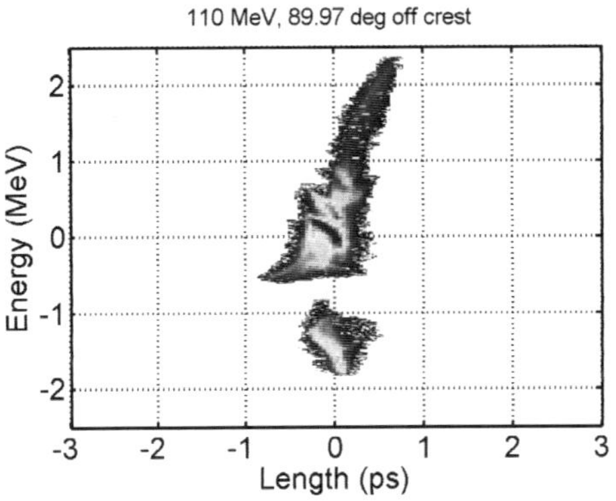

Figure 3. Measured Longitudinal Phase Space (LPS) for a four-pulses comb beam at the maximum compression phase; the title reports the average energy and the compression phase (200 pC total charge).

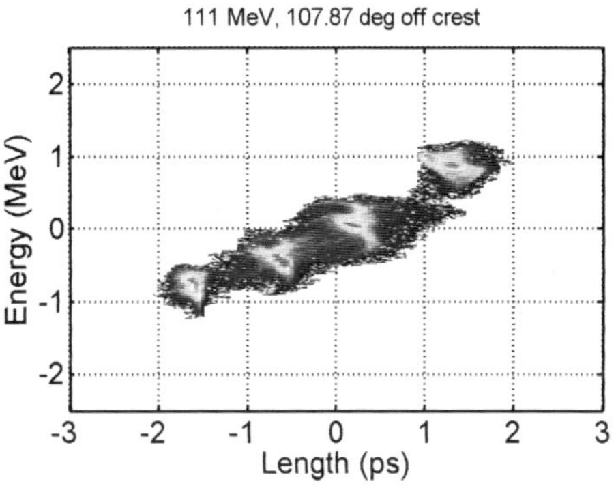

Figure 4. Measured LPS in the deep over-compression regime; the title reports the average energy and the compression phase (200 pC total charge).

in Fig.3, at the maximum compression phase (-97 deg) the four pulses are completely overlapped in time. As far as we move from the maximum compression phase, along the compression curve, the pulses get further and more modulated in time (Fig.4). This region is called deep over compression region.

3. Experimental Setup

A schematic layout of SPARC is shown in Fig.5. SPARC is based on a 1.6 cell S-band RF gun operating at 2.856 GHz (BNL/UCLA/SLAC type) and with high peak field (up to 130 MV/m) on the incorporated copper photo-cathode, generating up to 6 MeV electron beam. The beam is then focused and matched into three traveling wave (TW) accelerating sections, which boost its energy up to 180 MeV [14]. At the end of the linac a diagnostics section allows to fully characterize the accelerated beam by measuring transverse emittance, longitudinal profile and slice emittance through a RF deflector [15]. The beam can then be sent straight to the undulator, for FEL experiments or to a by-pass line, which hosts the experiment for the generation of THz radiation. The source is Coherent Transition Radiation (CTR) from an aluminum coated silicon screen. The screen is placed in the vacuum pipe at the end of the by-pass, at 45° with respect to the electron beam direction. TR is extracted at 90° with respect to the beam axis through a z-cut quartz window and then collected by a 90° off-axis paraboloid mirror. The parallel beam is then reflected down to a flat mirror at 45° which reflects radiation horizontally. Figure 6 shows the two schemes foreseen: one for interferometer measurements and one for integrated CTR measurements with the possibility of selecting custom band pass filters in the THz range [16].

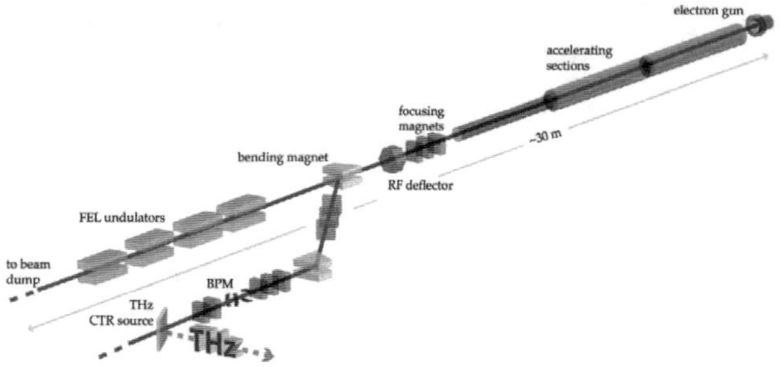

Figure 5. Layout of the SPARC accelerator, with the THz source placed at the end of the by-pass line.

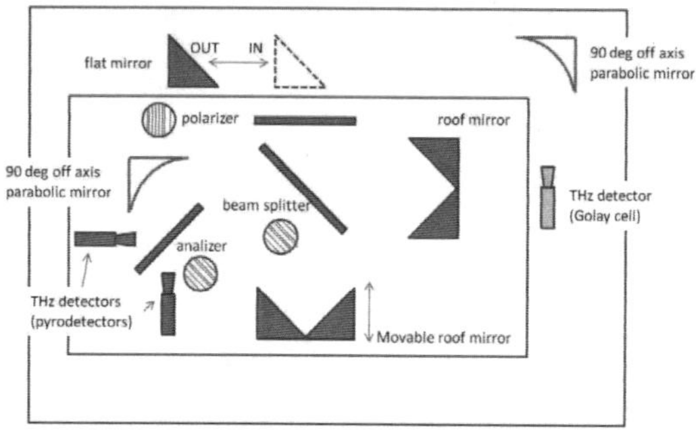

Figure 6. Drawing of the experimental layout for detection of THz radiation (top view).

A Martin-Puplett interferometer [17] is used for frequency domain measurements. The main advantages with respect to a Michelson one derive from the beam splitter properties. The beam splitter is a wire grid whose reflectivity coefficients depend on wires diameter and spacing. Furthermore, it splits polarizations, thus measuring the intensity of both polarizations, the difference interferogram can be normalized to the sum. Correlated fluctuations due to electron beam instabilities are then canceled. The main limitation is however due to low and high frequencies suppression introduced by the experimental setup.

The alternative branch allows to collect radiation to the 90° off-axis paraboloid mirror to be focused onto a THz calibrated detector. Band-pass filters can be inserted in front of the detector to select the THz wavelength.

4. Results

Broad band THz radiation source up to 5 THz has been measured at SPARC with a 500 pC, 500 fs electron beam. Details can be found in [1].

Narrow band THz radiation has been produced and characterized by studying the rotation of the longitudinal phase space of a four-pulses train under VB regime. The relevant regime for THz applications is the the so-called deep over compression regime. In this regime the sub-bunches are well separated and their distance, thus the THz emission frequency, can be tuned changing the compression phase. Interferometer measurements have been performed mainly in this regime. The comb current profile, shown in Fig.7, has been measured at the beginning of the by-pass line, i.e. after the first bending magnet in Fig.5. The corresponding LPS is shown in Fig.4. The four-pulses comb train has been then transported along the by-pass line down to

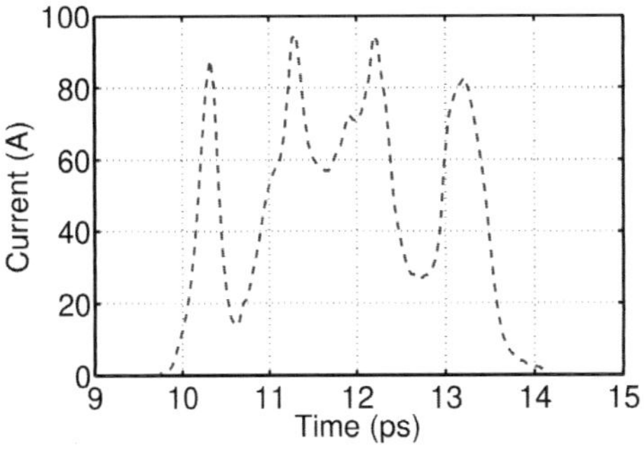

Figure 7. Comb current profile as measured at the beginning of the by-pass line. Corresponding LPS is depicted in Fig.4.

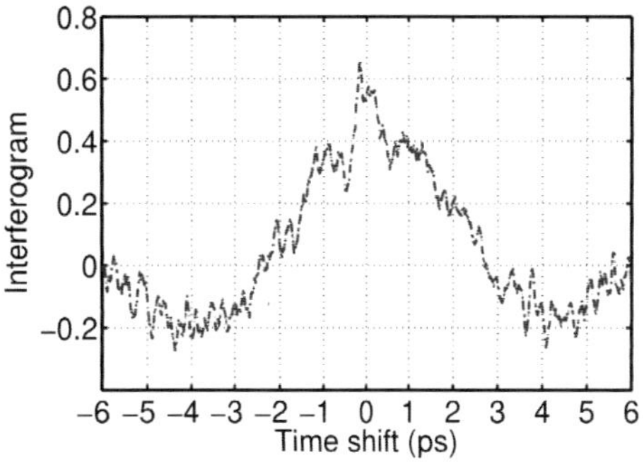

Figure 8. Measured interferogram (blue curve) and expected one from the modulated measured beam shown in Fig.7.

the TR screen. The measured autocorrelation function of the modulated current profile shows clearly five (Fig.8, blue curve) out of the seven peaks expected from a four-pulses comb train (Fig.8, red curve) [18]. In Fig.9 the form factor retrieved from the measured interferogram (blue curve) is compared to the one (red curve) retrieved from the current profile shown in Fig.7. Both the number of peaks in the measured autocorrelation function and the shift of emission frequency, shown in Fig.9, can be explained by considering the effect of the beam transport through the by-pass line. Indeed, under some conditions, it may act as a decompressing section. An increase of the pulse length and pulses separation might occur, causing two sub-pulses to merge into a single one so that the emission will then be driven by a three-pulses comb beam, and the shift of the emission frequency in the sub-THz range.

The measured CTR energy per pulse has then been retrieved (Fig.10, blue curve) taking into account the single particle TR spectrum and the transfer function of the experimental system. A well defined narrow band peak centered around 0.8 THz is visible, as expected from the interferogram peaks distance.

5. Conclusion

SPARC is becoming a test bench for advanced high brightness beam applications, in particular coherent THz generation. Frequency domain measurements in the THz regime are currently in operation at SPARC by means of both custom band-pass mesh filters and Martin-Puplett interferometer. Broad band, up to 5 THz, high peak power THz radiation has been produced and characterized for different beam parameters, e.g. charge and bunch length. Narrow band and tunable THz radiation has been generated combining comb beams and velocity bunching regime to tune the THz emission bandwidth. The multi-peaked autocorrelation function, measured with a Martin-Puplett interferometer, is a clear signature of the longitudinal modulation at THz repetition rate.

Acknowledgments

Authors wish to acknowledge F. Anelli, L. Cacciotti, S. Fioravanti and R. Sorchetti for their technical support.

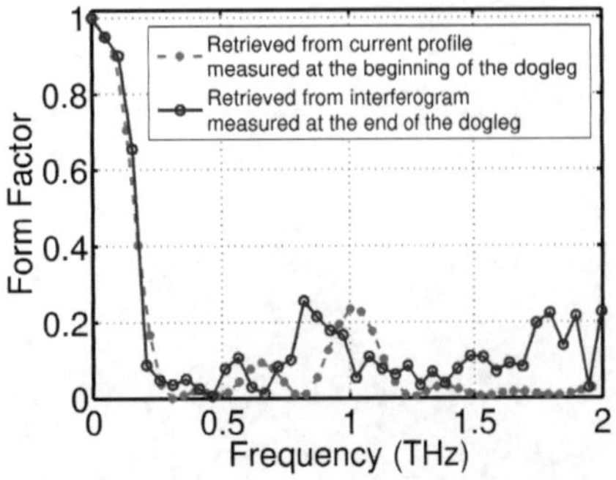

Figure 9. Form factor retrieved from current profile measured at the beginning of the dogleg (red curve), in comparison with the one retrieved by measuring the autocorrelation function of the CTR produced by the four-pulses comb beam.

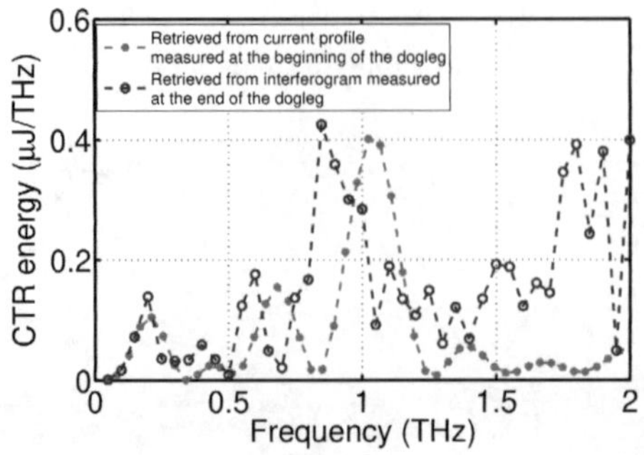

Figure 10. CTR energy per pulse from the measured interferogram (blue curve) and as expected (red curve) from the four-pulses comb beam measured at the beginning of the dogleg. Both curves are obtained by taking into account the single particle TR spectrum in far field approximation and for a finite target size. The experimental system transfer function is also considered.

References

[1] E. Chiadroni et al., submitted to Journal of Synchrotron Radiation.
[2] M.S. Sherwin and al., *Opportunities in THz Science*, Report of a DOE-NSF-NIH Workshop (2004).
[3] E. Chiadroni, *Bunch Length Characterization at the TTF VUV-FEL*, TESLA-FEL 2006-09 (2006).
[4] M. Ferrario et al., Phys. Rev. Lett. **99**, 234801 (2007).
[5] M. Ferrario et al., Phys. Rev. Lett. **104**, 054801 (2010).
[6] L. Giannessi et al., Phys. Rev. ST Accel. Beams **14**, 060712 (2011).
[7] L. Giannessi et al., Phis. Rev. Lett. **106**, 144801 (2011).
[8] M. Labat et al., Phys. Rev. Lett. **107**, 224801 (2011).
[9] J.S.Nodvick and D.S. Saxon, Phys.Rev. **96**, 180 (1954).
[10] A. P. Potylitsyn et al., Diffraction Radiation from Relativistic Particles, Springer Ed. (2010).
[11] L. Serafini and M. Ferrario, AIP Conf. Proc. 581, 87 (2001).
[12] M. Ferrario et al., Nucl. Instrum. and Meth. A **637**, S43-S46 (2011).
[13] Y. Shen et al., Phys. Rev. Lett. **107**, 204801 (2011).
[14] D. Alesini et al., *Technical Design Report for the SPARC Advanced Photo-Injector*, L. Palumbo and J.B. Rosenzweig (Eds.), (2004).
[15] D. Alesini et al., Nucl. Instrum. and Meth. A **568**, 488-502 (2006).
[16] N. I. Landy, IEEEXplore Digital Library 1-4244-1448-2, 151-156 (2007).
[17] D. H. Martin and E. Puplett, Infrared Physics, vol. 10, pp. 105-109 (1970).
[18] E. Chiadroni et al., E. Chiadroni et al., Proceedings of IPAC 2011, THPS101, San Sebastian (2011).

Journal of Physics: Conference Series **357** (2012) 012035 doi:10.1088/1742-6596/357/1/012035

Compact X-ray source at STF (Super Conducting Accelerator Test Facility)

J Urakawa
KEK, 1-1 Oho, Tsukuba, Ibaraki 305-0801, Japan

E-mail: junji.urakawa@kek.jp

Abstract. KEK-STF is a super conducting linear accelerator test facility for developing accelerator technologies for the ILC (International Linear Collider). We are supported in developing advanced accelerator technologies using STF by Japanese Ministry (MEXT) for Compact high brightness X-ray source development. Since we are required to demonstrate the generation of high brightness X-ray based on inverse Compton scattering using super conducting linear accelerator and laser storage cavity technologies by October of next year (2012), the design has been fixed and the installation of accelerator components is under way. The necessary technology developments and the planned experiment are explained.

1. Introduction

Evolution of photon source, which is an X-ray source based on accelerator and laser technologies, indicates the necessity of a stronger and brighter photon beam according to following equation:

$$Brightness = \frac{photons/\sec}{(mrad)^2(mm^2(source-area))(0.1\%spectrum-width)}. \quad (1.1)$$

The 10 μm size of photon source is considered, which means 1.0 mm mrad normalized emittance of the 40 MeV electron beam at the interaction point between electron and laser beams. Also, a 1mrad angular spread collimation to achieve a 1% energy spread X-ray means small energy spread of the electron beam.

The main purpose of the test facility for the Quantum Beam Technology Program (QBTP) is to develop technologies for next generation compact high brightness X-ray sources using super conducting RF acceleration techniques. The characteristics of this facility are its compact size of less than 10 m total length, monochromatic energy spread less than 1 %, higher flux of 100 times compared to compact normal conducting X-ray sources which can generate a flux of 10^{10} photons/sec/1 % band width, high brightness corresponding to 2.5 generation photon factory which generates 10^{17} photons/sec mrad^2 mm^2 0.1 % bandwidth X-ray, and ultra-short pulse X-ray of about 40 fs using 30 fs TW laser pulse and inverse Compton scattering with 90 degrees crossing angle [1].

2. Purpose

There are many applications for this compact high brightness X-ray source, including genome structural analysis, evaluation of nano-material and precise fine X-ray imaging.

Journal of Physics: Conference Series **357** (2012) 012035 doi:10.1088/1742-6596/357/1/012035

The impacts of compact high brightness X-ray are as follows:

1) The performance of a second-generation photon factory is obtainable within a 10 m laboratory.

2) Sub-psec X-rays can be obtained in 10m laboratory, which makes research for fast dynamic phenomena possible. Chemical reactions in solvents, functions of proteins, the destruction by impulses and the phase transition induced by photons will all be researched using this X-ray source.

3) To solve energy and environmental problems it is necessary to detect isotopes in radioactive nuclear waste. If we can use 300 MeV ERL (energy recovery linac) or a small storage ring with a ring circumference of about 30 m, r -ray with a high brightness of 1 to 5 MeV can be generated to diagnose precise components of nuclear waste [2].

The purpose of this project is to demonstrate the feasibility of the above applications by developing advanced accelerator and laser technologies.

The test accelerator for QBTP is under construction at STF. The three dimensional layout for this test facility is shown in figure1. It will be completed by March of 2012. The beam operation will start from spring in 2012. However, since the design for the collision scheme for inverse Compton scattering was recently changed from a 10 degrees crossing angle to a head-on collision, the set-up of the laser system will be delayed by about three months. Yields of X-ray photons will be increased by factor of about 5 compared to the former collision schemes and the beam collision and the setup of the detector system for X-ray imaging will become easier.

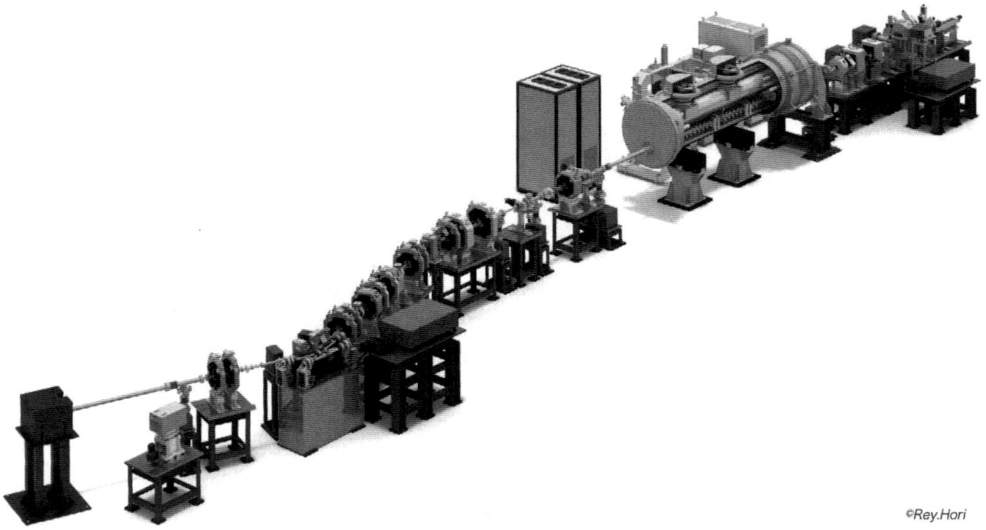

Figure 1. Layout of QBTP Test Accelerator.

3. Status of basic technologies

Generation of an ultra low emittance electron beam by photo-cathode DC electron gun is under development. So, we decided to use an L-band photo-cathode RF gun for the 5 Hz 1ms pulse beam operation in 2012. We manufactured two sets of the L-band photo-cathode RF gun, one is the old type of DESY Flash and other is modified one of the water cooling channels [3-5].

A compact RF Source has been developed as the Distributed RF Source (DRFS) for ILC (International Linear Collider) application with synergy of ILC group. ILC Baseline RF unit consists of one modulator, 10 MW klystron, PDS (power distributed system), three cryo-modules that include 26 9-cell SC (Super Conducting) cavities. We at KEK proposed a DRFS scheme that consists of one small modulator, 800kW klystron and two 9-cell SC cavities. A compact RF source has to have high reliability for ILC. Four manufactured 9-cell super conducting cavities have been operated at the highest accelerating gradient of 35 MV/m without the electron beam. We are assuming an accelerating gradient of 30MV/m with the beam for next year's operation.

We will use a 4 mirror optical cavity to accumulate the laser pulse and make the laser focus. Recently, we found a new method of fast switching for circularly polarized X-rays when three dimensional 4 mirror optical cavity is applied for the laser accumulation [6]. Also, we found a new laser storage scheme, so called 'self-starting oscillator scheme', and are developing a fiber laser system. A planar 4-mirror optical cavity will be made for QBTP test experiment because of its simplicity. The laser evolution is same in the tangential and sagittal plane of new planar 4 mirror optical cavity. Two laser pulses circulate with a spacing of 6.15ns in this ring optical cavity.

Figure 2. Layout of the inverse Compton scattering area for QBTP.

WFS (Wave Front Sensing) feedback will be used for the stabilization of this 4 mirror cavity. Figure 2 shows the area near the beam collision point for QBTP, which has a beam monitor screen, two sets of 20 degree bending magnets and a planar 4-mirror cavity system.

The laser system development for the laser oscillator and amplification is based on the fiber laser and LD (Laser Diode) burst amplifications. The injected laser power into the optical cavity will be about 1kW average power in 1ms pulse period. The 200W average power of the mode locked laser will be developed for the compact ERL project by the synergy with KEK-ERL group.

4. Perspective

We started the R&D for 30 MV/m 9-cell SC cavity from September 2008. Subsequently, the R&D for the stable operation of the SC cavity and the compact RF source system (DRFS) was started from October 2009. We established the technology on 30 MV/m 9-cell SC cavity by December of 2010 and the final fabrication for two 9-cell SC cavity system was ordered. Now, we have completed the final gradient test and two cavities have been installed into the cryo-module for the beam acceleration test at STF. We are aiming for the stable pulsed beam operation for the generation of high brightness X-ray at STF in 2012. Table 1 shows the comparison of the operation parameters for QBTP and ILC-STF

Phase2. After 5 Hz pulsed operation for the high brightness X-ray generation will be finished, the laser system including the optical cavity will move to the compact ERL facility in 2013 assuming the following the electron beam conditions will be achieved: beam energy: 35-65 MeV; average current: 10 mA; emittance: 1.0 π mm mrad.

Table 1. The comparison of the operation parameters for QBTP and ILC-STF Phase2.

	Operation Parameter for Quantum Beam	Operation Parameter for ILC-STF Phase2
Pulse length	1 ms	0.9 ms
Repetition rate	5 Hz	5 Hz
Bunch Spacing	6.15 ns (162.5MHz)	369.27 ns (2.708MHz)
Number of bunch/pulse	162500	2437
Bunch charge	62 pC	3.2 nC
Total charge /pulse	10,000 nC	7,798 nC
Beam current	10 mA	8.7 mA
Bunch length	12 ps (Laser, FWHM)	10 ps (Laser, FWHM)
Max. beam energy	50 MeV	21.5 MeV
Beam power	2.5 kW (50 MeV beam) Usually 2.0 kW (40MeV)	0.8 kW (21.5 MeV beam)

Acknowledgements
This work is supported by the Quantum Beam Technology Program of the Japanese Ministry of Education, Culture, Sports, Science, and Technology (MEXT). We would like to express our gratitude to all members of the ILC and ERL R&D's group at KEK for their helpful support.

References

[1] Urakawa Junji 2011 *Nucl. Instrum. and Meth. A* Development of a compact X-ray source based on Compton scattering using a 1.3 GHz superconducting RF accelerating linac and a new laser storage cavity **637** S47-S50
[2] Bulyak E, Gladkikh P, Omori T, Skomorokhov V, Urakawa J 2010 *Nucl. Instrum. and Meth. A* Compton ring for nuclear waste management **621** 105–110
[3] Deshpande A, Fukuda M, Sakaue K, Terunuma N, Urakawa J 2009 *Nucl. Instrum. and Meth. A* Design of a mode separated RF photo cathode gun **600** 361-366
[4] Terunuma N, Fukuda M, Sakaue K, Urakawa J 2009 *Nucl. Instrum. and Meth. A Scientific Instruments* Improvement of an S-band RF gun with a Cs2Te photocathode for the KEK-ATF **80** 123304-1-7
[5] Deshpande Abhay, Araki Sakae, Fukuda Masafumi, Sakaue Kazuyuki, Terunuma Nobuhiro, Urakawa Junji, Washio Masakazu 2011 *Phys. Rev. ST Accel. Beams* Experimental results of an rf gun and the generation of a multibunch beam **14** 063501-1-9
[6] Honda Y, Shimizu H, Fukuda M, Omori T, Urakawa J, Sakaue K, Sakai H, Sasao N 2009 *Optics Communications* Stabilization of a non-planar optical cavity using its polarization property **282** 3108-3112

Journal of Physics: Conference Series **357** (2012) 012036 doi:10.1088/1742-6596/357/1/012036

Estimation of grains size in mosaic crystals by means of radiation of fast electrons in these samples

D.A. Baklanov, S.A. Laktionova, R.A. Shatokhin, I.E. Vnukov
Belgorod State University, 14 Studencheskaya str., 308007 Belgorod, Russia
E-mail: vnukov@bsu.edu.ru

Abstract. There were analyzed the possibilities of the previously suggested method for determining the characteristic dimensions of grains in mosaic crystals of diffraction suppression of yield of the bremsstrahlung with the fixed energy and calculated results. There were defined the bounds of the method applicability and also there were taken into account more accurately the effect of the absorption process of X-rays on the reflectivity.

1. Introduction

Ordered arrangement of atoms in condensed matter leads to the appearance of orientation and interference effects in the yield of the secondary processes that occur while passing through it fast charged particles. The existence of connection between the internal structure of the target and the yield of the secondary processes allows us to raise a question about the analysis of the internal structure of the target in accordance with the results of the measurements. For example, by output of backscattered channeled ions we could see the location of impurities in the crystal lattice, and by emission spectra of fast channeled electrons we could specify the form of potential, electron density, the amplitude of thermal vibrations of lattice atoms and the like, see, for example, [1, 2] and references there in.

In the same row there is the task of analyzing the quality of the internal structure of crystalline samples, that is, determining the presence of outside inclusions and the mosaic blocks in them, also the distribution of the grains along the angle of misalignment regarding to the basic direction and the sizes according to the characteristics of hard electromagnetic radiation generated by passing through them fast electrons. The advantages of this approach include the high penetration ability of such radiation and clarity of interpretation. In the clearest form these benefits are realized in the case of analysis of the microstructure of the thick samples and the task of estimation of the characteristic dimensions of the blocks in which the use of traditional X-ray analysis can not be provided the quality control of internal structure.

It is well known (see [3] and references therein) that according to the degree of perfection, crystals can be classified by two factors: by the size of regular blocks or sections in the crystal, and by the extent of their mutual disorientation. According to the first factor all crystals can be divided into two classes a and b. In crystals of class a the separate sections are large enough to demonstrated a marked influence of the effect of primary extinction, that is, their linear size is comparable to the primary extinction length l_{ex}. In crystals of class b the size of right units is small, so the effect of primary extinction is almost not observed. According to the second factor the crystals can also be divided into two classes - α and β. In crystals of class α the blocs are almost parallel to each other, their mutual disorientation is small, so the contribution of a secondary extinction effect is large. In crystals of class β blocks are distributed irregularly, so that the contribution of the secondary extinction effect is small. The limit of the class $a\alpha$, in case when disorientation of grains is less than the region of the total reflection of X-rays $\Delta\Theta$ (Darwin "table"), is the perfect crystal, when the class bβ is the ideal mosaic crystal.

The most challenging of the above mentioned is the identification of common block sizes, as from them it often depends the possibilities of use of crystals for applied purposes. In particular, to extract beams from accelerators of high energy we need perfect crystals or, in extreme cases, the crystals of class $a\alpha$, and as for the use in medicine, also for X-ray and gamma-ray astronomy and for the obtaining of intensive quasi-monochromatic neutron beams we are required mosaic crystals of class $b\alpha$. It should be noted that belonging of crystal to a particular class is not given once and for all, because it is determined by l_{ex}, which, in turn, depends on the order of reflection and the photon energy [3]. That is, for different orders of reflection or significantly different photon energies one and the same sample may relate to different classes [4].

The direct measurement of the size of grains using X-ray radiation beams is a difficult experimental problem, and it can be only implemented for the analysis of surface layers [5]. Methods of electron microscopy can resolve this problem as applied to thin polycrystalline and crystalline samples, in the case when the angles of disorientation of the nearby blocks and their sizes are more than divergence and linear dimensions of electron beam on the target. If you can't get thin crystalline structure without disrupting the target as, for example, in case with ductile metallic crystals, the use of electronic microscopy cannot provide the required information.

When using fast electrons by varying the angle of observation [4] and the energy of the detected photons [6] we can get more qualitative information about the microstructure of crystals of large thickness, in particular, we can estimate the characteristic dimensions of grains instead of using X-rays with fixed-wavelength or the method of electron microscopy. Therefore, the problem of estimation of the quality of the internal structure of crystals and the identification of common grains sizes by radiation generated while passing through them fast electrons, and achievable parameters is important and relevant.

2. Formulation of the problem

Our method of estimating of the typical sizes of grains in crystals of class $a\alpha$ is based on the results of an experiment [7] in searching and researching of parametric X-ray radiation at small angles to the velocity of a particle in a tungsten crystal. From the point of view of our problem the most important outcome of the experiment was the detection of diffraction suppression of the bremsstrahlung yield i.e. the presence of minima of the orientational dependences (OD) of yield of high-energy photons with $\omega \simeq \gamma\omega_p$, where γ- is Lorentz factor, ω_p - is plasma frequency of the medium (see Figure 1), which position is consistent with kinematic conditions of the diffraction of photons with this energy, the error of which is not worse than 1%.

While calculating the orientational dependences of x-ray yield for the conditions of an experiment [7] we took into account the efficiency of the spectrometers and contribution of transition radiation from the output face of the crystal fulfilled in the work[6] and used the approximate method that took into account the diffraction of the bremsstrahlung in perfect crystals [9]. The calculation showed that the position of confidently manifested minima coordinate greatly with the calculation including those for the weaker reflecting planes (112) and (220). But depth of the dips in the calculated orientation dependences for the type of planes (110) and the photon energies 67 keV and 96 keV, respectively, $\sim$ 2.5% and $\sim$1.5% was almost five times smaller than the experimental ones $\sim$ 15% and $\sim$ 10%.

Based on these results and the fact that in the experiment [7] it was firstly observed the dynamic effects in the emission of fast electrons in crystals - so to say parametric X-rays along the particle velocity (forward PXR), which supposes the perfection of the structure by omission, the authors [6] classified the crystal used in this experiment as a crystal of class $a\alpha$ and suggested two methods of estimation of the sizes of the blocks in such crystals: the first method is estimated by the degree of manifestation of forward PXR in the region of the photon energy $\omega < \gamma\omega_p$ and diffraction of bremsstrahlung in the region of energies $\omega > \gamma\omega_p$; the second one is estimated by ratio of experimentally measured values of the diffraction suppression of the yield of photons with fixed energy in a mosaic crystal with the calculation or the result of measurements for the perfect crystal. The first method requires the accelerator with the energy ~1 GeV and above for its implementation but it is not always justified. Therefore, more interesting and promising is the second one, because such measurements can be made to by means of less energy accelerators.

3. Estimation of the size of blocks according to the diffraction suppression yield of bremsstrahlung

Method of estimating the typical sizes of grains in mosaic crystals according to the diffraction suppression of the yield of bremsstrahlung [6] is based on two assumptions. The first is that the investigated crystal is a mosaic crystal of class $a\alpha$. That is the sizes of grains in it is significantly longer than the primary extinction l_{ex} for a given order of reflection and the photon energy ω. The second one assumes that the probability of a repeated reflection of the diffracted photons in other grains is negligible.

For the crystal used in the experiment [7], validity of the first assumption is justified by the fact that in the crystal it was for the first time registered forward PXR. The presence of minimum in yield of high-energy photons, the depth of which significantly exceeds the contribution of the diffraction suppression of release of hard photons in a perfect crystal, is a clear indication of blocking structure along the direction of the photon beam, because otherwise we can't provide the experimentally observed magnitude of suppression. The same is with the width of the minimum which is larger than it is predicted by the calculation for the perfect crystal. For any other crystal, the implementation of this conditions must necessarily be checked.

The surface mosaicity of tungsten crystal, used in the experiments [7], is $\sigma_m \leq 0.2$ mrad. In fact, this value corresponds to the sensitivity of the methodology used certification of crystal surface, that is the true meaning of the interior mosaicity can be either smaller of the this value or bigger. This value is comparable with the width of the Darvin Table for photon energy ω=67 keV $\Delta\Theta \sim 0.03$ mrad. Therefore, to determine the bounds of applicability of techniques [6] we are required additional study of the effect multiple reflection on the recorded values of diffraction suppression.

For the exact solution of the contribution of repeated reflections we need unknown information about the distribution of grains along the angle of misalignment and the sizes which, strictly speaking, we need to find by the results of the measurements. So let's analyze the influence of this factor on the results of measurements of the photons yield with fixed energy in the case of the crystal belonging to the class $b\alpha$ and by using techniques [10].

Figure 1 shows results in comparison of the measured photon OD with ω = 67 keV of a tungsten crystal in the spectral range, clipping crystal-diffraction spectrometer in the experiment [7] and calculation results for three crystals of tungsten with a thickness of 0.41 mm and different microstructure. They are as follows: a perfect crystal by the method of [9]; mosaic crystal with σ_m=0.2 mrad and σ_m=0.5 mrad by the method [10], that is there is the assumption that the characteristic length of the blocks in it is smaller than the primary extinction length of the photons of this energy $l_{ex} \approx 3.1$ μm, and the number of blocks on the absorption length $l_a \approx$ 183 μm $\geq$ 50-100, depending respectively 1-3. The calculation takes into account the efficiency and angular capture of crystal-diffraction spectrometer. The experimental errors are statistical and do not include normalization errors and definitions efficiency of the NaI(Tl) detector.

The figure shows that the presence of a mosaic leads to a broadening of a dip in yield of photons in comparison with the case of a perfect crystal and of its much more depth. As we

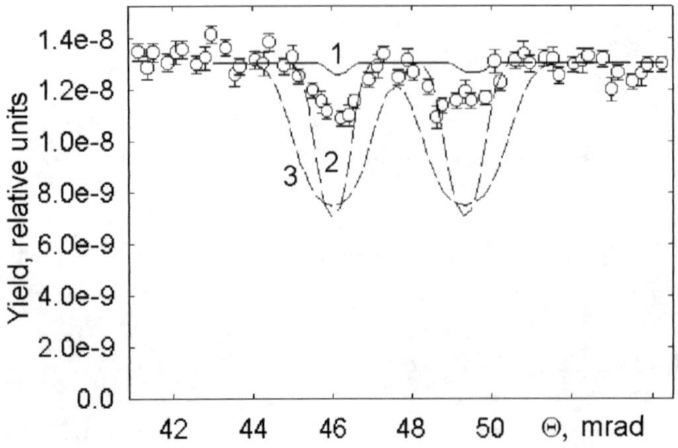

Figure 1. OD yield of photons with ω=67 keV. Points - the experiment [7]; Curves: 1 - a perfect crystal, 2 - σ_m=0.2 mrad, 3 -σ_m=0.5 mrad.

expected, none of the dependencies describes the results of measurements, because the used in the experiment [7] tungsten crystal is a crystal of class $a\alpha$, and not $b\alpha$. Nevertheless, its mosaicity value is close to the estimation of $\sigma_m \approx 0.2$ mrad since the dependence of 2 with σ_m=0.2 mrad is closer to the experimental one than the dependence of 3 for $\sigma_m = 0.5$ mrad.

As can be seen from the figure, the difference in the magnitude of σ_m didn't practically count to the magnitude of the diffraction suppression. In both cases it is close to 45%. For mosaic crystals of class b the total reflectivity of X-ray radiation is $Q \sim z^2$, see, for example, [3], and the probability of the reflection is $W \sim Q/\sigma_m^2$ [10], so for small values of σ_m multipath probability increases dramatically. In order to improve abovementioned in Figure 2 we found the estimated value of W_n which is probability for a photon to experience at least n reflections before it absorption in the crystal, or departure from it, averaged according to the energy resolution of the spectrometer. The value of n varies from 1 to 17-18, depending on the value of σ_m. The calculation is made for the experimental conditions [7] and for four values of σ_m.

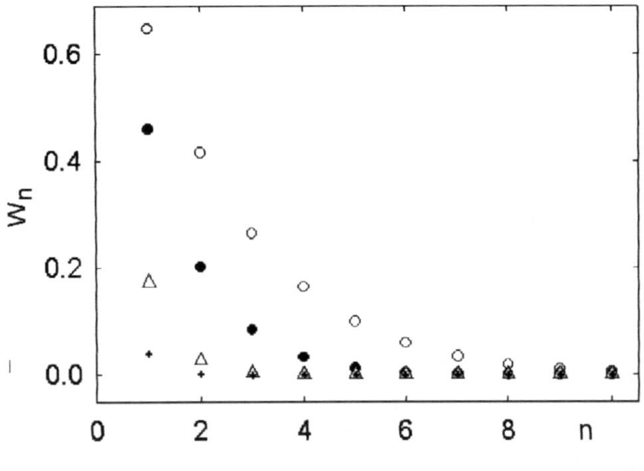

Figure 2. The dependence of W_n on the crystal mosaicity.
○ - σ_m=0.2 mrad; ● - σ_m=0.5 mrad; △ - σ_m=2 mrad and + - σ_m=10 mrad.

The figure shows that in the crystal with σ_m=0.2 mrad the probability of one or more reflections of $\sim$ 65% is more than in the crystal with σ_m=0.5 mrad ($\sim$ 46%) and it exceeds the value of diffraction suppression of the yield of photons ($\sim$ 45%). The resulting suppression of the bremsstrahlung yield is determined by the ratio of a number of odd and even reflections. The odd reflections decrease the recorded yield, while the even ones partially compensate this loss (see Figure 3).

As can be seen from Figure 3, for σ_m=0.2 mrad the fraction of photons that have experienced

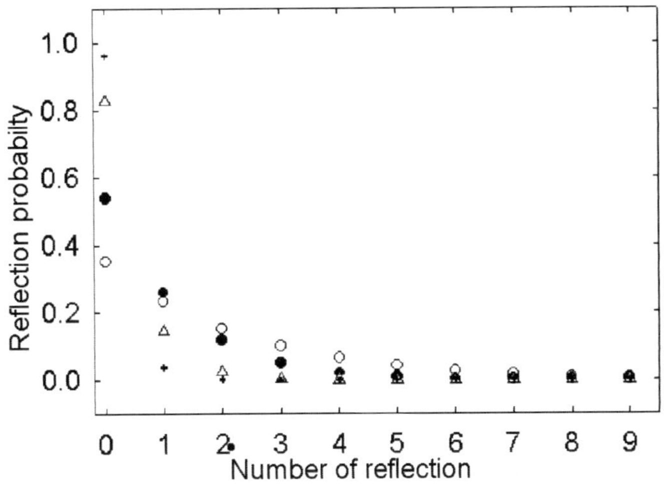

Figure 3. The probability distribution of reflections as a function of σ_m. ○ - σ_m=0.2 mrad; • - σ_m=0.5 mrad; △ - σ_m=2 mrad and + - σ_m=10 mrad

only a single reflection, is responsible for the observed value of the diffraction suppression, $\sim$ 23% and is significantly smaller than the diffraction suppression of $\sim$ 45% and slightly less than for a crystal with $\sigma_m = 0.5$ mrad ($\sim$ 26%). While the probability of all subsequent reflections in a crystal with lower value of σ_m is significantly higher. As a result, the net suppression of yield for these two crystals is close enough (See Fig. 1, curves 2 and 3). With the increase of the characteristic angle of mosaicity the probability of multiple reflections decreases sharply. For photon energy ω=67 keV, the probability of re-reflection is reduced by 10 times in comparison with the probability of a single reflection, but for a crystal with $\sigma_m \sim 3$ mrad. That is, for large values of σ_m and for the photon energy we may not take into account the effect of multiple reflections. It is necessary to emphasize that the reflectivity for crystals of class b is $Q \sim \lambda^2$, so that when the photon energy changes the boundary value of σ_m, in which the effect of re-reflections may be ignored, should be determined again.

The analysis prove that multiple reflections significantly change the observed dependence of the diffraction suppression of the radiation yield in comparison with the case in which this effect can be neglected. For crystals of class $a\alpha$ the situation is significantly more complicated, since there is no information about the distribution of blocks in size, so that the accurate statistical analysis is impossible without it. This problem seems to be solved by successive approximations using an additional information about the distribution of blocks in the corners of misalignment and size. If this information is missing, the methodology of estimating the number and characteristic sizes of blocks by means of the magnitude of the diffraction suppression of photon yield, proposed in [6], can only be used if value of registered suppression is much smaller than for the crystal class of $b\alpha$ with the same value σ_m. If it is wrong and they are comparable, the estimated number of grains obtained in this way, would be understated, but their characteristic size, on the contrary, would be significantly overstated.

This condition is satisfied for the tungsten crystal used in the experiment [7], and for photon energy ω=67 keV. The recorded magnitude of diffraction suppression of radiation yield $\sim$ 17% (see Fig. 1) is significantly less than calculated magnitude for the crystal class of $b\alpha$ with the same thickness and characteristic mosaicity angle ($\sim$ 45%), and therefore we can use the ratio of the depths of experimental and calculated minima in the radiation yield as an estimate of the number of blocks along the path of the photons in the crystal $N_{bl} \approx 5.4$.

As we noted above, the magnitude of the diffraction suppression for a perfect crystal (Fig. 1, curve 1) is calculated by using the approximation about absence of influence of the photon absorption on the process of diffraction [9]. While comparing this approach with a more precise one [11] we found out that for unidirectional and monochromatic beam of photons the difference in the diffraction suppression, calculated by means of these approaches for blocks with thickness up to 50 μm is not more than 10%. That's why the use of results obtained with the help of a simple technique [9], to estimate the number of blocks is justified. The value $\sigma_m \approx 0.2$ mrad is almost an order larger than the characteristic angle of total reflection $\Delta\Theta \sim 3 \cdot 10^{-5}$ mrad for photon energies ω=67 keV. Therefore, if there is a photon on the way of $\sim$ 5 blocks, the probability of re-reflection of the once diffracted photon in the first approximation cannot be taken into account.

The formation of recorded yield of photons with the fixed energy in the present experiment depends not only on their decreasing because of diffraction processes and absorption, but also because of the photons generation by electrons passing through the investigated crystal. Therefore, as a characteristic distance over which the recorded radiation yield is formed it is better to take not the length of photon absorption l_a, as it is done in [6], but the effective path length of photons in crystal, that is the average distance traveled by the photons which were emitted from crystal into the solid angle, which was overlained by the spectrometer.

$$< l_{ph} >= \frac{\int_0^T (T-t)dt \int d\omega \int \frac{d^2I^*}{d\omega d\Omega} S(\omega,\vec{n},t)d\Omega}{\int_0^T dt \int d\omega \int \frac{d^2I^*}{d\omega d\Omega} S(\omega,\vec{n},t)d\Omega},$$

where $\frac{d^2I^*}{d\omega d\Omega}$ is a spectral and angular distribution of bremsstrahlung intensity, if we take into account multiple scattering of electrons in the target [9], and $S(\omega,\vec{n},t)$ is a coefficient, which takes into account the absorption of photons with the direction of $\vec{n}$ and energy ω in the target and the efficiency of crystal-diffraction spectrometer, as for T it means thickness of crystal. The integration takes place all over the angles of emission and photon energies taking into account angular and energy capture of the spectrometer. For photon energies ω=67 keV and the crystal thickness of tungsten is 0.41 mm, $< l_{ph} >$=236 μm $> l_a \approx 184$ μm. Therefore, the characteristic size of the blocks in the crystal of tungsten, which was used in the experiment [7], is $l_{bl} =< l_{ph} > /N_{bl} \sim$40 μm instead of 30 μm [6].

In order to measure by assess of characteristic dimensions of grains in crystals of class $a\alpha$ it is not necessary to use an accelerator with energy of $\sim$ 1 GeV as in the experiment [7]. The measurements may be conducted with the help of relatively cheap and more common electron accelerators with energies of 30-50 MeV and with a sufficiently large database that would give an opportunity to implement the method of allocation of emission with the fixed energy by means of crystal diffraction spectrometers.

The additional advantage of low-energy electrons is the lack of forward PXR in X-ray frequencies, which can mask diffraction suppression of radiation yield [7, 6]. The analyzed method for estimating the characteristic dimensions of grains is based on its registration. This allows scientists to conduct measurements with less energy photons thus limiting the size of the viewing area of the crystal and the number of blocks contributing to the suppression of the recorded yield of photons.

From the viewpoint of the testing method for assessing the quality of the internal structure of crystals and the characteristic dimensions of grains, one of the important differences of medium-energy accelerators from accelerators with higher energies is significantly larger size of the electron beam on the target $\sim 5 \times 10$ mm^2 and more, see, for example, [12]. Resolution and efficiency of the crystal-diffraction spectrometer depends on the angular distribution of the radiation hitting on the analyzer crystal, and on the distances between the target, where investigated radiation is generated, and the crystal-analyzer together with the detector, registering the diffracted radiation. The change of the size of the emitting section leads to a change of the angular distribution of radiation on crystal-analyzer and it can also lead to a deterioration of monochromatic of the detected radiation and, consequently, to a deterioration of sensitivity of the proposed method.

In order to test the influence of this factor on the resolution of the spectrometer by means of the method [10] it was carried out the simulation of efficiency of crystal diffraction spectrometer based on a mosaic crystal of pyrolytic graphite with the width of 2.5 mm for different transverse dimensions of electron beam hitting on a crystalline sample under study. Figure 4 shows the simulation results for the electron and photons energies of 50 MeV and 67 keV, respectively. The remaining conditions are identical with experimental conditions [7] for the photon energy. In the first case (∘) the size of the electron beam were not taken into account (approaching a point of the beam), while the second (•) were 5×16 mm^2 horizontally and vertically, respectively.

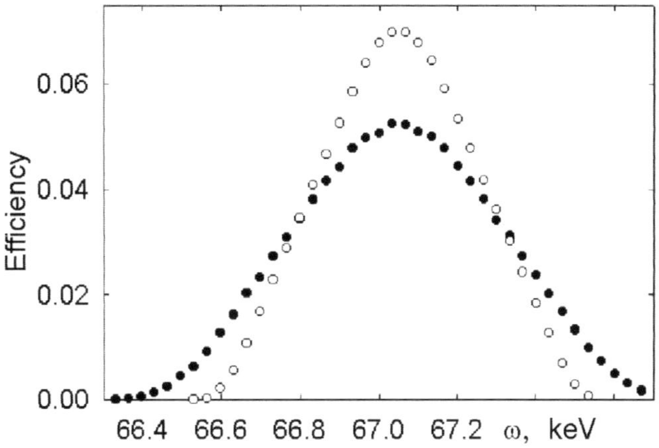

Figure 4. The efficiency of crystal diffraction spectrometer. E_e=50 MeV, ω=67 keV. ∘ - spot beam; • - with a beam size of 5×16 mm^2.

The figure shows that increasing of the size of the electron beam on the target practically did not affect the characteristics of the spectrometer. The area under the two dependencies (total efficiency of the spectrometer) has changed less than 2%, while resolution deteriorated by ∼ 30%. Therefore we can expect that approximately the same may take place with the change of depth and width of a minimum in OD caused by diffraction suppression of the yield of bremsstrahlung in a perfect crystal. However, it should be noted that for the experimental conditions [7] OD of the yield of radiation calculated for the same size of an electron beam on the tungsten crystal, were completely identical.

Another feature of such common electron accelerators of medium-energy as microtron or a linear accelerator is a short time of acceleration cycle ∼ 6-10 μs, which is comparable with a typical duration of event for traditional X-ray detectors as NaI(Tl) or silicon detector (∼ 1-10 μs), depending on the type of a detector. Therefore, to avoid overlap and separation of the photons only one order of reflection with a help of a differential discriminator as it was done in the experiment [7], we should use faster and more expensive detectors as CeF_3 for example or the current of the accelerator must be maintained in the way that it would be recorded no more than 0.2-0.4 event during a single cycle of acceleration [13].

For an NaI detector in order to obtain the statistical error of 1-2% even when the frequency of the accelerator is 50 Hz we need at least 200 seconds for each crystal orientation. In order to identify without doubts the effects of diffraction suppression it is necessary to make measurement of the yield of radiation for at least 200-500 points in increments of ∼ 0.1 mrad that requires 10-20 hours of non-stop operation of the accelerator. We need a comparable time, first to adjust the orientation of the crystal-analyzer spectrometer, and second to search the area of diffraction suppression for each plane of interest in the investigated crystal.

The inclusion of the detector may become the way out of the situation. The detector will record the diffracted radiation in the integral mode. As it is shown in [14] such mode of an NaI detector located in an accelerator hall, allows us to carry out measurement of radiation yield from the OD number points of ~ 500 for 30-40 minutes with an accuracy of ~ 1-2%. When the detector is well protected from the background, the contribution of photons, which are not related with the process of diffraction, doesn't exceed 3-5%, see for example, [6]. Simultaneous recording of several orders of reflection, but not one as in [7], can be accounted during data processing, the more so as for the crystal pyrolytic graphite reflectivity for the second one allowed reflection order is of at least twenty times smaller than the first one [8]. If necessary, the contribution of second and higher orders of reflection can be further reduced by choosing a detector with low detection efficiency for these photon energies.

One of the advantages of the proposed approach over traditional ones is in the fact that this method is useful for the thickness of crystals, which is much higher than absorption length of photons of the characteristic X-rays material of the anode x-ray tubes used in X-ray analysis. With the help of this approach you can see the mutual misalignment of blocks in the volume sample at an angle ~ 10^{-5} rad, which is practically impossible to make using traditional methods.

4. Conclusion

The results of the analysis of the method of estimation of the internal microstructure of crystals of large thickness according to the diffraction suppression of the bremsstrahlung yield with the fixed energy can be briefly summarized as follows:

1) Belonging of the crystal to the class $b\alpha$ drastically changes the appearance of orientation dependence of the diffraction suppression in comparison with the measurements [7].

2) Measurement of the diffraction suppression of the X-ray radiation yield with the fixed energy of fast electrons in mosaic crystals and its comparison with the calculated or measured ones for a perfect crystal allows us to determine the characteristic dimensions of grains in these crystals under condition that the magnitude of the recorded suppression is significantly less than for the crystal of class $b\alpha$ with the same value of the characteristic mosaicity angle.

3) Methodology to estimate the characteristic dimensions of grains in such crystals according to diffraction suppression of the yield of the X-ray radiation with the fixed energy possesses great capacity for work not only on high-energy accelerators, but also on medium energy electron accelerators, since the size of an electron beam target does not affect on the resolution of crystal diffraction spectrometers based.

The authors thank the co-authors [7, 8, 9] for participation in the development and implementation of techniques used in the process of research, and assistance during the measurements. This work was supported partly by internal grants program of BelSU and the federal program "Academic and Teaching Staff of innovation Russia", the government contract 16.740.11.0147 dated 02.09.2010.

References

[1] Ryabov V A 1994 *Channelling effect* (Moscow: Energoatomizdat) (in Russian)
[2] Datz S *et al* 1986 *NIM* B **13** 19
[3] James R *The optical principles of the diffraction of X- rays* 1958 (London: G. Bell and Sons)
[4] Baklanov D A *et al* 2010 *Journal of surface investigation. X-ray, synchrotron and neutron techniques* **4** 295
[5] Ohler M, Baruchel J, Moore A W, Galez Ph, Freund A 1997 *NIM* B **129** 257
[6] Baklanov D A *et al* 2011 *IL Nuovo Cimento* C **34** 125
[7] Aleynik A N *et al* 2004 *JETP Letters* **80** 393
[8] Vnukov I E *et al* 2001 *RUSSIAN PHYSICS JOURNAL* **44** 281
[9] Baldin A N, Vnukov I E, Kalinin B N, Karataeva E A 2006 *Poverkhnost'* No. 4, 72 (in Russian)
[10] Baklanov D A *et al* 2011 *IL Nuovo Cimento* C **34** 439
[11] Pinsker Z. 1984 *Dynamical Scattering of X-rays in Crystals* (Berlin: Springer)
[12] Shvedunov V I *et al* 2001 *Proceedings of the 2001 Particle Accelerator Conference, Chicago* 2596
[13] Shchagin A V, Pristupa V I, Khizhnyak N A 1990 *Phys. Lett.* A **148** 485
[14] Baklanov D A *et al* 2010 *Journal of surface investigation. X-ray, synchrotron and neutron techniques* **4** 201

Journal of Physics: Conference Series **357** (2012) 012037 doi:10.1088/1742-6596/357/1/012037

Microbunch Instability Observations from a THz Detector at Diamond Light Source

W Shields[1], R Bartolini[2,3], G Boorman[1], P Karataev[1], A Lyapin[1], J Puntree[3], G Rehm[2]

[1] John Adams Institute, Royal Holloway, University of London, Egham, TW20 0EX, UK
[2] Diamond Light Source, Oxfordshire, UK
[3] John Adams Institute, University of Oxford, OX1 3RH, UK

E-mail: `william.shields.2010@live.rhul.ac.uk`

Abstract. Diamond Light source is a third generation synchrotron facility dedicated to producing radiation of outstanding brightness, ranging from infra-red to x-rays. The short electron bunches that are accelerated around the storage ring are susceptible to the phenomenon of microbunching instabilities when the bunch charge exceeds a threshold. The primary feature of the microbunch instabilities is the onset of bursts of radiation in the THz range. The high frequencies involved in the emissions make detection and analysis challenging. A 60-90 GHz Schottky Barrier Diode detector was installed to investigate turn by turn evolution of the instabilities.

1. Introduction

The phenomenon of microbunch instabilities has been observed at many facilities around the world including Diamond Light Source [1, 2], UVSOR-II [3], ALS [4], BESSY-II [5, 6], the VUV ring at BNL [7], and many other facilities [8]. The key feature of the instability is the onset of bursts of coherent synchrotron radiation (CSR) in the microwave range. The wavelength of these radiation bursts are shorter than the bunch length, $\lambda < \sigma_z$. The onset of the bursts only occurs above a threshold current, where the bursts appear quasiperiodic. As the current increases, the bursting occurs in a random manner.

An exact explanation for the cause of the instability is an active research topic. The mechanism which drives the microbunching instability is not clear. Evidence has shown that they can be driven by a collective force which is self generated by the CSR [9, 10], or by broadband resonator wake fields [11]. We present the result of a series of measurements taken with a Schottky Barrier Diode detector at Diamond Light Source, investigating the onset and evolution of the CSR bursts as a function of bunch current. We were able to clearly detect the mm wave emissions and obtain a preliminary classification of the various emission patterns.

2. Theory

The frequency of radiation we are interested in is that of the order of the size of the microbunches. If we take a bunch of N ultrarelativistic electrons with a Lorentz factor $\gamma \gg 1$, then the radiation spectrum from the bunch can be expressed as

$$\left(\frac{dU}{d\lambda}\right)_{Bunch} = \left(\frac{dU}{d\lambda}\right)_e \left(N + N(N-1)\left|G(\lambda)\right|^2\right) \tag{1}$$

where $\left(\frac{dU}{d\lambda}\right)_e$ is the emission from a single electron. The first term, proportional to N, describes the incoherent emissions. The second term is the coherent part, which for $N \gg 1$ can be written as $N(N-1) \approx N^2$, meaning the coherent part will dominate at those wavelengths where $G(\lambda)$ is non zero. $G(\lambda)$ is the longitudinal form factor, given by:

$$G(\lambda) = \int_{-\infty}^{\infty} f(z) e^{2\pi i z/\lambda}\, \mathrm{d}z \tag{2}$$

where $f(z)$ is the longitudinal particle distribution function and is normalised such that $\int_{-\infty}^{\infty} f(z)dz = 1$. The onset of the instability is determined by the number of electrons in the bunch being sufficiently high to pass the current threshold. The onset of the instability is characterised by the current threshold, and is given by:

$$(\sigma_0 c)^{7/3} = \frac{c^2 Z_0}{2\pi F 3^{1/3}} \frac{I \rho^{1/3}}{V f_{rf} f_{rev}} \tag{3}$$

where σ_0 is the zero current bunch length, I is the single bunch current, c is the speed of light, $Z_0 = 377\Omega$ is the free space impedance, $F = 7.456$ [14], ρ is the dipole bending radius, V is the RF cavity voltage, f_{rf} is the RF frequency, and f_{rev} is the revolution frequency [14].

3. Experimental setup

Diamond Light Source is an electron synchrotron designed specifically to generate high intensity radiation. Basic parameters are shown in table 1. The primary source of the radiation is dipoles, however undulators and wigglers are also present in the Diamond lattice as insertion devices. Diamond's design means that it can operate in numerous fill pattern modes, including a mode which consists of a single bunch. For this experiment, single bunch mode was chosen to ensure monitoring of the evolution of the same bunch. To record the signal appearance as a function of current, the beam current was manually ramped up to its highest value. A collimator was moved closer to the beam to introduce a rapid decay. To keep the decay rate quasi-linear, the collimator was moved closer to beam at numerous stages throughout the experiment.

Table 1. Parameters of Diamond Light Source.

Parameter	Value
Energy	3 GeV
Circumference	561.6 m
RF Voltage	2.4 MV
RF Frequency	499.7 MHz
Bunch Length	3 mm rms (10 ps)
Dipole bending radius	7.13m
Synchrotron frequency	1.7 kHz

The detector that was employed was a Schottky Barrier Diode (DXP-12) from Millitech, USA, along with a standard gain horn antenna (SGH-12). Figure 1 shows the experimental setup of the detector at the beam port. The detector is off centre due to a mirror inside the vacuum vessel reflecting light out of the port beneath the antenna [15] which is used in optical diagnostics. The antenna was aligned to receive horizontally polarised radiation. General properties of the detector are shown in table 2. The detector was installed at the port at Diamond light source, which was then amplified by 40 dB by a Femto HSA-X-1-40 amplifier before being read by an Agilent N9020A Signal Analyser. The output from the analyser was displayed in the Diamond control room via remote desktop connection, using Agilent 89601B Vector Spectrum Analyser (VSA) software. The data was recorded separately in both the time and frequency domains.

Table 2. Properties of the Schottky Barrier Diode detector and Standard Gain Horn antenna.

Element	Style
Frequency Range	60 - 90 GHz
Wavelength	3.33 - 5 mm
Video Sensitivity	20 mV/mW
Response Time	250 ps
Antenna Gain	24 dB
Input Aperture	30 × 23 mm

Figure 1. Photo of the 60-90 GHz Schottky barrier diode detector and horn gain antenna.

4. Microbunch Instability Observations

Four experiments were conducted to investigate the observations of the instability. Three of these were performed at different RF voltages of 2, 3, and 3.5 MV in order to establish the instability current threshold and also to observe the relative intensity of the detected frequencies. The quoted RF voltage values that were investigated are the sum of two RF cavities in the Diamond storage ring. The fourth trial was to investigate the bandwidth of the signal for an RF voltage of 3.5MV. The frequency domain data was recorded as a function of time, and the beam current was also recorded in a separate file in three second intervals, again as a function of time of day. As a result, the current data has to be interpolated to match the frequency data. The interpolated current function for the experiment conducted at 3.5MV RF voltage is shown in figure 2. As mentioned in section 3, the decay is near linear, and can easily be correlated with the frequency data.

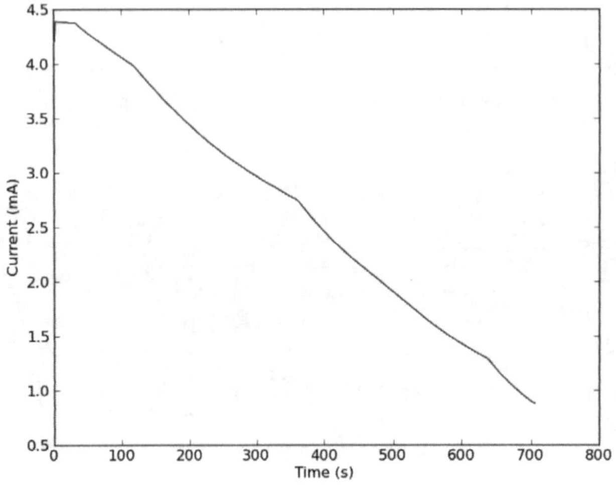

Figure 2. Interpolated plot of beam current decay throughout the frequency detection experiment at 3.5MV RF voltage.

Figure 3 shows the time and frequency domain signals for several currents for an RF voltage of 3.5MV. The signal at 0 kHz in figures 3-6 corresponds to the revolution frequency to which our signal is modulated. The signal at 1.001 mA shows there are three frequencies present before the instability starts. These are presumed to be an artefact of a component in the RF system and remain present and unchanged throughout the experiment. The instability starts at 1.5mA and is characterised by the sudden appearance of several frequency components, with two prominent spectral features at 6.8 and 13.6 kHz. These two signals are multiples of the synchrotron frequency of 1.7 kHz, an effect which has previously been observed [6]. At higher currents, these two frequencies shift and develop sidebands, and a quasiperiodic structure appears in the time domain signal. The periodicity of this structure varies significantly with current, and at 3.150 mA, can be seen at a much higher peak voltage. As a result, there is an increase in overall power in the background. At the peak current of 4.210 mA, there is very little time domain structure and the bursts appear chaotic, as observed in previous experiments [1, 2]. Another noticeable effect throughout the current increase is the drifting of the prominent signals to higher frequencies.

Equation 3 can be used to predict a threshold for the onset of the instability. Using some of the parameters listed in table 1, an RF voltage of 3.5 MV, a revolution frequency of 534 kHz and a zero current bunch length of 3mm rms, the estimated threshold current is 1.26 mA. This is slightly smaller than our observed value of 1.5 mA. Figure 4 shows the signal at the instability threshold for the tests conducted for the lower RF voltages. The results also show that the threshold is higher for the lower RF voltages. At 3MV, the threshold current is similar to the value of 1.98 mA obtained in a previous experiment for the same RF voltage [1] . A higher threshold is unexpected as the threshold current is directly proportional to the RF voltage, shown in equation 3. The reason for these discrepancies is unknown, additional experiments would be required to deduce any further conclusions and provide an explanation.

The two tests also revealed that the two prominent frequencies that are produced at the threshold change with RF voltage. For 2 MV, the frequencies are at 6.9 and 13.8 kHz, and for 3 MV, the frequencies are at 7.2 and 14.4 kHz. This can be partially explained by the change in synchrotron frequency that results from an increase in RF voltage. Unfortunately, our results do

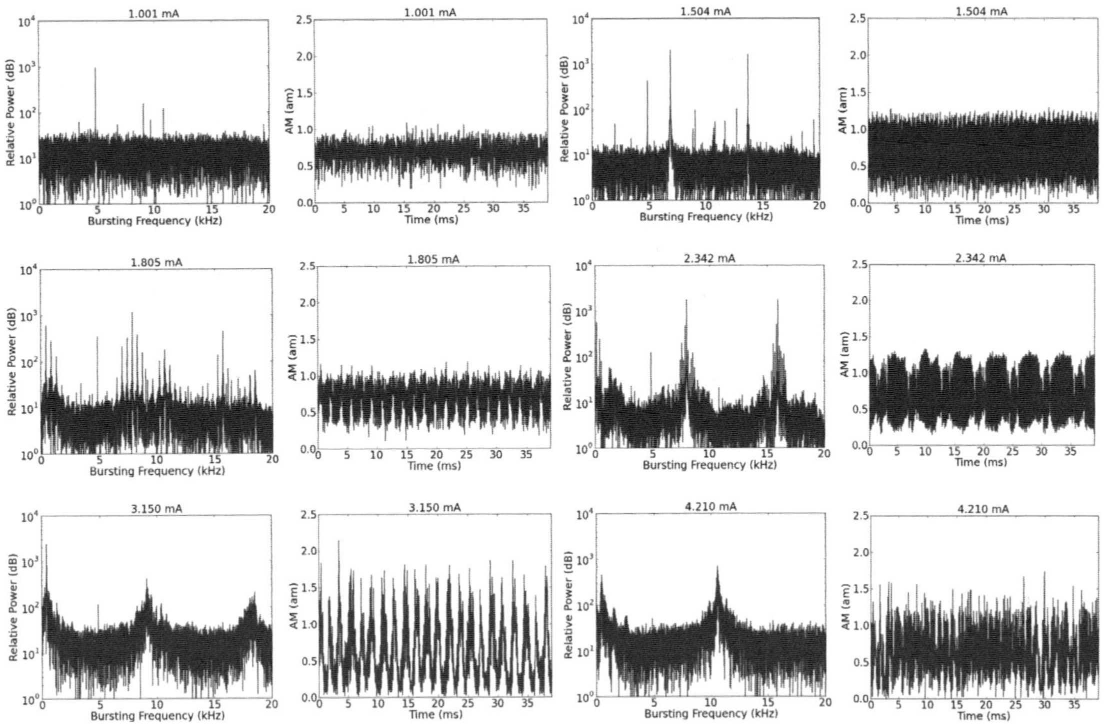

Figure 3. Frequency and time domain signals showing the onset and evolution of the instability for an RF voltage of 3.5 MV. The time domain graphs show the normalised amplitude modulation of the signal.

not match this relationship as the frequencies at 3.5 MV are lower than those at 3 MV. This is possibly due to the two RF cavities using different voltages, whereas the voltages were identical in the other two tests, however future research is necessary to provide an explanation for this phenomenon.

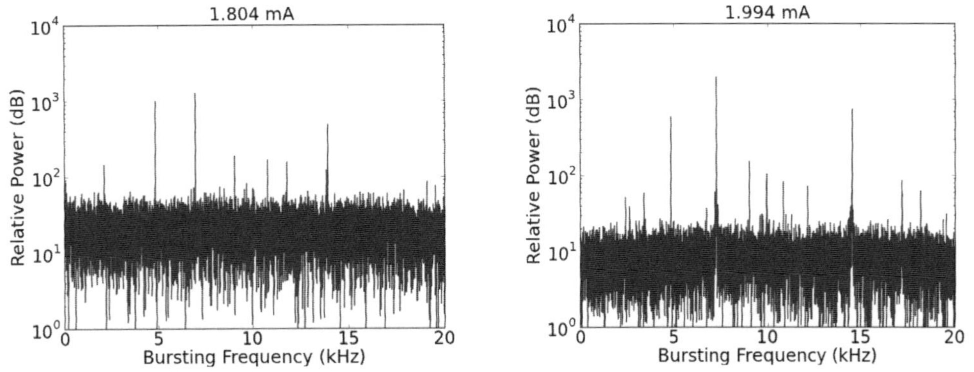

(a) (b)

Figure 4. Frequency domain signals at instability threshold for RF voltages of 2 MV (a) and 3 MV (b).

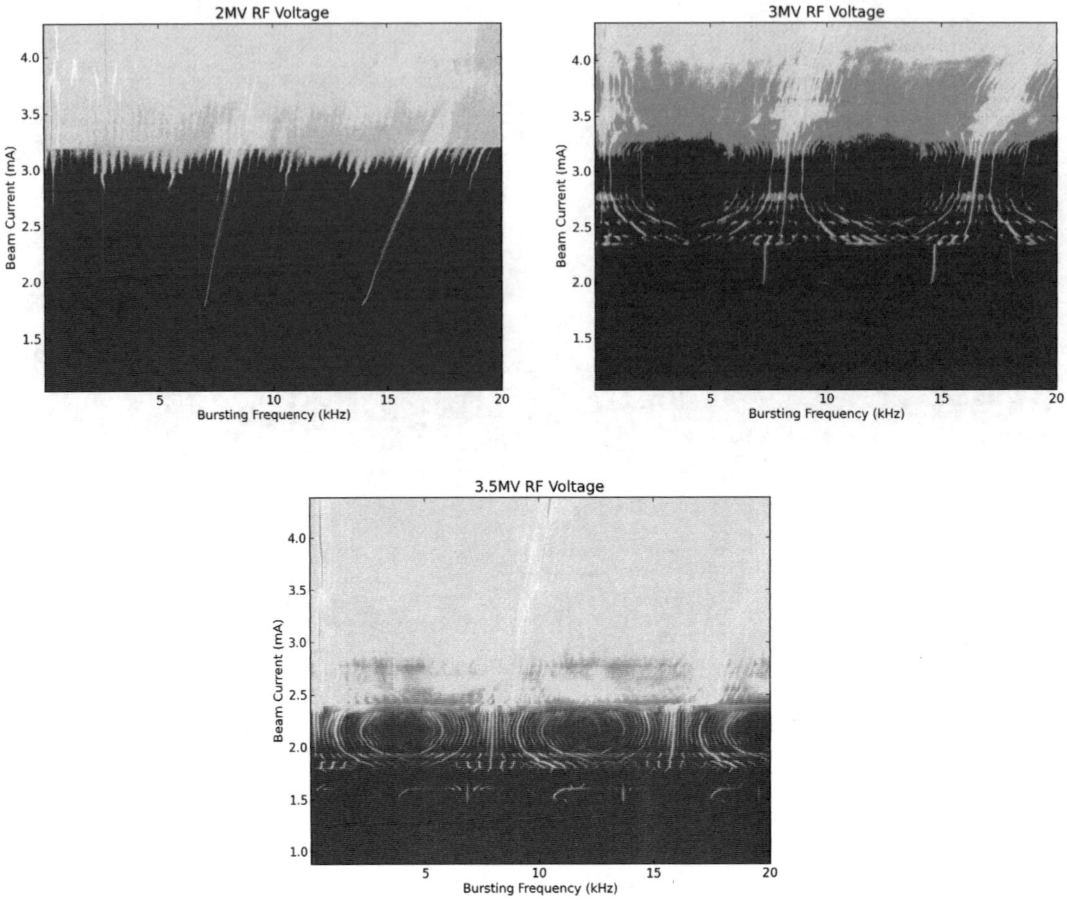

Figure 5. Spectral displays of frequency domain signals as a function of single bunch current for varying RF voltages.

Figure 5 shows the spectral data across across the whole current range for all three trials. The colour applied to the plots represents the power of the signal at that frequency with blue being the background power without CSR signals. There are clear differences in the signals observed as expected due to the variations in bunch length. In all tests, we observed the fundamental and the first harmonic of the bursting signal at the onset of the instability. In the 2 MV test, these slowly increase in frequency until around 2.8 mA where the signals develop sidebands, however in the 3 MV test, the sidebands appear at a lower current of approximately 2.3 mA after a sudden jump of the prominent signals to higher frequencies. These sidebands also change frequency and are much more distinguished. In both these tests, there is a further jump at approximately 3.2 mA after which there is an increase in background signal over all frequencies. These features were also evident in the 3.5 MV test. Here, the sidebands around the two main signals are very prominent over a wide current range. There is also the unusual feature of a region above the threshold where the bursts are not observed. Another test was conducted at 3.5 MV where the CSR bursts were measured over a larger bandwidth, shown in figure 6. It is clear that there are additional harmonics of the synchrotron frequency where the CSR signal is observed, up to approximately 50 kHz.

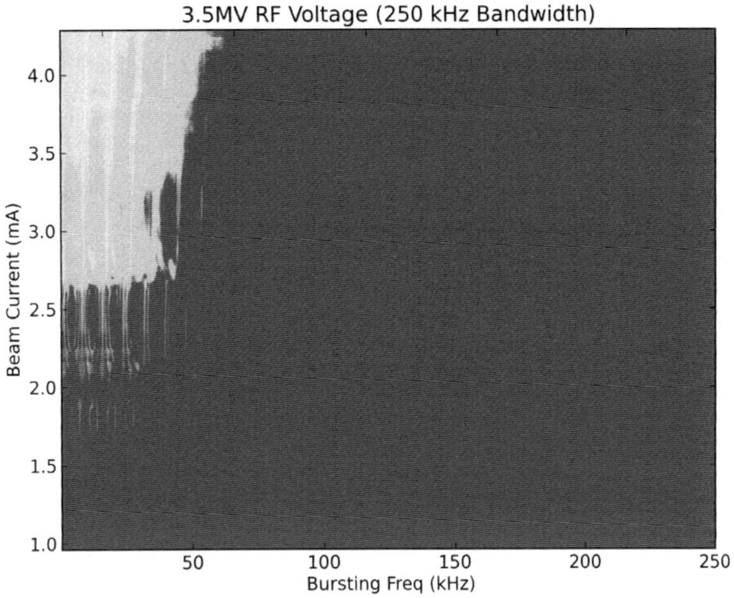

Figure 6. Full bandwidth spectral display of CSR signal as a function of bunch current at a 3.5 MV RF voltage.

5. Conclusions and Future Work

We have gathered further data on the microbunching instability using a 60-90 GHz Schottky barrier diode detector. The results have shown that there are significant levels of CSR bursts produced from a single bunch. There is a variation in observed signals from different RF voltages, with a higher voltage producing more prominent features and higher power bursts due to a shortened bunch length. The threshold for the instability also varies with RF voltage which can possibly be attributed to bunch length variations, however further experiments are planned to in which additional data on the instability thresholds will be obtained and firmer conclusions can be drawn. The evolution of the CSR bursts has also been observed. Strong signals are produced at harmonics of the synchrotron frequency which slowly increase in frequency with bunch current. Highly visible sidebands about these signals are also produced which exhibit unusual behaviour of larger changes of frequency. Furthermore, rapid frequency jumps in the signal are also seen, after which there is an increase in signal power over the whole 20 kHz bandwidth rather than the discrete signals at lower currents. There are no firm conclusions that can be drawn from our experiments about an explanation for the appearance of the instability and its evolution during changes in bunch current, resulting in the production of these CSR bursts. In the future, a dedicated port for experiments using mm-waves will be built where we plan to install additional detectors which operate at higher frequencies. We also intend to build a Michelson or Martin Puplett interferometer to measure the spectral content of the CSR bursts.

References

[1] G. Rehm, A.F Morgan, R. Bartolini, I.P. Martin, and P. Karataev, 2009, Ultra-fast mm-wave Detectors for Observation of Microbunching Instabilities in the Diamond Storage Ring, *Proc. 9th European Workshop on Beam Diagnostics and Instrumentation for Particle Accelerators (DIPAC 09)*, Basel, 25-27 May.

[2] R. Bartolini, G. Rehm and P. Karataev, 2009, Observation of Longitudinal Microbunching Instabilities in the Diamond Storage Ring, *Proc. 23rd Particle Accelerator Conference (PAC09)*, Vancouver, 4-8 May.

[3] A. Mochihashi, M. Hosaka, M. Katoh, M. Shimada and S. Kimura, 2006, Observation of THz Synchrotron Radiation Burst in UVSOR-II Electron Storage Ring, *Proc. 10th European Particle Accelerator Conference (EPAC 06)*, Edinburgh, 26-30 June.

[4] J. M. Byrd et al. 2002, Observation of broadband self-amplified spontaneous coherent terahertz synchrotron radiation in a storage ring, *Phys. Rev. Lett.* **89** 224801.

[5] M. Ako-Bakr, J. Feikes, K Holldack and G. Wüstefeld, 2002, Steady state far-infrared coherent synchrotron radiation detected at BESSY II, *Phys. Rev. Lett.* **88** 254801.

[6] P.Kuske, 2009, Investigation of the temporal structure of csr-bursts at BESSY-II, *Proc. 23rd Particle Accelerator Conference (PAC09)*, Vancouver, 4-8 May.

[7] B. Podobedov, G.L Carr, S.L. Kramer and J.B. Murphy, 2001, Longitudinal density modulation of unstable bunches emitting coherent IR , *19th Particle Accelerator Conference (PAC 2001)*, Chicago, IL, 18-22 June.

[8] G. Wüstefeld, 2008, Short Bunches in Electron Storage Rings and Coherent Synchrotron Radiation, *Proc. 11th European Particle Accelerator Conference (EPAC 08)*, Genoa, 23-27 June.

[9] L.V. Iogansen and M.S. Rabinovich, 1960, *Sov. Phys. JETP* **37** 83.

[10] J-M Wang, 1998, Microwave instability as a coherent light source, *Phys. Rev. E* **58** 984.

[11] S. Heifets and G. Stupakov, 2001, Beam instability and microbunching due to coherent synchrotron radiation, *19th Particle Accelerator Conference (PAC 2001)*, Chicago, IL, USA, 18-22 June.

[12] O. Grimm and P. Schmüser, 2006, Principles of longitudinal beam diagnostics with coherent radiation, *TESLA FEL 2006-03*

[13] G. Stupakov, 2003, Theory and Observations of Microbunching Instability in Electron Machines, *Proc. 2003 Particle Accelerator Conference (PAC03)*, Portland, OH, USA, 12-16 May.

[14] G. Wüstefeld, J. Feikes, M. v. Hatrott, M.Ries, A. Hoehl, R. Klein, R. Müller, A. Serdyukov and G. Ulm, 2010, Coherent THz Measurements at the Metrology Light Source, *Proc, 2010 International Particle Accelerator Conference (IPAC 10)*, Kyoto, Japan, 23-28 May.

[15] N.P. Hammond, K. Kay and G. Rehm, 2008, Developments in the Diamond storage ring, *Proc. 2008 Mechanical Engineering Design of Synchrotron Radiation Equipment and Instrumentation (MEDSI) workshop*, Saskatoon, Canada, 10-13 June.

Several hundred femto-second soft X-ray source in compact several meter facility

J Urakawa
KEK, 1-1 Oho, Tsukuba, Ibaraki 305-0801, Japan

Abstract. A compact MW peak power THz source is of high interest and possible to construct in a small accelerator facility (~2 m) by using pre-bunched FEL scheme. High gradient photo-cathode RF gun and few tens of femto-second laser system can directly generate a pre-bunched electron beam of a few hundred femto-seconds. Furthermore, small undulator can coherently generate MW THz photons in a few hundred femto-seconds. We are considering the reflection of the THz photons to make a collision with the pre-bunched electron beam and to generate a few hundred femto-second soft X-ray. A multi-bunch photo-cathode RF gun to generate ~10 MeV electron beam is under development. The design for the soft X-ray source is presented.

1. Introduction

The short pulse high-brightness coherent THz light source of ~0.3 to 10 THz in the intermediate zone between radio and light waves has caused a breakthrough in the rapidly expanding field of THz photon science. A photocathode is irradiated with a fs (femoto-second) laser pulse train of ~10 pulses, and an fs electron bunch train (Comb beam) is accelerated by a radio-frequency (RF) accelerating field of more than ~50 MV/m. In this way, the Comb beam is carried on a single RF accelerating field, enabling it to be accelerated to 5 MeV in a 7.5 cm RF gun. When the Comb beam is passed through a small wiggler (30cm), super-radiance in the THz region arises. The objective is to develop and apply an ultra compact high-brightness coherent THz light source, with short pulses of ~10 MW which can vary between 0.3 and 10 THz with a radiation of ~10 MJ /pulse.

The peak power of ~10 MW is about 100 times that of earlier THz light sources, and comparable to the intensity of THz light generated by the 10m facilities using advanced accelerator technologies that are being developed around the world. With this light source, it is possible to substantially reduce THz time-domain spectroscopy (THz-TDS) measurement time, and greatly improve the accuracy of measurement. In addition, it enables the fs timescale and multi-photon absorption nonlinear science phenomena to be captured with high precision. Applied experiments using the developed device will be carried out from 2014.

The THz-FEL is a good candidate due to its characteristics of high peak brightness, short duration, and tunable wavelength. However, the need for a facility in excess of ~10 m and for substantial funds has limited THz-FEL development. Two important goals of the proposed facility are to make the THz-FEL facility compact and to increase its output radiation power. Furthermore, the compact accelerator facility for the THz photon source may have additional possibilities as a novel and interesting photon source, making this compact facility very attractive. For this reason a several hundred fs soft X-ray source facility that can be installed into a ~5m area is proposed. Figure 1 shows the proposed compact THz source using pre-bunched beam train and edge-focusing wiggler [1], which can be installed on a

2 m long table. Later a short accelerator to generate a relatively high bright and fs soft X-ray will be added to the table-top THz light source.

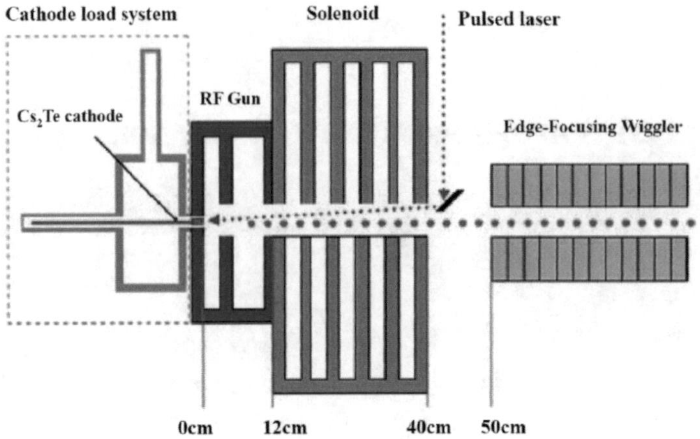

Figure 1. Table-Top Short Pulse and Coherent THz Light Source.

2. Research Methods

In order to maintain the time structure of the photoelectrons generated in the RF gun, it is necessary to engage them with an RF phase that overcomes the Coulomb repulsion and in which bunch compression arises dynamically. The cathode end plate of the RF gun is fixed in the position where the high electric field of the cavity arises. If the accelerating field is an increasing phase (20 degrees) and the cathode is irradiated with 100 fs micro-pulses, the S-band (2856 MHz) RF accelerating field (130 MV/m) is changed from 44.46 to 44.68 MV/m. The subsequent photoelectrons gain a slightly larger acceleration or a dynamic bunch compression while at the same time receiving rapid acceleration. By coming close to relativistic energy, the Coulomb repulsion and Lorentz force reach equilibrium. The time difference between the beginning and end of the Comb beam is about 8 ps. The accelerating field increases to 61.03 MV/m, therefore the 8ps bunch receives about 30 % bunch compression at the RF gun exit. The electron micro-bunch structure predicted in the simulation is confirmed by CDR (Coherent Diffraction Radiation) measurement.

3. Basic Technologies

3.1. High Gradient S-band RF Gun

A laser-driven RF gun with a Cs_2Te photocathode has been in development at KEK since 2002. This gun has been operated as an electron source for the ATF (Accelerator Test Facility) and generates a beam with an operational intensity of up to 2×10^{10} electrons per bunch. In 2008, a new gun incorporating all of the earlier modifications was produced for the ATF. Tests have confirmed a significant improvement of the Q value of the latest model. A typical transverse emittance of 1.3 π mm·mrad at 80 MeV has been obtained under the following conditions: solenoid field of 0.18 T, beam intensity of 1×10^{10} electrons per bunch, and RF power of 9 MW. The usual maximum accelerating gradient is 120 MeV/m for the multi-bunch beam generation [2].

We are also operating an approximately 10m long normal conducting accelerator facility [3] LUCX (Laser Undulator Compact X-ray source) for developing the technologies on multi-bunch electron beam generation, pulse laser accumulation system and precise multi-collision. Imaging experiments for X-ray applications will be attempted and the possibilities for their applications in this facility will be confirmed in early 2012.

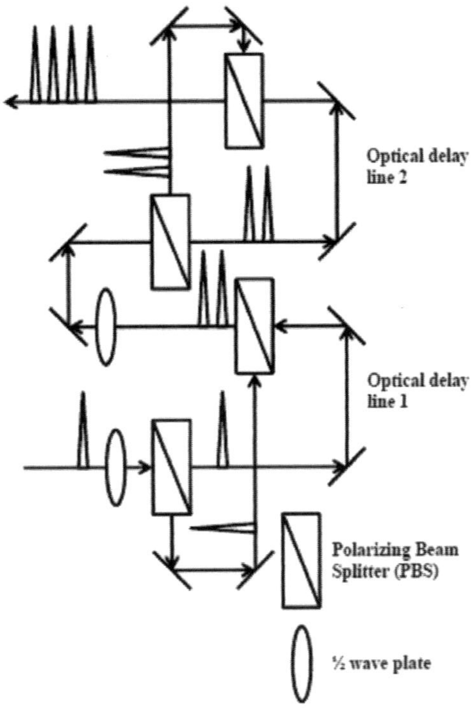

Figure 2. The method for generating multi-pulse laser train.

3.2. Multi fs Laser Pulse Train

CPA (Chirped Pulse Amplification) Titanium-Sapphire laser system will be used for this proposed experimental plan as it can generate ~10 mJ/pulse at 800 nm in the 30 to 100 fs pulse duration range. For example, the Trident laser system of Amplitude Technologies generates a 30 fs and 10mJ single pulse. By following pulse splitting and time delay methods [4], we can generate a multi-pulse laser train with pulse separation of THz wave period due to sufficient laser power. A half-wave plate rotates the S-wave by 45 degrees. PBS (Polarizing Beam Splitter) makes S-waves and P-waves by reflection and transmission. When repeated 4 times with a delay of about 500 fs, a 16 micro-bunched laser train within 7.5 psec was obtained. Figure 2 shows the generation of the multi-pulse laser train in the case of 4.

Subsequent micro-bunches create the bunching of former micro-bunches in a resonated wiggler. We need detailed research on beam loading effects due to multi micro-bunch and tuning techniques in a wiggler field created by pole-gap that makes the FEL resonate. If we accept low micro-bunch charge, say 100 pC or less, and not many micro-bunches, say 16 or less, the above research can confirm the generation of the comb beam. Also, we assume the time response of the Cs_2Te cathode is same as a Cu cathode and a minimum 0.2 % QE (Quantum Efficiency). If we try to generate a total single bunch charge of 50 pC assuming 5 % conversion efficiency from 800 nm to 266 nm by nonlinear crystal, the necessary laser pulse energy is about 250 μJ .

If we assume the peak RF field gradient at the cathode surface is 100 MV/m, 200 pC total charge in a micro-bunch train and the laser injection phase 20 degrees [5], the bunching factor for 2 THz radiation is still high, 0.446 at the wiggler entrance, which is predicted by beam tracking simulation. Our simulation results indicate the necessity of a higher gradient acceleration, lower total charge and a laser injection phase of less than 20 degrees to maintain a higher bunching factor. For example, we assume the peak field gradient at the cathode surface is 120 MV/m and the laser injection phase is 20 degrees, the bunching factor is above 0.6 and the beam energy is 5.68 MeV. Also, we are considering a wiggler period length of 30 mm and 2 THz radiation (wave length 150 μm), uniform laser size on cathode 1.0 mm ϕ , and a total charge of 25 pC. Since a higher gradient acceleration (100 MV/m→140 MV/m) produces a shorter bunch length and an earlier laser injection phase (20→10→1) produces a high bunching factor, the THz peak power of 100 MW will be generated by optimum tuning of the photo-cathode RF gun.

4. Soft X-ray Generation

To make the compact THz source more attractive, a booster linac and microwave cavity to generate the fs soft X-ray were added. Figure 3 shows the layout of the proposed facility. The S-band 12 cell booster linac is under manufacture and the microwave cavity is under research. Estimated soft X-ray flux is 7.5×10^{6} photons/sec 10% bandwidth with 2 THz microwave.

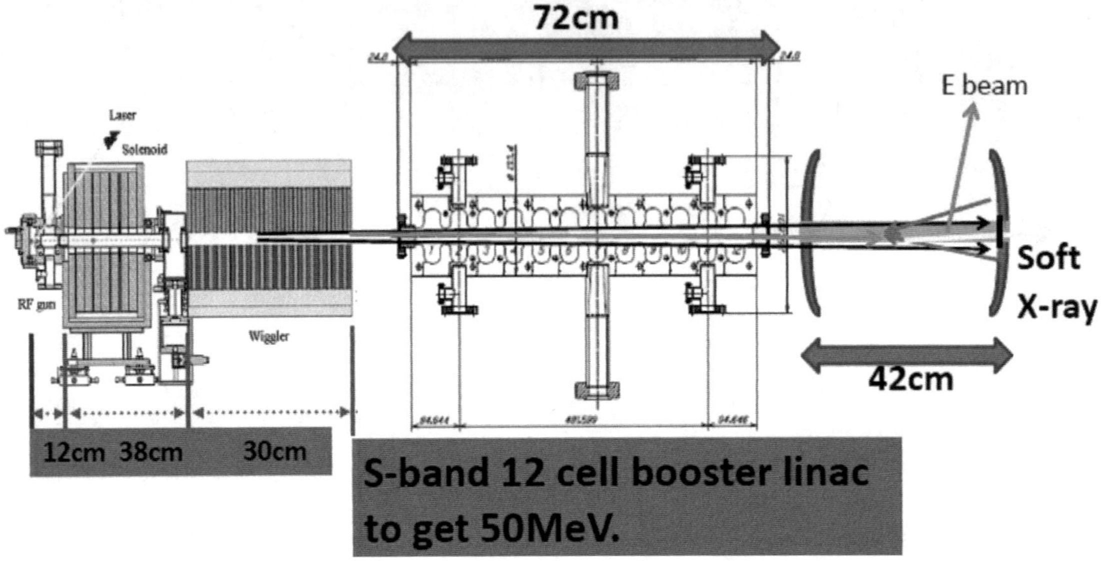

Figure 3. The layout of the compact accelerator for fs THz and soft X-ray generation.

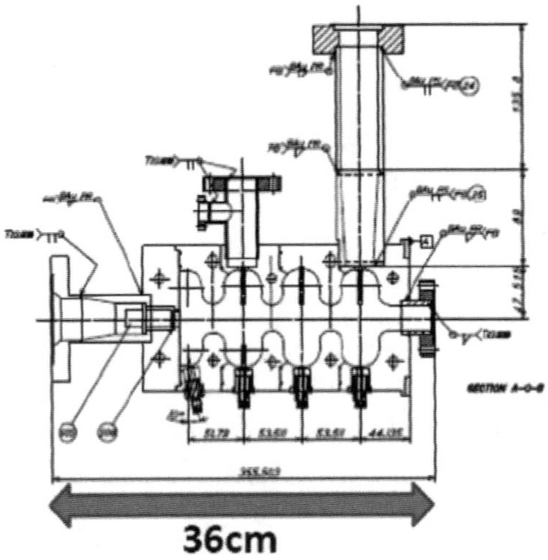

Figure 4. 3.5 cell S-band RF Gun.

In order to increase the photon flux of the soft X-rays, we have to increase the beam energy. So, we are manufacturing the 3.5 cell photo-cathode RF gun to get a 10MeV micro-bunch train beam. Figure 4 shows the design. After the completion of the gun, we will replace the 1.6 cell RF gun with this gun.

5. Summary

By tuning the micro-bunch spacing to the THz wavelength, it is possible to generate a narrowband coherent THz wave. Depending on whether an ideal Comb beam can be formed in the RF gun by high gradient acceleration, super-radiance peak power from the small wiggler (30 cm) reaches ~100 MW. Thus innovative THz light source applications can be developed. Also, by adding the booster linac and the microwave cavity, we can generate the fs soft X-rays for biological research applications.

Acknowledgements

The work is supported by the Quantum Beam Technology Program of the Japanese Ministry of Education, Culture, Sports, Science, and Technology (MEXT). Author would like to thank Dr. Shengguang Liu for his initiative suggestion to this project.

References

[1] Kashiwagi S et al 2007 *Proceedings of PAC07* THPMN037
[2] Terunuma N et al 2010 *Nucl. Instr. and Meth.in Phys. Res.* Improvement of an S-band RF gun with a Cs2Te photocathode for the KEK-ATF" **A613** 1
[3] Deshpande Abhay et al 2011 Experimental results of an rf gun and the generation of a multibunch beam *Phys. Rev. ST Accel. Beams* **14** 063501-1
[4] Liu S and Huang Y 2011 *Nucl. Instr. and Meth.in Phys. Res.* **A637** S172
[5] Yang Jinfeng et al 2011 *Nucl. Instr. and Meth.in Phys. Res.* Femtosecond pulse radiolysis and femtosecond electron diffraction **A637** S24

Journal of Physics: Conference Series **357** (2012) 012039 doi:10.1088/1742-6596/357/1/012039

The TPLUS project : a table-top tunable parametric UV radiation source

J -M André[a1], G Turk[a], K Le Guen[a], P Jonnard[a], F Delmotte[b], S de Rossi[b], A Jérôme[b], A -L Coutrot[b]
[a] Laboratoire Chimie Physique – Matière Rayonnement, UPMC Univ Paris 6, CNRS UMR 7614, 11 rue Pierre et Marie Curie, F-75231 Paris Cedex 5, France
[b] Laboratoire Charles Fabry, Institut d'Optique, CNRS, Univ Paris-Sud, 2 avenue Augustin Fresnel, F-91127 Palaiseau Cedex, France

E-mail: jean-michel.andre1@upmc.fr

Abstract. We present the project of a parametric radiation source, the Tunable Parametric Laboratory Ultraviolet Source (TPLUS). We aim at building a prototype of such a source with an electron gun of moderate energy, 100 keV, interacting with a periodic multilayer within a high vacuum experiment chamber. The objective of this table-top facility is to provide users with a cost-effective, simple-handling mean to provide a tunable and intense radiation source in the extreme UV and soft X-ray ranges.

1. Introduction

The objective of the TPLUS (Tunable Parametric Laboratory Ultraviolet Source or T+) project is to prove that it is possible to build, at a relatively low cost, a laboratory tunable extreme UV (EUV) source based on the emission of parametric radiation (PXR) by a periodic multilayer, working in continuous mode operation, which can be easily implemented in academic and industrial centres for various purposes: characterization of optics, detectors and spectrometers for the EUV, chemical analysis, source of EUV radiation for physical chemistry studies. Currently, no EUV source using non-relativistic electrons interaction with multilayer targets and exploiting the transition (TR) or parametric radiation has been reported in the literature. Nevertheless several successful experiments have been carried out in the soft X-ray domain using relativistic electron beams supplied by LINAC or storage rings by a Japanese team [1] or a US-Russian collaboration [2].

In the hard X-ray domain, some PR sources using crystal targets and relativistic electrons supplied by medium energy LINAC have been developed and used for scientific applications especially in Germany and Japan [3]. It is worth mentioning the PR-based source [4] in operation in Japan at the Nihon University. It produces x-rays around 15–20 keV and is dedicated to perform phase contrast imaging. It is based on a 100 MeV pulsed LINAC whose electrons interact with a silicon crystal. By rotating the crystal it is possible to change the emitted wavelength. This source works in pulsed mode (2 Hz) that induces long acquisition times.

Different EUV sources have been developed in the context of the EUV lithography [5]. They are mostly based on laser-produced plasma using liquid target or discharges in gas. They can offer an important power but their emission is limited to a narrow spectral range around 13.5 nm. A drawback

[1] To whom any correspondence should be addressed.

for metrology application is the uneven "shape" of their spectrum and their relative lack of pulse-to-pulse stability. Another system for EUV lithography is the "pocket synchrotron" Mirrorcle developed in Japan [6, 7]. In these 1 to 2 m large apparatus, the electrons are accelerated up to 4 or 20 MeV and produce mainly infrared (IR) radiations. EUV radiation can be obtained by inserting a thin foil in the storage ring, giving rise to an intense TR emission. Let us also note the CEMOX (Centrale d'Elaboration et de Métrologie des Optiques X) instrument working with a plasma source and used to characterize new multilayer coatings [8]. However it presents two main limitations: the available spectral range is limited due to region just near absorption edge of available filters and the pulsed operation (at 1 Hz) induces limited signal-to-noise ratio.

Except the synchrotron radiation or the free electron lasers, only available in large facilities, there is at present no tunable source of EUV radiation. Its use is quickly developing, mainly because of the implementation of the EUV lithography dedicated to the manufacturing of the next generation of electronic chips [9], of research in solar and extra-solar imaging from spatial telescopes [10] and of "attosecond" physics [11]. In addition, most of the private or institutional laboratories elaborating EUV optical components generally characterize their products by means of X-ray diffractometers operated in the hard x-ray range and complete their measurements using synchrotron radiation at the wavelength of application. The availability of a laboratory tunable EUV source would be invaluable for such purposes.

2. Physical principle

Our EUV source is based on the PXR emission in the case of Bragg resonant transition radiation, occurring when fast electrons propagate through a periodic medium (crystal, multilayer interferential mirror) close to the Bragg conditions. It can be seen as TR undergoing Bragg diffraction in a periodic lattice, as shown in figure 1. When electrons impinge a stratified structure of period d at a glancing incidence θ, radiation of wavelength λ is produced in the direction close to 2θ with respect to the incident electron beam direction, as given in first approximation by the Bragg law. By the coherent superposition of diffracted waves, this radiation is monochromatic, intense and directional.

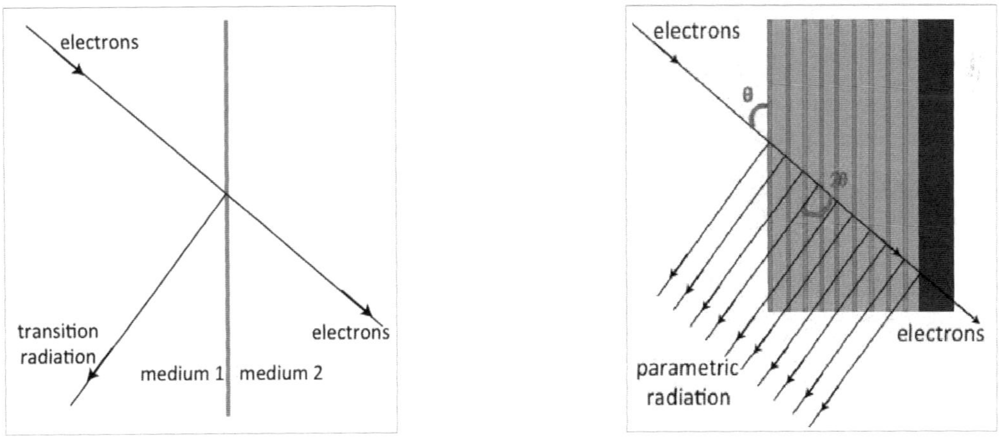

Figure 1. Principle of the transition (left) and parametric (right) radiations.

When high-energy electrons of velocity v are bombarding the periodic multilayer under a glancing angle equal to θ, PXR is emitted close to the Bragg conditions. The nth order PXR frequency is given by the following relation in which γ is the Lorentz factor of the relativistic electrons [12]:

$$\omega^{PR} = \frac{2\pi nc}{2d\sin\theta} = 1 - \frac{1}{2\gamma^2\sin^2\theta} \quad \text{with } \gamma = \frac{1}{\sqrt{1-\frac{v^2}{c^2}}}.$$

The frequency is related to the diffraction of virtual photons associated with the Coulomb field of a fast electron. In the case of ultra-relativistic electrons impinging on the periodic multilayer structure under a θ glancing angle, the previous expression of the nth order PR frequency is simplified and given by:

$$\omega^{PR} = \frac{2\pi nc}{2d\sin\theta}.$$

The radiation is collimated within an emission angle of about $1/\gamma$ with respect to the specular direction of electron trajectory, that is to say, the emission is very anisotropic: the intensity decreases very rapidly when one gets away from this well defined direction.

In the non-relativistic case the spatial distribution of the PR is more complicated. The general expression of the intensity and yield of the radiated energy is developed into details in reference [13–15] in the PARAMETRIX calculation code working within the framework of the classical theory of electromagnetism in continuous media. N N Nasonov *et al.* have published recently theoretical calculations of the PXR spectral and angular distributions emitted from a multilayer irradiated by 100 keV electrons [16]. Upon a 45° incidence angle, photons of 23.5 nm wavelength are emitted in the case of Si/Nb multilayer target having a 10 nm period. Their calculations demonstrate the feasibility of such a monochromatic source with an efficiency that is large enough for practical applications. They predict a yield of approximately 10^{-5} photon per electron with 100 keV impinging electrons the multilayer. In this case, with an electron gun providing 1 mA current, we can predict an order of magnitude estimate intensity of about 5.10^{9} photons.eV^{-1}.s^{-1}.sr^{-1} at the energy of 100 eV considering a spectral width of 1 eV. This value is lower than the 10^{15} photons.s^{-1}.sr^{-1} for both He I (58.4 nm) and He II (30.4 nm) lines supplied by a He-plasma, generated with the electron cyclotron resonance technique. Nevertheless this He source can emit only a few intense lines.

3. Proposed apparatus

The device under construction consists in the source by itself plus the characterization tools of the electron and PR beams. The scheme of the system is shown in figure 2. The main components of our apparatus are :

- a 100 keV Kimball Physics electron gun, which will deliver electrons under 1 mA current, focused on a 1 mm diameter spot on the surface of multilayer;
- a multilayer target, *i.e.* a periodic stack of nanometric layers deposited on a very thin silicon nitride membrane, and its four axes sample holder (3 translations, 1 rotation);
- as elements of the source, and
- a Roper back-illuminated CCD camera whose pixel size is 20 µm x 20 µm for the diagnosis of the EUV radiation
- a YAG scintillator coupled with a visible camera, and Bergoz wire vibrating monitors for the diagnosis of the electron beam.

The layers of the target will be made of light materials, to minimize the *Bremsstrahlung* radiation yield, proportional to the square of the atomic number, and preserve the spectral purity of the source. The materials have to resist to the electron beam and to an eventual heating. Monte Carlo simulations have shown that if the thickness of the stack and the membrane does not exceed 600 nm, then less than 5 % of the incident beam energy will be deposited in the target. The option to produce a self-standing multilayer will also be considered.

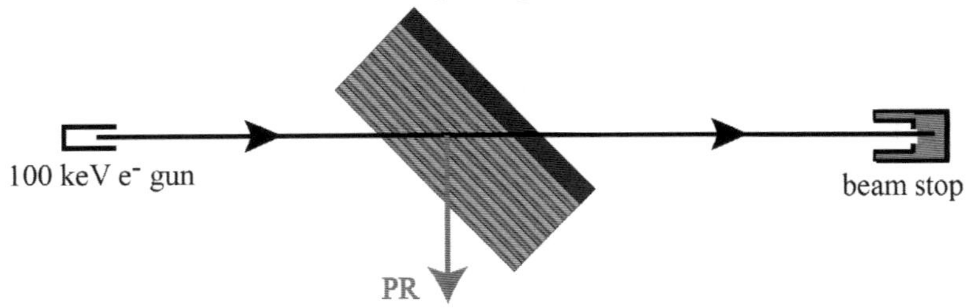

Figure 2. Scheme of the T+ apparatus. The PXR will be detected by a CCD camera.

The energy of the emitted PXR can be tuned either by changing the glancing angle or the period of the multilayer. However, it is preferable to obtain a source with a fixed exit. Indeed, the fixed exit of the source is a valuable feature because it makes easy its implementation for most of its uses. So we plan to use laterally d-graded multilayer targets [17]. These particular multilayers present a lateral gradient of period, see figure 3. Then, by translating the target under the electron beam, the multilayer period changes and so the PXR energy.

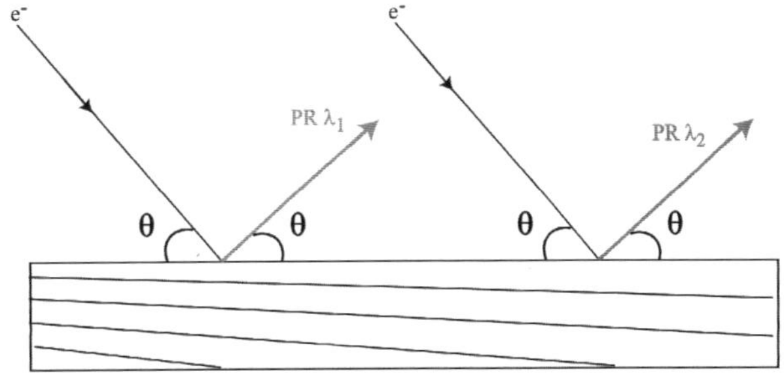

Figure 3. Wavelength tunability with the laterally d-graded multilayer target.

We first plan to use the well-known Mo/Si multilayer to obtain a PXR source at 92 eV (13.5 nm) in the following conditions. The multilayer is designed with 50 bilayers, with 4 nm-thick Mo layers and 6 nm-thick Si layers. The PARAMETRIX code is used considering 100 keV electrons impinging the target at incidence of 45° and with a current of 1 mA, and a square 100 mm x 100 mm plane detector located at 40 mm from the centroid source point. The result of this simulation giving the spectral power density is shown in figure 4.

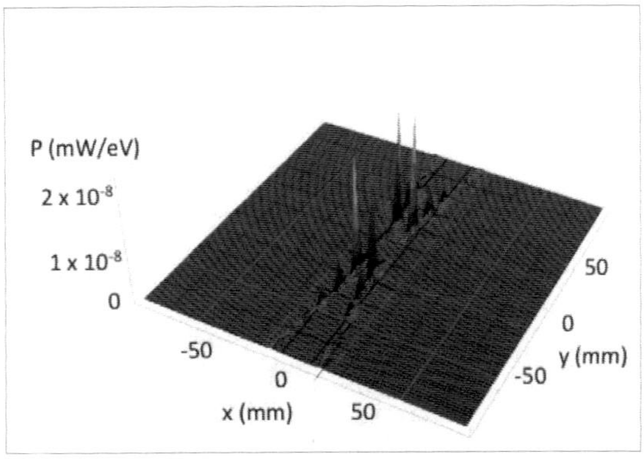

Figure 4 : Spectral power density of PXR emitted by a Mo/Si multilayer irradiated at 45° by 100 keV, 1 mA electrons and measured on a square 100 mm x 100 mm plane detector located at 40 mm from the source point.

The scale of intensity of emitted radiation is given in $mW \cdot eV^{-1}$, which corresponds to the power in 1 eV spectral bandwidth. Given the power value of about 10^{-8} $mW \cdot eV^{-1}$ at the energy of 92 eV, this leads to about 10^5 $photons.s^{-1} \cdot eV^{-1} \cdot sr^{-1}$. This is in agreement with the value predicted by Nasonov et al. [16]. It is observed that the spatial distribution of the emitted PXR is complicated and presents some intense and narrow symmetric peaks. We expect that the experimentally observed distribution will result from the convolution of the simulated distribution by the spatial, energy and angular distributions of the electron beam impiging the target.

Acknowledgments
The project is funded by Agence Nationale de la Recherche, granting N°2010BLAN092401. We greatly thank Mr. R Bartolo for his active work in control electronics and informatics, and Pr. C Bonnelle for helpful discussions.

References

[1] Yamada K, Hosokawa T and Takenaka H 1999 *Phys. Rev. A* **59** 3673
[2] Nasonov N N, Kaplin V V, Uglov S R, Piestrup M A and Gary C K 2003 *Phys. Rev. E* **68** 036504
[3] Inagaki M, Hayakawa Y, Nogami K, Tanaka T, Hayakawa K, Sakai T, Nakao K and Sato I 2008 *Japanese Journal of Applied Physics* **47** 8081
[4] Hayakawa Y, Hayakawa K, Inagaki M, Kuwada T, Nakao K, Nogami K, Sakae T, Sakai T, Sato I, Takahashi Y and Tanaka T 2008 *Nuclear Instruments and Methods in Physics Research Section B: Beam Interactions with Materials and Atoms* **266** 3758
[5] Robert E, Gonthiez T, Sarroukh O, Thomann A, Viladrosa R, Fleurier C, Pouvesle J and Cachoncinlle C 2002 *Emerging Lithographic Technologies Vi, Pts 1 and 2* vol 4688, éd R Englestad (Bellingham: Spie-Int Soc Optical Engineering) p 672
[6] Toyosugi N, Yamada H, Minkov D, Morita M, Yamaguchi T and Imai S 2007 *Journal of Synchrotron Radiation* **14** 212
[7] Yamada H, Minkov D, Toyosugi N, Morita M, Hasegawa D, Moon A and Okoye E K 2008 *Journal of Micro/Nanolithography, MEMS and MOEMS* **7** 043004
[8] Defise J-M, Halain J-P, Berghmans D, Denis F, Mazy E, Thibert T, Lecat J-H, Rochus P, Nicula B, De Groof A, Hochedez J-F, Schühle U, Ravet M-F and Delmotte F 2007 *Proc. SPIE* **6689** 66890S
[9] Montcalm C, Bajt S, Mikarimi P B, Weber F J and Folta J A 1998 *Proc. SPIE* **3331** 42
[10] Anon SECCHI - Solar Physics Branch - Naval Research Laboratory, http://secchi.nrl.navy.mil/
[11] Paul P M, Toma E S, Breger P, Mullot G, Augé F, Balcou P, Muller H G and Agostini P 2001 *Science* **292** 1689
[12] André J-M, Bonnelle C and Pardo B 2003 *Nuclear Instruments and Methods in Physics Research Section A: Accelerators, Spectrometers, Detectors and Associated Equipment* **498** 532
[13] André J-M, Pardo B and Bonnelle C 1999 *Phys. Rev. E* **60** 968
[14] Pardo B and André J-M 2000 *Phys. Rev. E* **63** 016613
[15] Pardo B and André J-M 2002 *Phys. Rev. E* **65** 036501
[16] Nasonov N N, Kubankin A S, Zhukova P N, Goldstein M, Williams D L, Piestrup M A and Park H 2007 *Nuclear Instruments and Methods in Physics Research Section B: Beam Interactions with Materials and Atoms* **254** 259
[17] Stemmler P, Bizeuil C, Marmoret R, André J-M, Barchewitz R et Rivoira R 1999 *Journal of Electron Spectroscopy and Related Phenomena* **101-103** 919

AUTHOR INDEX

AUTHOR INDEX